Springer Textbooks in Earth Sciences, Geography and Environment

The Springer Textbooks series publishes a broad portfolio of textbooks on Earth Sciences, Geography and Environmental Science. Springer textbooks provide comprehensive introductions as well as in-depth knowledge for advanced studies. A clear, reader-friendly layout and features such as end-of-chapter summaries, work examples, exercises, and glossaries help the reader to access the subject. Springer textbooks are essential for students, researchers and applied scientists.

More information about this series at ▶ https://link.springer.com/bookseries/15201

Wolfgang Frisch · Martin Meschede · Ronald C. Blakey

Plate Tectonics

Continental Drift and Mountain Building

Second Edition

 Springer

Wolfgang Frisch
Gomaringen, Baden-Württemberg, Germany

Martin Meschede
Institute of Geography and Geology
University of Greifswald
Greifswald, Germany

Ronald C. Blakey
Colorado Plateau Geosystems
Phoenix, AZ, USA

Department of Geology,
Center for EnvironmentalSciences
and Education
Northern Arizona University
Flagstaff, AZ, USA

ISSN 2510-1307   ISSN 2510-1315 (electronic)
Springer Textbooks in Earth Sciences, Geography and Environment
ISBN 978-3-030-89001-8   ISBN 978-3-030-88999-9 (eBook)
https://doi.org/10.1007/978-3-030-88999-9

Preface to the First Edition

In the late 1960s, Wegener's theory of continental drift, originally conceived fifty years earlier, was merged with the theory of plate tectonics and the concepts gained global acceptance among geoscientists. For the first time, a unifying concept, plate tectonics, could reconcile and unify all phenomena in geoscience into a common synthesis. The basic tenants have not changed since the 1960s, although many corrections and refinements have been added regarding questions of detail. The impact of the theory of plate tectonics cannot be overestimated—its importance to geoscience is as basic as Darwin's theory of evolution is to the biological sciences.

Earth is subject to steady change. This can impressively be realized in volcanic bursts and earthquakes around the Pacific Ocean or in the grandeur of young mountain ranges. Whilst new parts of plates are created along the mid-ocean ridges, older parts of plates disappear in subduction zones. Mountain chains arise from the collisions of plates. These dynamic processes are driven by the heat that is released from the interior of the Earth; this is what keeps the dynamo running. Continental drift, mountain building, volcanism, earthquakes, and in consequence of quakes, sea waves like the devastating tsunami of 26th December 2004 in the Indian Ocean—all this is the expression of the dynamics of the Earth.

This book presents an introduction in the wide field of plate tectonics and is dedicated to a broad audience interested in natural sciences, and students and professionals in geosciences and related subjects. Expertise in tectonics is not a pre-condition; the technical terms and geological processes are explained and reviewed in adequate detail. Many terms are defined in the glossary at the end of the book and a keyword index enables quick location of topics and terms in the text. Figures at the inside front-end and back-end covers present concise overviews of the geologic timescale and critical events in Earth history as well as the classification of magmatic and metamorphic rocks. More than 200 references enable the studious reader to gain deeper insight into special topics.

The book begins with an historical introduction concerning early ideas of continental drift and Earth dynamics that leads into discussion and consideration of plate motions and geometry. This is followed by several chapters that define, describe in detail, and illustrate the various features, processes, and setting that comprise the plate-tectonic realm: graben structures, passive continental margins, ocean basins, mid-ocean ridges, subduction zones, and transform faults. The remaining chapters deal with mountain-building processes as a consequence of plate tectonics and the collision of terranes and large continents. These chapters illuminate plate-tectonic processes from the early history of the Earth to the present. We examine how plate tectonics played a role in the construction of ancient Archean continents and then built large volumes of Proterozoic crust via "modern" plate-tectonic processes. Following the late Proterozoic break-up of the supercontinent Rodinia, we follow the plate-tectonic events that generated older and deeply eroded Paleozoic mountains and the formation of Pangaea to the more recent Mesozoic and Cenozoic mountains that remain the pinnacles of our modern planet. Examples from across the Earth are presented, including the young and lofty mountain ranges of the Himalayas, the Alps, and the North American Cordilleras. We relate the concepts, processes, and examples from the earlier chapters of the book to the plate-tectonic evolution discussed in the latter part of the book. In this way the book not only describes the plate-tectonic phenomena, but also focuses on the processes behind them and how they have worked in concert to produce the present plate configuration. Throughout the book, we strive to communicate to the reader an understanding that the Earth is a body in constant motion and change—a tectonic machine.

The present edition has evolved from an earlier German edition, Plattentektonik (Wissenschaftliche Buchgesellschaft, Darmstadt) by Wolfgang Frisch and Martin

Meschede (2005). Although largely a translation of the German edition, it also contains a number of new contributions by Ron Blakey, who especially enriched the text with geological examples from North America and upgraded the English text from an earlier translation.

Wolfgang Frisch
Tübingen, Germany

Ronald C. Blakey
Sedona, USA

Martin Meschede
Greifswald, Germany
May 2010

Contents

Contractional Theory, Continental Drift and Plate Tectonics

Contents

© The Author(s), under exclusive license to Springer Nature Switzerland AG 2022
W. Frisch et al., *Plate Tectonics*,
Springer Textbooks in Earth Sciences, Geography and Environment,
https://doi.org/10.1007/978-3-030-88999-9_1

1.1 Plate Tectonics—A Change in the Paradigm of the Geosciences

Earth's tectonic system concerns the movement of the lithosphere, the relatively brittle outermost solid Earth, which consists of a mosaic of independent plates. The boundaries of these plates are the most dynamic areas in the world and are the locations of most orogeneses, a word from the classical Greek meaning mountain building. The concepts embraced above define the field of geodynamics—energy, forces, and motion of a changing Planet Earth; mountains are the most obvious results of this system. Although nearly all geologically recent and many ancient mountain ranges are at obvious plate boundaries, all continental crust was at some time generated at plate margins, an observation that clearly relates the global phenomenon of mountain building and plate tectonics.

The formation and evolution of mountain ranges has been a major concern of geologists since the inception of the field of geology. Insights from hundreds of years of investigations are involved in the paradigm of plate tectonics. However, Wegener's theory of continental drift, which was published in 1915 and is regarded by many as the direct precursor of modern plate tectonics, contributed astonishingly little to the understanding of mountain building processes. Only the theory of plate tectonics was capable of combining all dynamic phenomenons to one unifying and explanatory theory. This major change in paradigm was formulated in the 1960s, revolutionized the geosciences, and blazed the trail towards a process-oriented consideration of geodynamic processes.

1.2 Early History of Geodynamic Thought

The French philosopher and naturalist René Descartes (1596–1650) was one of the first scientists to consider the composition of the interior of the Earth. In his "Principia philosophiae" (1644) he proposed that Earth contains a core with a liquid similar to that of the sun and wrapped by layers of rock, metal, water and air (Bonatti 1994). The Danish naturalist Niels Stensen, alias Nicolaus Steno (1638–1686), recognized that rocks are deformable and that the original position of deformed rocks can be reconstructed (Steno 1669).

Peter Simon Pallas (1777), James Hutton (1795), and Leopold von Buch (1824), who together can be regarded as the founders of tectonics as an independent branch of science, considered the forces of rising magmatic rocks as the main cause of mountain uplift. They noted that granitic rocks are common along the central axis of many mountain ranges. The opposing theory of horizontal forces, which suggested that resulting compression and folding of the crust created mountains, was championed by numerous geologists and also had several variations. De Saussure (1796) and Hall (1815) were the first to propose that horizontal forces were the prime drivers of mountain building. The existence of strong horizontal forces has later been verified by the discovery of large nappe thrusts in the Alps.

Horizontal forces generally were considered to be a consequence of Earth's contraction and the results were a compression of the Earth's crust (Élie de Beaumont 1852). The contractional hypothesis was based on the concept of an original liquid Earth followed by long-term cooling and shrinking. A variation of contractional tectonics is manifested in the geosynclinal theory of Dana (1873), which assumed that sedimentary rocks, now folded in a mountain range, were deposited in large, linear subsiding marine troughs, the so-called geosynclines. The sedimentary accumulation in a geosyncline is typically several kilometers thick and is many times thicker than sedimentary accumulations of the same age deposited on cratons. Dana considered the subsidence of the troughs as well as the later folding and uplifting to be the result of a shrinking of the Earth. The geosynclinal theory was later widely extended (e. g., Stille 1913) and a large number of different types of geosynclines was defined. The contractional theory, which was supported far into the twentieth century, is not accepted today because it conflicts with most modern hypotheses that concern the origin and early history of Earth. In fact, instead of shrinking, the diameter is actually slowly increasing due to tidal friction that slows down the velocity of rotation, which at present, is about 16 millionths of a second per year. At the beginning of the Cambrian approximately one-half billion years ago, a day was two and a quarter hours shorter and one year had 400 days.

The concept that mountain building occurs in phases, which globally act at the same time, was supported by Élie de Beaumont (1852) although Lyell (1833) had previously argued resolutely against it. This theory assumes that tectonic events that deform rocks and lead to folding in mountain ranges occur globally in temporally very limited phases. It was insistently advocated and improved by Stille (1913 and later publications). The concept of tectonic phases is not accepted today in its original, rigid form; however, it is well known that plate–tectonic events tend to create widespread, relatively synchronous mountain-building events.

The scientific dispute as to whether vertical or horizontal forces are the primary drivers of mountain building is today clearly settled in favor of the horizontal forces. However, these forces have their origin in the dynamics of plate tectonics and not in a shrinking Earth. The uplift of a mountain range is a secondary process that is induced by the horizontal movement of crustal blocks.

At the beginning of the twentieth century, a significant advancement in tectonics developed through studies of large geologic structures in the Alps (Lugeon 1902; Termier 1904). Outcrops displayed rock patterns with kilometers to tens of kilometers of overthrusting that had formed during mountain building processes. These overthrust units are called nappes and are characteristic of nearly all mountain ranges. This proved that mountain ranges are zones of extreme compression and crustal shortening; furthermore, results from geophysical investigations indicate that the continental crust beneath mountain ranges is significantly thickened and shortened. Because this thickened stack of continental crust below the mountains is significantly less dense than the displaced mantle below, according to the principle of isostasy, a buoyancy develops that leads to increase in topographical elevation. The relations between crustal shortening and thickening are complex and critical to the understanding of the tectonics of mountain building.

1.3 From Continental Drift to Plate Tectonics

Since the end of the sixteenth century, naturalists have noted the similarity of the coastlines on both sides of the Atlantic and concluded an original unity followed by later drift of the continents. The English philosopher Francis Bacon (1561–1626), who had access to the first accurate maps of the continents, was the first to make this point. Early explanations were sought, but at this point were in vain. The Flemish cartographer Abraham Ortelius (1527–1598) stated in 1596 that America was torn off from Europe and Africa by earthquakes and floods (Braun and Marquardt 2001)—an interpretation, though without evidence when Ortelius wrote it, that stands today. In 1756 the German theologian Feodor Lilienthal found the biblical confirmation of this observation: "And unto Eber were born two sons: the name of one was Peleg; for in his days was the Earth divided" (First Book of Moses, 10: 25).

From 1910 until his early death in 1930 the German meteorologist Alfred Wegener published works that attempted to add credibility to his theory of continental drift (◘ Fig. 1.1; Wegener 1912, 1915, 1929). Although his observations were supported by geographic fit, fossil evidence, and geologic patterns across the Atlantic, he was never able to convince a skeptical audience who demanded more proof—and especially an explanation for the mechanism that propelled the continents long distances. But he kept trying; his publications suggested that the continents were composed of lighter, less dense materials that he termed sial—an acronym that reflects the prevailing elements silicon and aluminum. The lighter continents drifted across the denser material of the Earth's mantle and the ocean floor, his sima—an acronym for silicon and magnesium. Sial was able to drift or plow through sima. He proposed that the driving forces behind continental drift were derived from known forces such as the rotation of the Earth, precession (a small conical rotation) of the Earth's axis, or tidal friction. The rotation of the Earth would pro-

◘ **Fig. 1.1** Reconstruction of the supercontinent Pangaea after Wegener (1915). The diverse geological and climatological data from different continents fit like a jig-saw puzzle on this reconstruction. Panthalassa was the giant ocean that stood opposite to Pangaea

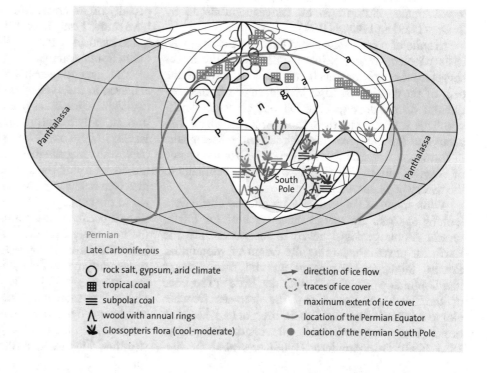

Permian

Late Carboniferous

○ rock salt, gypsum, arid climate
▦ tropical coal
≡ subpolar coal
Λ wood with annual rings
🌿 Glossopteris flora (cool-moderate)

→ direction of ice flow
⊂⊃ traces of ice cover
    maximum extent of ice cover
‒ location of the Permian Equator
● location of the Permian South Pole

1

duce polar escape, the slow drifting of the continents away from the poles, and the westward drift of the continents. Wegener used these forces to explain specific mountain ranges and the folding associated with them. Polar escape formed the mountain ranges that extend from the Alpine–Mediterranean area across the Iranides and into the Himalayas and Southeast Asia; they formed due to the convergence of Eurasia, which was drifting southward from the North Pole, and the southern continents Africa and India. The western drift from the Earth's rotation formed the high mountain ranges along the western coast of the Americas by frontal compression. Coincidentally and independently, similar ideas were expressed by the American geologist Taylor (1910).

Wegener's theory of continental drift could explain a number of problems that existed during his time: (1) the fit of the coastlines along both sides of the Atlantic, (2) the linear and narrow shape of the mountain ranges (the contractional theory in fact should produce much wider orogens), (3) the dominance of two levels of global altitude, the deep-sea abyssal plains and the broad continental lowlands (each reflects the two types of crustal material—oceanic and continental; ◘ Fig. 1.2), (4) the otherwise inexplicable appearance and disappearance of land bridges that were necessary to explain the faunal exchange between continents now separated by large oceanic areas. Continental drift could also explain why indicators of warm climates, such as Carboniferous coals, are today found near the poles. He postulated that all large continental blocks were unified at the end of the Paleozoic and beginning of the Mesozoic into one large continent that he named *Pangaea* (*Greek* all the land). His paleogeography followed similar illustrations of Pangaea published by Snider (1859) and Baker (1911).

In spite of a seemingly comprehensive theory, physicists rejected Wegener's ideas stating that the forces suggested by him are much too weak to explain the drift of the continents. After Wegener's death, the continental drift theory quickly fell out of scientific favor. This was particularly true in America where Wegener had aroused great scientific hostility. Meanwhile in Europe, a few scientists continued to examine some of Wegener's ideas—ideas that would lead eventually to the theory of plate tectonics.

While many of the geologists mentioned above were looking at rocks exposed at the surface of the Earth, several Alpine geologist began to look deep into the Earth for answers regarding the origin of mountains. Eduard Suess, in his magnificent and revolutionary multi-volume work "Das Antlitz der Erde" (The Face of the Earth), suggested that the deep-sea trenches along the border of the Pacific are zones where the ocean floor plunges beneath the continents (Suess, 1885–1909). Otto Ampferer (1906) presented the the-

ory of undercurrents (*German: Unterströmungstheorie*) which postulated that compression and nappe transport were forced by mass currents beneath the mountain ranges. Robert Schwinner (1920) took these ideas to develop a more far-reaching theory that assumed that currents in the Earth's interior are produced by convective heat transport. Subduction replaced the older German term "Verschluckung" that was used by Ampferer; Amstutz (1951) used the term subduction in conjunction with the development of tectonic nappes in the Swiss Alps. The term was later adopted and used by plate tectonicists (White et al. 1970). Many of the above concepts were amalgamated into a mobilistic model of geodynamics by the British geologist Holmes (1931, 1944), who proposed that convection currents were the driver of the Earth's tectonic system.

The mechanisms proposed by Ampferer, Schwinner, and Holmes suggested that rising currents underneath continental extensional structures and oceanic ridges, mountain building above descending branches of currents and continental drift on top of the horizontal parts of the currents drove the Earth's tectonic engine. Their ideas are not far from those of modern plate tectonics. Interestingly, if Schwinner and Wegener, who were both professors at the university of Graz (Austria) in the 1920s, would have communicated, they could have unified the drift theory with the correct theory of driving mechanism thus accelerating the development of the theory of plate tectonics. Meanwhile, German research vessels discovered the large, long-stretched submarine mountain ranges that today are known as midocean or mid-oceanic ridges and are a key to the concept of modern plate tectonics. But scepticism to all of this persisted well into the 1960s, especially in America, where a geologist is purported to have said: "I only accept the theory of continental drift if the head of a fossil will be found in Africa and the tail in South America".

The general acceptance of continental drift was achieved in the 1960s with the development of the theory of plate tectonics (for compilations of the most important early literature on plate tectonics see: Bird and Isacks 1972; Cox 1973). Modern investigations of the previously inaccessible ocean floor and in particular the discovery of the striped pattern of magnetic polarities (see below) at both sides of the mid-ocean ridges led to the concept of " sea-floor spreading", the dispersion of ocean floor from the mid-ocean ridges. Now for the first time, data supported a firm base on which to build the theory of plate tectonics and geologists had a single unifying model, which as it developed in the following years, could unify all basic geological and geophysical phenomena. The vast accumulation of data and knowledge since the inception of the basic model, have tweaked the original, but the basic tenants remain steadfast 50 years later. Plate tectonics remains the first

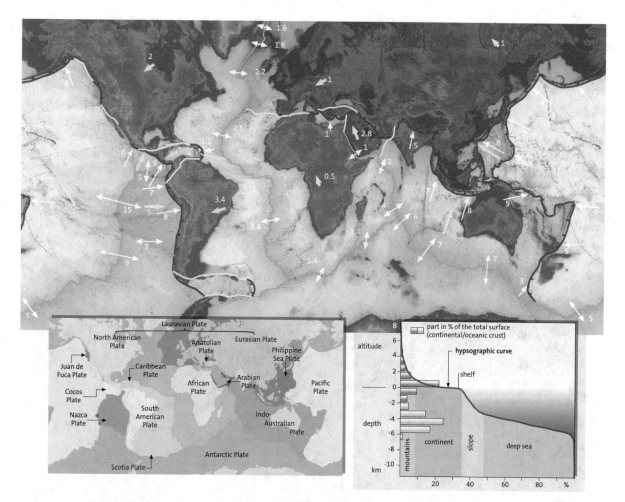

Fig. 1.2 Map showing the present plate configurations and plate motions on Earth; plate names shown on insert. White arrows indicate relative motion along plate boundaries (see ▶ Chap. 2) and yellow arrows indicate absolute plate motions related to a reference system based on stationary hot spots (see ▶ Chap. 6; DeMets et al. 1994). Constructive plate boundaries (mid-ocean ridges) are marked by green lines, destructive plate boundaries (subduction zones) by red lines (teeth point towards the upper plate), and conservative plate boundaries (transform faults) by gray lines. The hypsographic curve (box in the lower right) indicates the percentage of different topographic levels on the continents and under the sea. Continental and oceanic crust are characterized by different topographic levels

and only global geodynamic theory which orchestrates all known tectonic phenomena including earthquake zones, mountain building, structural patterns, nature of sedimentary basins, magmatism, and metamorphism—plate tectonics is an elegant and comprehensive synthesis of Earth's geodynamics.

One of the strongest tenants of the theory of plate tectonics is that it is based on the tectonics of present Planet Earth (e. g., Hess 1962; Vine and Matthews 1963; Wilson 1965; Isacks et al. 1968; LePichon 1968; Morgan 1968; McKenzie and Morgan 1969; Dewey and Bird 1970). Therefore, plate tectonics is an actualistic model. Geologists attempt to apply this concept to mountain building throughout Earth history. This application has been successful in most cases over the last 2 billion years of Earth history. Of course numerous older mountain ranges have generated contrasting hypotheses regarding the details of their origin as

much original information has been destroyed during later mountain building processes. Also, all ocean crust older than ca. 180 million years has been destroyed so detailed reconstructions of oceans older than this are not directly possible. Furthermore, mountain building older than 2.5 Ga (*Giga anni*—billions of years ago) does not straightforwardly conform with modern plate tectonics because the outer layers of the Earth—the place where plate tectonics is manifested—were somewhat different back then (▶ Chap. 11).

1.4 The Plate Tectonic Concept

The outermost two layers of Planet Earth, the lithosphere and asthenosphere, are directly involved in plate tectonics and are the only parts of Earth described in detail here. The term plate tectonics takes its name from the rigid lithospheric plates that form the outermost shell around the entire Earth. The size of individual plates is significantly different (■ Fig. 1.2). The lithosphere (*Greek* shell of rock) ranges in thickness from 70 to 150 km or more, thicker below the continents, thinner below the oceans. Beneath some mountain ranges it may exceed 200 km in thickness. It consists of two components—the crust (oceanic or continental), and the lithospheric part of the mantle (■ Fig. 1.3). The lithospheric mantle behaves in brittle fashion and contrasts in this regard from the underlying "plastic-like" *asthenosphere* (*Greek* weak shell). The asthenosphere behaves in ductile fashion and locally contains pockets of molten rock. These attributes point out the significant difference between Wegener's theory of continental drift and the theory of plate tectonics—the former suggested that continental blocks somehow plowed through the oceans but the latter documents that rigid plates, containing both continental and oceanic crust, move on top of ductile mantle.

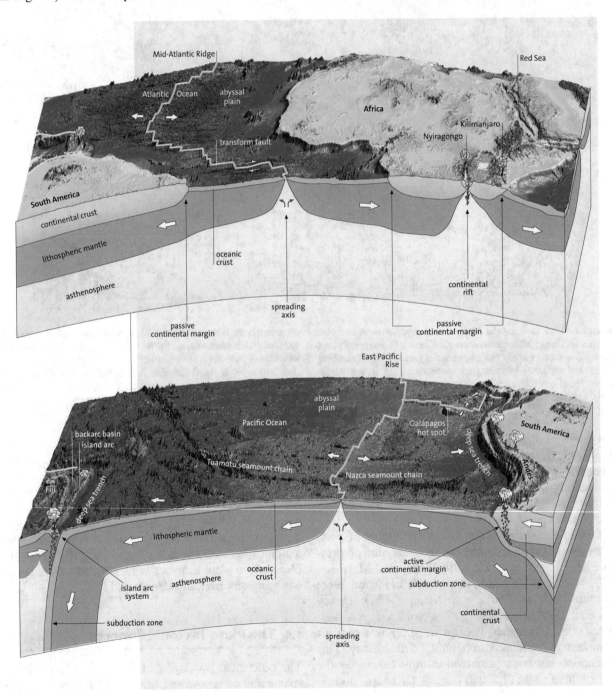

■ Fig. 1.3 Block diagrams of the outer shells of the Earth in the Atlantic and the Pacific region. Shown are the three types of plate boundaries, passive and active continental margins, island arcs, volcanic chains fed by hot-spot volcanism, and a graben system (strong vertical exaggeration). The plates consist of crust and lithospheric mantle. Relief data are from etopo30 (land surface) and gtopo2 data by Smith and Sandwell (1997), and etopol data by Amante and Eakins (2009)

Earth's crust, which forms the uppermost 5–60% of the plates, is generally divided into two types, continental and oceanic. Continental crust has an average thickness of 30–40 km but ranges up to 70 km under mountain ranges and high plateaus like the Andes or the Tibetan Plateau. Unlike the pure geographic definition, continents consist of both land and their adjacent shelf areas covered by shallow seas. Oceanic crust is markedly thinner with typical thicknesses of 5–8 km and forms the ocean floor. The top of oceanic crust lies an average 4–5 km deeper than that of the continental crust (◻ Fig. 1.2). This stunning bimodal distribution of Earth's surface defines the first-order aspect of topography and controls the primary pattern of land versus sea-most land and continental shelf areas lie within several hundred meters of sea level and most ocean floor lies 5 km below sea level. In fact, the fit of the continents across the Atlantic, the prime piece of evidence to Wegener's continental drift, is nearly perfect at the 500 fathom (~1000 m) contour—the famous Bullard fit (see ◻ Fig. 4.7; Bullard et al. 1965).

The bimodal elevation of Earth is a direct reflection of the difference in composition and density between continental and oceanic crust. Continental crust consists of relatively light (less dense) material that consists of acidic (rich in silicic acid, > 65 weight percent SiO_2), granitic and meta-morphic rocks (granites, granodiorites, gneisses, schists); hence the expression that continental crust is granitic. Primary mineral components include potassium and sodium feldspar, quartz, and mica, especially in the upper portions of the crust. In deeper parts of continental crust, the amount of basic (poorer in SiO_2) minerals such as hornblende and amphibole increase and rocks types include diorite and gabbro. The average density of the continental crust is 2.7–2.8 g/cm^3, and the average chemical composition is that of an andesite or diorite, magmatic rocks with an intermediate content on SiO_2 (about 60% SiO_2). Oceanic crust consists of basic basaltic rocks (~50% SiO_2), mostly basalts and gabbros; density averages about 3.0 g/cm^3, and calcium-rich feldspar and pyroxene are the most important minerals. The mantle lithosphere is composed of ultrabasic peridotites (~42–45% SiO_2) that have a density of 3.2–3.3 g/cm^3; the main mineral components are olivine and pyroxene.

The lithospheric plates move against and towards each other with different velocities in different directions (◻ Fig. 1.2). This raises the question of how a balance is possible within a closed system of plates across the sphere of the Earth. According to the Euler theorem of 1770 the movement of an object at the surface of a sphere occurs by rotation around an axis that passes through the center of the sphere. Therefore, all plate motions are defined by a rotation around such an axis plus an angular velocity (see ▶ Chap. 2). Plate motion results in three types of plate boundaries: constructive, destructive, and conservative (◻ Fig. 1.4).

Constructive plate boundaries are characterized by diverging plates—commonly referred to as divergent plate boundaries. The developing gap along the line of separation is immediately filled by newly formed lithospheric material including oceanic crust; hence the term constructive plate boundary. Constructive plate boundaries are represented by the *mid-ocean ridges*—the under sea mountain ranges that circle the Earth like seams on a baseball. Ridges are formed by rising basaltic melts generated from the mantle asthenosphere, that solidify to brittle ocean crust. The topographic mid-ocean ridge (spreading axis in ◻ Figs. 1.3 and 1.5) reflects the warm, less dense lithosphere that underlies it (older oceanic lithosphere is more dense and expresses the low topography of the abyssal ocean plains). The ocean floor spreads from constructive plate boundaries, hence "sea-floor spreading".

Destructive plate boundaries are characterized by converging plates. Where two plates move towards each other, the denser plate is bent and pulled beneath the less dense plate, eventually plunging downward at an angle into the depths of the sub-lithospheric mantle. Such areas are called *subduction zones*. Eventually the subducted plates become recylced into the mantle and thus destroyed. At convergent boundaries as they are commonly termed, only dense, oceanic lithosphere can be diverted into the sub-lithospheric mantle in large quantities; thicker, less dense continental lithosphere cannot subduct very deeply—this explains why old continental crust, billions of years old exists today while no ocean crust greater than ca. 180 Ma (*Mega anni*—millions of years ago) is present—older ocean

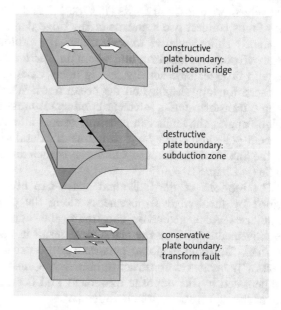

◻ **Fig. 1.4** Block diagrams showing the three types of plate boundaries

constructive plate boundary: mid-oceanic ridge

destructive plate boundary: subduction zone

conservative plate boundary: transform fault

1

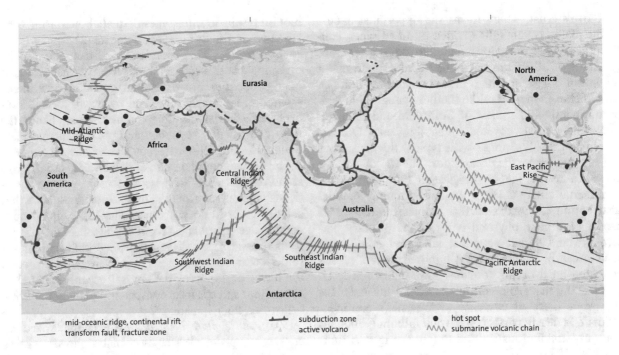

□ Fig. 1.5   First-order tectonic elements of Earth. Each of the present plates is readily discernable

crust has all been recycled. Surface expression of subduction zones is manifested in the deep-sea trenches, common features around the Pacific Ocean (□ Figs. 1.3 and 1.5).

Conservative plate boundaries occur where two plates slip past each and little or no crust is created or destroyed in the process. These boundaries are also characterized by strike slip or *transform faults* and are commonly called transform margins. Transform faults are rare in the purely continental realm. Mid-ocean ridges on the other hand are cut by numerous, mostly relatively short transform faults (□ Figs. 1.3 and 1.5). Such faults connect two segments of the ridge that are apparently shifted relative to each other. The prolongation of oceanic transform faults contain *fracture zones* with little tectonic activity that in many cases can be traced for long distance on the ocean floor. Where oceanic transform faults intercept continental crust at oblique angles, the faults can penetrate deeply into the adjacent plate and create large distances of lateral offset; the San Andreas Fault of California is such an example (▶ Chap. 8).

The motions of the individual plates can be described by their relative movements along the plate boundaries. As a geometric constraint, the sum of the movements of all of the plates must result in zero (▶ Chap. 2). From a global perspective, the drifting apart of the plates at constructive boundaries must be compensated by the opposite movement and destruction of lithosphere at destructive boundaries.

1.5   The Pattern of Magnetic Polarity Stripes

The discovery and interpretation of the striped pattern of magnetic polarities (□ Figs. 1.6 and 1.7) led to the concept of sea-floor spreading. Although generally ascribed to Vine and Matthews (1963), L. W. Morley submitted a paper a year before that was rejected because the reviewers considered the idea absurd (Cox 1973).

Minerals and the rocks in which they are contained acquire a magnetic signature as a given mineral cools below a certain temperature, its Curie temperature. Below the Curie temperature, named after the physicist Pierre Curie, a given mineral acquires the magnetic signature of the Earth's magnetic field that was present at that time. As an example, magnetite has a Curie temperature of 580 °C. Three signatures of magnetism are generally infused into magnetic minerals: inclination, which reflects latitude; declination, which reflects direction to the poles; and normal or reversed polarity, which indicates magnetic reversals (by convention, the current situation is defined as "normal"). Magnetic signatures in minerals are maintained for hundreds of millions of years, although some overprinting from subsequent geologic events does occur so that samples must be "cleaned" to eliminate younger events. Also, the perturbing effect of the current magnetic field must be compensated for during the analysis of the sample.

The magnetic pole moves around the geographic pole (the rotational pole of the Earth) in an irregular, sinu-

Fig. 1.6 **a** Stripe pattern of magnetic polarities on the ocean floor at the Reykjanes Ridge, part of the Mid-Atlantic Ridge southwest of Iceland (Heirtzler et al. 1966). **b** Curves representing the magnetic field strength measured along the track of ships crossing the ridge. Normal (in colors) and reverse magnetization can be obtained from these curves. **c** Graph showing detailed magnetic stripe pattern for the last 4.5 Ma. By comparison with measured profiles, the ocean floor can be dated

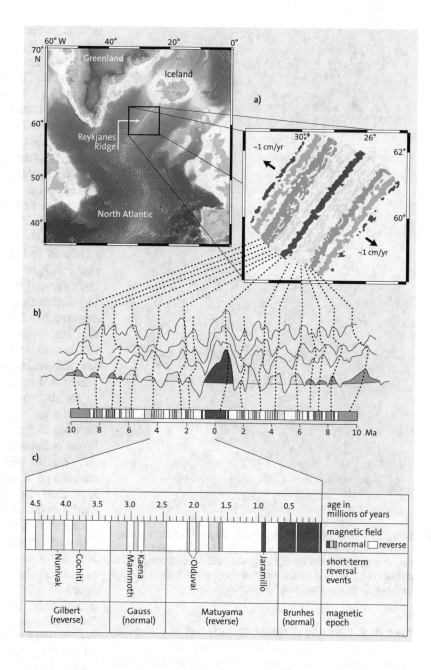

ous manner to produce what is called secular variation. However, averaged over a period of several thousand years the two poles coincide. Therefore, the orientation of earlier geographic poles can be detected using paleomagnetics if the mean value is calculated from enough samples. The present magnetic South Pole is located near the geographic north pole. This has not alwas been the case. At very irregular intervals over periods of variable duration, the polarity reverses and the earlier South Pole becomes the North Pole and the other way round.

Using dated basaltic and other rocks with magnetic signatures on land, a magnetic time scale has been defined that displays the periods and epochs with normal (like today) and reversed magnetization. These patterns of magnetization are found parallel and sym-

metrically aligned to the oceanic ridges (■ Fig. 1.6). Based on the characteristic patterns of normal and reversed magnetization, the stripes can be dated by comparing them with known sequences. This is very strong proof for sea-floor spreading because the method shows that variable magnetic stripes of oceanic crust are formed parallel to the ridges and that they become older with increasing distance to the ridge (■ Fig. 1.7). It was the discovery of this symmetric pattern parallel to the ridges that proved in the early 1960s the concept of sea-floor spreading and associated drifting of continents, two of the most basic tenants of plate tectonics. Magnetic reversals in oceanic rocks only yield data back to approximately 180 Ma, the Early Jurassic (see ■ Fig. 2.12)—all older oceanic crust has been sub-

1

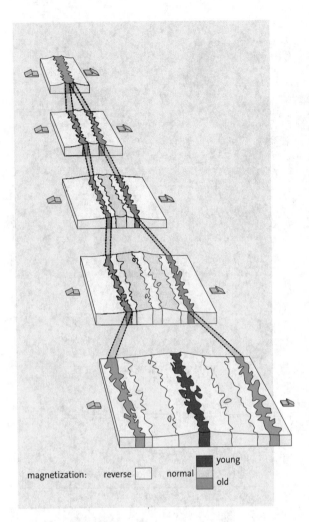

magnetization: reverse ☐   normal   ☐ young
                                          ☐ old

▢ Fig. 1.7 Simplified sketch showing development of the magnetic stripe pattern along the spreading axis. The pattern is caused by repeated reversals of the Earth's magnetic field. Irregularities of the stripes are caused by submarine extrusion of basaltic lavas that adapt to the existing, commonly rough topography

ducted. A paramount reason for this fact is that older ocean crust is colder and more dense, and therefore subducts more readily; for example, if 20 Ma ocean crust and 150 Ma ocean crust collide, the older will be subducted (▶ Chap. 4).

1.6 Plate Motions and Earthquake Zones

Convectional currents in the sub-lithospheric mantle are interrelated to overlying plate motions. Based on the propagation behavior of earthquake waves, it is known that the Earth's mantle is primarily in a solid state. Nevertheless, it is able to flow on the order of several centimeters per year; this value is close to the velocity of plate motion. The flow motion is facilitated by gliding processes along mineral grain boundaries, a condition accentuated by the high temperature con-

ditions in the Earth's mantle. The Earth's mantle contains relatively small but important areas where molten material forms a thin film around and separates the solid mineral grains. The mobile asthenosphere directly beneath the lithosphere is assumed to contain a few percent of molten material.

The pattern of convection cells movement in the Earth's mantle is extremely complex and has even been "photographed" using seismic tomography, which is a technique based on methods used in the medical industry (▶ Chap. 2). Probably, the outermost system of convection cells in the upper mantle (down to about 700 km depth) is separated from a second system in the lower mantle; both systems, however, are strongly interrelated and induce and influence each other. Tomography suggests that rising and descending currents in both parts of the mantle commonly have the same spatial distribution. The Earth's core, which consists predominantly of iron and nickel, has an outer liquid shell that surrounds an inner solid sphere. The relations of the core to convection cell currents and plate motions are still under investigation. However, it can be assumed that interactions and material transfer occur.

Relative movements of the plates along the plate boundaries induce earthquakes. The gliding process between the plates produces great but variable stress. Stress is induced along slip planes within rocks, which to a certain degree are elastically deformable, and is released in a jerky movement when a limiting value is reached. Looking at the distribution map of earthquake epicenters (points at the Earth's surface directly above the earthquake center) it impressively shows that earthquakes are mainly restricted to narrow zones around the globe (▢ Fig. 1.8), the present plate boundaries. Distribution of earthquakes varies at different types of plate boundaries. Deep earthquake centers occur only along subduction zones, whereas shallow earthquakes occur at all plate boundaries. Moreover, distributed earthquake centers can be found elsewhere indicating that the plates are not free of deformation in their interior parts that may be cut by large fault zones. The rates of movement at intra-plate fault zones are generally less than a few millimeters per year when averaged over long geologic periods of time; this tends to be an order of magnitude less than rates at plate boundaries.

Globally, earthquakes are strongly concentrated along destructive plate boundaries and are especially notable ringing the Pacific Ocean. Zones of epicenters are relatively wide (▢ Fig. 1.8) because the subducting plates that produce the earthquakes plunge obliquely into the mantle. Subduction zones can be traced downward, using earthquake foci (the location and depth of the earthquake) to depths of approximately 700 km. The map view of the plate boundary is located at the trench side of the earthquake belt where the centers are

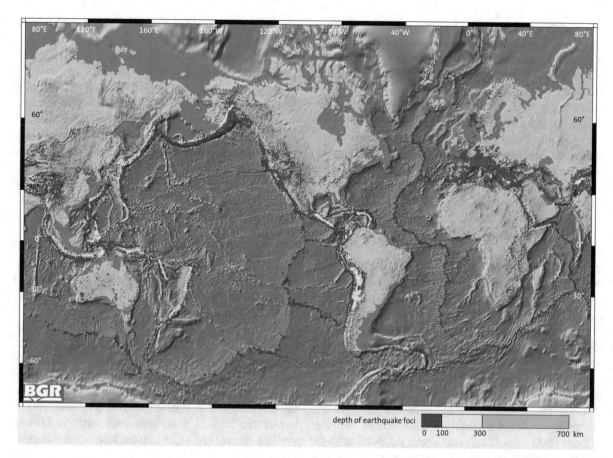

☐ Fig. 1.8 Global distribution of earthquake centers with epicenters mapped according to their depth. Note how the epicenters define plate boundaries (☐ Fig. 1.5); also note that most earthquakes occur within 100 km of the surface except along subduction zones where they deepen under the upper plate (produced with the kind support of Mrs. Agneta Schick, Federal Institute of Geosciences and Resources, BGR, Hannover, Germany)

at a shallow depth; the boundary plunges downward at different angles. Earthquake foci at shallow depths, with epicenters mostly near the surface of the plate boundary, may have devastating consequences—these are the locations of the Earth's most destructive earthquakes.

Along transform faults the epicenters of earthquakes are much more concentrated near the surficial trace of the plate boundary because the fault zones are vertical. Where transform faults cut through continental crust, they may also cause devastating earthquakes similar to the shallow earthquakes at subduction zones. Friction is generated by the thick, stiff plates and is dependent upon the motion velocity between the plates. Examples of continental transform faults include the San Andreas Fault in California and the North Anatolian Fault in Asia Minor.

Earthquake activity is much less at mid-ocean ridges. Uprising currents transport molten rock material to the Earth's surface and the stiff shell that accumulates and releases stresses is quite thin. Hot and recently solidified rock material is more likely to deform plastically. Therefore, only small, shallow earthquakes occur. Nevertheless, the constructive plate boundaries are also clearly visible on the earthquake map (☐ Fig. 1.8).

Young mountain ranges like the Alps–Himalaya belt or the Andes–Cordillera belt, which are still tectonically active, are also characterized by frequent earthquakes. Because of the diffuse collision of large continental masses, wide zones of deformation with numerous slip planes develop. Therefore, exceptionally wide belts of shallow earthquakes occur in these zones (☐ Fig. 1.8). Occasional deeper earthquakes testify to the preceding subduction activity.

1.7 Two Kinds of Continental Margins

Nearly all of the presently existing plates contain areas with both continental and oceanic crust. A good example includes the large plates on either side of the Mid-Atlantic Ridge, one of Earth's most prominent plate boundaries. This plate boundary separates the

1

two American plates from the Eurasian and African plates, all of which have large amounts of both continental and oceanic crust (◻ Figs. 1.2 and 1.5). The Indo-Australian, the Antarctic and many smaller plates also contain both crustal types. In contrast, the huge Pacific Plate which extends westward from the East Pacific Rise to the eastern Asian island arc systems, contains only very small amounts of continental crust, mostly in California and New Zealand. The Philippines, Cocos, and Nacza plates—smaller plates that surround the Pacific Plate—only contain oceanic crust.

The fact that most plates contain both crustal types means that some boundaries between oceans and continents occur within a given plate; hence, two types of continental margins exist. Where continental crust merges with oceanic crust, shelf areas generally slope towards the abyssal plains—thus the ocean-continent boundary is an intra-plate feature. Continental and oceanic crust belong to the same plate. Such continental margins are widespread around the Atlantic Ocean. Here only slight (mostly vertical) movements occur; therefore, they are commonly called *passive continental margins* (◻ Fig. 1.3 upper part). Passive continental margins do not represent plate boundaries.

On the other hand, *active continental margins* are those margins where a plate boundary exists between continent and ocean. Two types occur-subduction margins and transform margins. At subduction margins, a part of a plate with oceanic crust is being subducted beneath the continental crust. At transform margins, the oceanic plate slides laterally along the continental margin. A deep sea trench forms along subduction zone plate boundaries. This type of continental margin is today prominent along the Andes (◻ Fig. 1.3 lower part) and along numerous subduction zones around the Pacific Ocean that are characterized by island arc systems. The margin of the upper plate in these cases is characterized by chains of volcanic arcs, built either on continental crust or on continental pieces that were separated from the neighboring continent.

1.8 Magmatism and Plate Tectonics

Magmatic belts as well as earthquake activity are closely related to plate boundaries. The average yearly production of magmatic (volcanic and plutonic) rocks formed at destructive plate margins is slightly less than 10 km^3 (Schmincke 2004). The melting that produces magmatism is caused by complex interrelations between the asthenosphere and the subducting plates plunging into it. These melts, which are marked by specific chemical characteristics, intrude into the upper plate and feed volcanic chains above subduction zones (◻ Figs. 1.3 and 1.5) to produce subduction re-

lated magmatism. Modern examples include the eastern Asian island arcs (island arc magmatism) and the Andes (magmatism at an active continental margin).

Mid-ocean ridges are the location of major production of basic magmatites, namely basalts and gabbros. High temperature and pressure release beneath the ridges combine to generate partial melting of up to~20% the rocks of the mantle (peridotite). Oceanic crust develops from these melts and annually more than 20 km^3 of new crust is formed (Schmincke 2004). Therefore, mid-ocean ridges generate more than twice the amount of melts than are generated above subduction zones. At transform faults significant melting does not occur so magmatic processes are unimportant.

Although constructive and destructive plate boundaries are responsible for the formation of most of Earth's magmatic rocks, annually approximately 4 km^3 of magmatic rocks are produced in intraplate settings. This intraplate magmatism is mostly related to hot spots (◻ Fig. 1.5). Hot spots are point-sources of magma caused by mantle diapirs and occur on either the continents or oceans. Diapirs are hot, finger-like zones of rising material within the mantle. When they reach the upper asthenosphere beneath the plates, melting is induced that creates volcanic eruptions and doming of the surface over long time periods. Hot spots are less commonly superimposed on constructive plate boundaries.

Modern continental hot spots include the Yellowstone volcanic field in North America, the French Central Massif and the volcanic Eifel Mountains in Europe, and the Tibesti Mountains and the Ahaggar (Hoggar) in North Africa (◻ Fig. 1.5). Modern oceanic hotspots include the active part of the Hawaiian Archipelago and the Canary Islands; Iceland is an example of a hotspot superimposed on a mid-ocean ridge. As plates drift over hot spots, long volcanic chains develop with the hot spot located at the active end; Hawaii is a good example of this (◻ Fig. 1.5). On continents, hot spots are commonly related to graben structures characterizd by extensive, deep fault systems that cut through the entire thickness of continents; the best known example are the volcanoes of the East African graben system. Graben structures are characterized by crustal extension and bordered by faults; such areas cause thinning of the lithosphere and provide the opportunity for magma to rise along fault zones. If extension continues, new ocean can be formed at these structures. An example for such a newly developing ocean is the northern part of the East African graben system (Afar) and the Red Sea (▶ Chap. 3). The so-called Afar triangle is also characterized by a hot spot. Graben structures may be transferred into constructive plate boundaries where the hot spot commonly plays an important role.

1.9 What Drives the Plates and What Slows Them Down?

The growing plate boundaries at mid-ocean ridges always forms new oceanic lithosphere because the basaltic/gabbroic oceanic crust is a product of partial melting directly from the mantle; on the other hand, continental crust forms by much more complicated melting and recycling processes above subduction zones (▶ Chap. 7). Only oceanic lithosphere can be completely reintegrated into the mantle at subduction zones whereas subducted continental crust experiences strong buoyancy and accretion to the overlying plate. Plate movements are thus mainly controlled by the formation of oceanic lithosphere at the oceanic ridges, and its subduction and reintegration into the Earth's mantle at destructive margins. Oceanic lithosphere, therefore, forms the conveyor belt of plate tectonics whereas the continental blocks go along for the ride.

In fact, the driving forces for the plate movement are to be found under the mid-ocean ridges and in the subduction zones—at the plate boundaries. Plate motion is thus orchestrated by rising magma at the mid-ocean ridges and sinking dense lithosphere at the subduction zones (Bott 1982). These processes are called "ridge push" and "slab pull". Ridge push is caused by the upward movement of hot and relatively light rock melts at the mid-ocean ridges where, in the area of newly forming lithosphere, the vertical movement is transferred into a horizontal vector that pushes the plates apart. Slab pull arises because of the higher density of cooled lithosphere with respect to the mantle underneath. Of the two driving forces this is the more important one. Mineral changes to denser species that, because of lower temperatures in the descending plate, occur at shallower depth as compared to that of the surrounding mantle, intensify this process. An earlier idea that suggested that the carrying of middle parts of plates on top of the horizontal currents of the asthenosphere may, however, not be important in plate motion and, in contrast, actually hinder the process in certain regions.

Ridge push and slab pull act in accord with the state of stress in the plate interiors. They produce compression near the mid-ocean ridge and extension near the deep sea trenches. If plates were actually carried by currents, the state of stress would be the other way round. Also, from a thermodynamic view, it is logical that rising hot and descending cold material supply the driving forces. These considerations are supported by the following observations. The velocities of plate movements are independent of the size of the plates, and plates with subduction borders move faster than those without subduction borders; this emphasizes the importance of the slab pull as the driving force. Plates with a large percentage of thick continental crust move more slowly, an observation that suggests that dragging at the bottom of the plates (like the keel of a boat on sand) negatively influences the movement (Kearey and Vine 1990).

1.10 Collision and Mountain Building

Subduction zones tend to form in locations with mature (older), cool, and thus denser lithosphere. These conditions exist at the edges of large oceanic basins like the present Pacific Ocean. If an oceanic basin is not bounded by subduction zones, it will widen as is the case with the present Atlantic Ocean. Spreading rates at middle oceanic ridges range from a low of 1 cm/year to a high of 15 cm/year (◻ Fig. 1.2). The rate of subduction (at present up to 9 cm/year—in the past probably faster) in an ocean basin can exceed the rate of new crust formation at a ridge, especially if opposite sides of the basin both contain subduction zones. In such a case, the oceanic basin shrinks and adjacent continental blocks move closer together. Continuing convergence finally leads to the collision of the continental blocks and the passive continental margin of the subducting (lower) plate is dragged beneath the active boundary of the upper plate. The low density of the subducted part of the continent prohibits extensive subduction and it cannot be dragged down to great depths. Rather, it buoys up on the surrounding denser mantle material and rises, following the principle of isostasy.

Buoyancy and strong frictional forces following collision of two continental blocks eventually brings the subduction of continental crust to a standstill. During this process, complex tectonic structures such as folds and nappes develop and rocks are deeply buried and heated; metamorphism and partial melting typically occur. After convergence ceases, the attached subducted oceanic lithosphere breaks off under its own weight ("slab breakoff"). The lessening of slab pull due to the release of the counterweight induces isostatic uplift of the continental crust, which has been thickened to double its normal thickness. The Alpine–Himalayan mountain chain was created by such processes and forms sharply protruding edifices on the surface of the Earth.

Collisional orogens are something like a Janus head: both an interior and an exterior sight, different, but closely related. The interior processes of mountain building or orogenesis consist of the collision of the continental parts of the plates where large blocks of crust are stacked and thrust upon each other, deformed, and metamorphosed. These processes occur at depth and are not necessarily immediately accompanied by the formation of a high mountain range. The exterior of a mountain range is characterized by topographic expression at the surface. Mountain building in this sense is the uplift and dissection of rock masses to form a high mountain range. Although the geological mountain building process in the Earth's crust causes the topographical uplift of a mountain range, the coupling, however, is complex and not always directly related in time. Uplift commonly follows with a considerable delay of several millions of years the internal mountain building process at depth.

The Alps are a product of the collision of the Adriatic Plate, which at the northern edge of the African Plate formed a protrusion, with Europe. During this collision, Europe was shoved beneath the Adriatic Plate along a southward-dipping subduction zone. The Himalayas formed by the collision of India, which previously separated from Africa as an independent plate, with Central Asia. But here, the subduction zone was dipping northward, thus pulling India beneath Asia. In both cases the collision happened in the early Tertiary at about 40–50 Ma. Ensuing uplift and the formation of the high mountain ranges, however, began millions of years later.

Not all mountain ranges form strictly through continental collision. Cordilleran-type mountain ranges form after immensely long periods of subduction of oceanic crust under continental crust. In the case of the Andes and the North American Cordillera, long-term subduction and accompanying volcanic and plutonic activity along the active margin caused crustal thickening, thrusting, metamorphism, and ultimately mountainous uplift. In fact, the North American Cordillera may be the world's greatest concentration of plutonic and volcanic rocks over the last 300 million years.

Subduction at complex plate tectonc junctions such as those currently in the western Pacific may induce the formation of several chains of island arcs. Collision of these island arcs, with each other and with adjacent continents, leads to the formation of mountain ranges and new continents.

Another variation at plate margins that results in mountain building occurs when dense, cold, old ocean crust is rapidly subducted. This results in a tearing away of the lower (subducted) plate from the upper plate. The locus of the trench moves away from or "rolls back" from the upper plate, much like slowly tearing masking tape off of a ceiling. The upper plate extends and thins to fill the space left by the trench roll-back. In one type of scenario, thin ribbon-like island arcs migrate with the trench as backarc spreading creates new crust to fill in the gap. In another common case, the trench roll-back occurs next to a continent and rifts and rafts a ribbon of continental crust from the adjacent continent. Backarc spreading and new oceanic crust occur in the rift zone. In both cases, the island arc or rifted continental ribbon tends to follow the rapidly rolling-back trench. These ribbon-like wedges eventually collide with other continents forming accreted terranes. Many ancient mountain ranges including the Appalachians, Alps, and Himalayas are replete with such accreted terranes. Each phase of terrane accretion formed another phase of orogeny. Presently, the rapidly eastward migrating Antilles arc and Caribbean plate provide an example.

Long lasting subduction and subduction related magma generation leads to the creation of new continental crust. Today about one third of the Earth's surface is covered by areas of continental crust. Continental crust has not grown at a continuous rate throughout the geological past. It was particularly rapid during the late Archean about 2.5–3 billion years ago (► Chap. 10). Since that time, continental crust has probably formed at a fairly steady rate.

Plate Movements and Their Geometric Relationships

Contents

© The Author(s), under exclusive license to Springer Nature Switzerland AG 2022
W. Frisch et al., *Plate Tectonics*,
Springer Textbooks in Earth Sciences, Geography and Environment,
https://doi.org/10.1007/978-3-030-88999-9_2

Three different types of plate boundaries were introduced in ► Chap. 1: constructive boundaries (mid-ocean ridges), conservative boundaries (transform faults), and destructive boundaries (subduction zones). We also introduced the concept that describes plate motion on the spherical surface of the Earth as the rotation of a given plate around an axis that passes through the center of the Earth. Therefore, every plate movement or relative movement between two plates can be described by the location of the pole of rotation and the angular velocity of the movement. Although this chapter uses geometry to explain the movement of plates, the major concepts presented here can be followed without using the equations with trigonometric functions.

There are several important rules and concepts of plate motion: (1) Relative movement between two plates is always parallel to transform faults. Though deviation from this rule can occur in nature at certain regions along transform faults, especially where they have a compressional or extensional component (► Chap. 8), these can be neglected when considering the theoretical concepts presented below. (2) Although oblique movement is possible, movement along divergent plate margins tends to be perpendicular to the plate boundary because this condition generates maximum stability of the spreading axis. (3) At convergent plate boundaries there is no such preferred orientation; subduction movement ranges from oblique to perpendicular to the plate boundary.

2.1 Helpful Transform Faults

Transform faults are very useful in the construction of plate movements. If a transform fault is precisely oriented, motion is in a pure strike slip sense and has a geometric relationship to the common pole of rotation of the two plates separated by the fault: the fault follows a small circle around the pole. Small circles form when the surface of the Earth's sphere is cut by planes that do not pass through the center of the Earth. In contrast, great circles, which are larger than small circles, are formed when the surface of the Earth's sphere is cut by a plane that passes through the center of the Earth (the largest possible circles on the surface of a sphere and the shortest distance between two points on the surface of a sphere). These geometries can be illustrated by examining Earth's latitude–longitude grid—all longitudes and the equator are great circles whereas all other latitudes are small circles (◻ Fig. 2.1). Like the arrangement of latitudes with respect to the geographic pole, transform faults are arranged concentrically around the common pole of rotation of two plates. Great circles vertical to transform faults meet in the common pole of rotation, as do the longitudes of

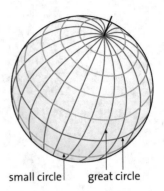

◻ **Fig. 2.1** Possible great circles (red) and small circles (green) on the surface of a sphere

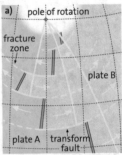

◻ **Fig. 2.2** Construction of the common pole of rotation of two diverging plates (Morgan, 1968). **a** Principle of construction: normals (deep yellow) to the transform faults meet in the pole of rotation; the transform faults follow concentric small circles around the pole of rotation. **b** Determination of the common pole of rotation of the South American and African plates using transform faults at the Mid-Atlantic Ridge between Brazil and Western Africa

the Earth in the geographic pole, i.e., the pole of rotation of the Earth. Following this concept, poles of rotation between two adjacent plates can be constructed (◻ Fig. 2.2).

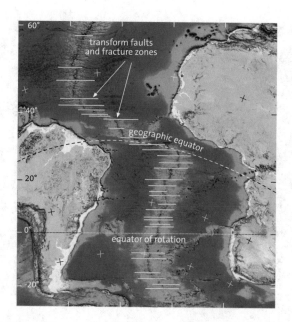

Fig. 2.3 Geometric relationships between the common pole of rotation and transform faults respectively fracture zones in their prolongation in the southern Atlantic (LePichon, 1968). The Mercator projection is related to the common pole of rotation between South American and African plates (58° N, 36° W, see ■ Fig. 2.2). In ideal cases, the transform faults run parallel to the equator of projection. The mid-ocean ridge tends to become orthogonal to this

The correct construction of the pole of rotation can be controlled by the Mercator projection of the Earth's surface. In a Mercator projection the poles are at infinity; therefore, it is not possible to display the polar regions in this projection. The grid is orthogonal, i.e., longitudes and latitudes are linear and vertical to each other. If we use the common pole of two plates as the pole of a Mercator projection, then all transform faults which separate these two plates must be parallel to the equator of this projection (■ Fig. 2.3). Irregularities cause small deviations from the ideal orientation of the transform faults. Mid-ocean ridges in this projection are exactly orthogonal to the transform faults if the divergence of the ocean floor is vertical to the spreading axes.

The poles of rotation of the two sides of both the Mid-Atlantic Ridge and the East Pacific Rise are located near the geographic pole (LePichon, 1968). Both ridges are more or less north–south oriented and their transform faults east–west oriented. The preferred north–south orientation of the mid-ocean ridges is probably a result of a long lasting relation between plate drift and the rotation of the Earth.

2.2 Relative Movements and Triple Junctions

Relative movements between two plates are different at different places along their common boundary, only the angular velocity remains the same. Along any given

transform fault, the movement is the same because all points along the fault are at the same distance to the common pole of rotation of the two plates. The spreading rate along the boundary between two diverging plates becomes smaller with decreasing distance from the pole of rotation; at the pole itself, it is reduced to zero (■ Fig. 2.4). Correspondingly the same is valid for rates of convergence between converging plates. The largest movement is at a distance of 90° from the common pole of rotation along the only possible transform fault which represents a great circle: the equator of rotation.

The decrease of velocity of the relative movement from the equator to the common pole of rotation follows the equation

$$V_\alpha = V_0 \cos \alpha$$

where v_α represents the velocity of the relative movement at a small circle which is at a distance of the angle a from the equator (■ Fig. 2.4). If the velocity at the equator of rotation is v_0, and is, for instance, at a distance of 30° from the equator ($\alpha = 30°$)

$$V_{30} = \frac{\sqrt{3}}{2} V_0 \text{ or } 0.866 \, V_0,$$

at a distance of 60° from the equator

$$V_{60} = \frac{1}{2} V_0 \text{ or } 0.5 \, V_0$$

and at the common pole of rotation $v_{90} = 0$.

The geometry of the plates on Earth contain a number of triple junctions, i.e., locations where three plates and three plate boundaries meet in a point. The nature of triple junctions ranges from simple to complex. The simplest case is a triple junction where three oceanic ridges meet; this kind of triple point is called an RRR triple point: R means (mid-ocean) ridge. If the three spreading axes have the same diverging rates, a stage of balance will evolve with three plate boundaries in an angle of 120° to each other (■ Fig. 2.5). A possible scenario would be a triple junction situated over a uniformly rising mantle diapir unless restraints of other plates cause a distortion.

Assuming that each single plate moves perpendicularly away from the ridge with the same velocity v/2 the geometric relation shown in ■ Fig. 2.5 will be the result. The vectors which are orthogonal to the spreading axes only display the relative movement between two adjacent plates. Looking at the triple junction at a fixed point, plate B moves with the velocity of v:√3 towards east–northeast (N60°E). Because the other two plates move in an analogous way, the movement of the two halves of a point (P′ und P″), which have been formed

2

■ **Fig. 2.4** Map showing different relative plate motion velocities along the destructive plate boundary between the Caribbean and Cocos plates. Velocity is dependent on the distance to the common pole of rotation of the two plates which is located northwesterly off the field of the map (DeMets et al., 1990). The sphere schematically indicates how relative plate velocity increases between of two diverging plates with increasing distance from the rotational pole

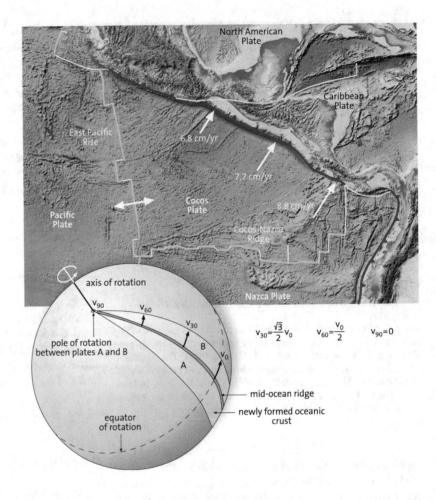

$$v_{30} = \frac{\sqrt{3}}{2} v_0 \qquad v_{60} = \frac{v_0}{2} \qquad v_{90} = 0$$

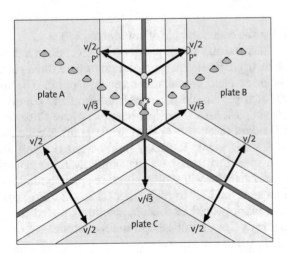

■ **Fig. 2.5** Movement direction and velocity of three plates at an RRR triple junction (three ridge systems meet in a point). The relative movements of two plates against each other do not coincide with the absolute plate movement that is indicated, e.g., by volcano chains formed at a (stationary) hot spot. Stripes of oceanic crust of the same age have the same colors

as a common point (P) at one of the ridges, appears as a relative movement orthogonal to the ridge. The effec-

tively covered distance, however, is larger than the distance to the ridge.

An example for such a situation exists in the Southern Atlantic, where the hot spot of Tristan da Cunha, located near the mid-ocean ridge, feeds volcanoes. The hot spot produced volcanoes on both the South American Plate and the African Plate as they drifted away from the spreading center on their respective plates (■ Fig. 2.6). Traces of these hot spots are thus visible in volcano chains oriented towards NW (Rio Grande Ridge) and NE (Walvis Ridge) respectively. If we consider the hot spot in this example to be point P, then volcanoes in a given distance from the hot spot are the points P′ und P″, which have the same age. The age of the volcanoes increases with distance from the hot spot. This example shows that the relative plate movement is rather close to east–west, as also indicated by the orientation of the transform faults and their traces. The movement relative to the hot spot, however, is NW directed on the South American Plate and NE directed on the African plate. If the hot spot is stationary, the Rio Grande Ridge and the Walvis Ridge, respectively, indicate the absolute movement of the plate. Today the hot spot is located at a certain distance to the mid-ocean ridge on the African Plate. Therefore, new sub-

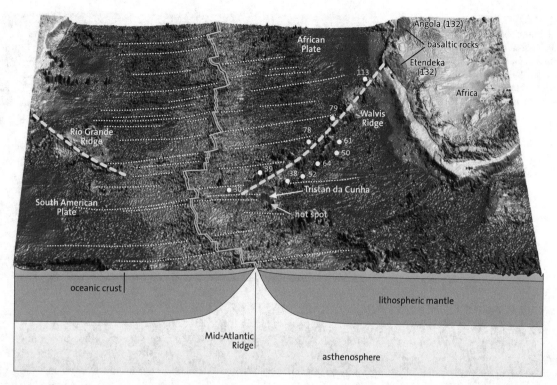

marine volcanoes develop only in prolongation of the
Walvis Ridge.

Relative movements of plates can be illustrated by
vector diagrams. ◘ Figure 2.7 follows the method of
McKenzie and Parker (1967). If an observer stands on
plate A, plate B moves away from him/her with velocity v towards the east ($_A v_B$). If s/he stands on plate B,
plate C moves with the same velocity towards the SSW,
standing on plate C plate A moves with the same velocity towards the NNW. If the observer returns to the
origin, the conditions of the vectorial equation are fulfilled. Using this diagram, relative movements of three
plates around a triple junction can be depicted. If two
vectors are known the third results from these.

$$_A v_B + _B v_C + _C v_A = 0$$

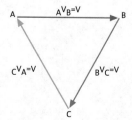

◘ **Fig. 2.7** Vectorial diagram used to determine the relative movement between three plates around a triple junction as illustrated in
◘ Fig. 2.5. The vector $_A v_B$ indicates the movement direction and velocity of plate B relative to plate A

An *RRR triple junction* exists in the Indian Ocean between African, Indo-Australian, and Antarctic plates
(◘ Fig. 2.8). Drifting of plates does not coincide with
the symmetrical model presented above, but the geometric relationships are fulfilled as can be demonstrated by the vectorial diagram.

Numerous other possible triple-junction geometries
exist, including where three subduction zones meet, as
can be observed in the western Pacific, where the Eurasian, the Pacific, and the Philippine Sea plates join at
a single point (◘ Fig. 1.2). However, not all theoretical triple-junction geometries are possible; for example,
three converging transform faults are geometrically impossible.

Two RTF Triple Junctions off North America

An RTF triple junction is a rather complex situation where a constructive (R stands for ridge), a destructive (T stands for trench), and a conservative
plate boundary (F stands for [transform] fault) meet.
◘ Figure 2.9a represents a theoretical case proposed
by McKenzie and Parker (1967). Plate A is being subducted towards the south beneath plate B with the velocity v (teeth along the subducting plate boundary
always point at the overriding plate). Plate C moves
away from B towards west with $2 \cdot v/2 = v$. The

2

■ **Fig. 2.8** Map of RRR triple junction between African (Af), Indo-Australian (In), and Antarctic (An) plates in the Indian Ocean. The vectorial diagram describes the different spreading rates and directions

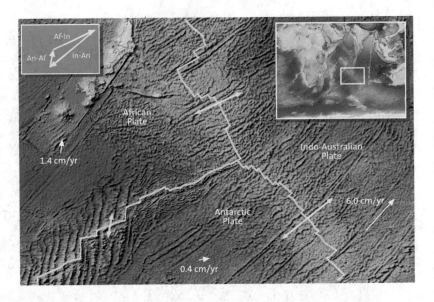

result is a dextral strike slip movement with the velocity of $v \cdot \sqrt{2}$ along the transform fault (dextral means that if an observer stands on one of the two plates and looks towards the other plate across the fault, then this plate moves towards the right).

The geometric relation illustrated in ■ Fig. 2.9a; however, is not stable and thus can only exist for a short time. New oceanic crust formed at the spreading axis between plates B and C at the triple junction (point P) must shift towards the NW parallel to the transform fault (P') if it belongs to plate C. Otherwise a gap would develop at this location. In order to move away in an eastward direction perpendicular to the common spreading axis (looking from the NW-ward moving point P'), the other half of P (P") must move by exactly the same amount towards the NE. Since plate B is not being subducted, P" remains

■ **Fig. 2.9** RTF triple junction and its related vectorial diagram (McKenzie and Parker, 1967). **a** Instable situation. **b** Stable constellation of the triple junction if plate A subducts beneath plate B. **c** Stable constellation if plate B subducts beneath plate A

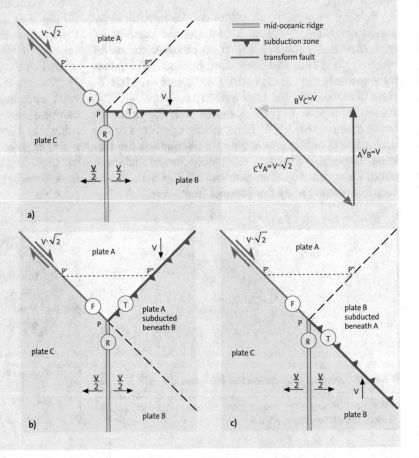

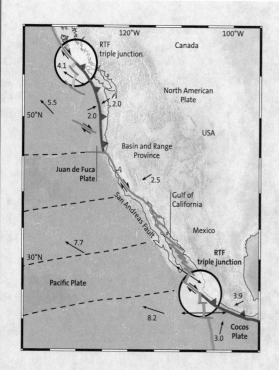

Fig. 2.10 Map of plate tectonic setting along the western coast of North America showing two RTF triple junctions. Both triple junction situations are in accordance with that of Fig. 2.9c. Motion velocities in cm/yr

at the surface controlling the course of the plate boundary along the subduction zone. The plate boundary A/B thus takes a SW–NE orientation. In this manner, the triple junction may remain stable over a longer period. Subduction proceeds in an angle of 45° oblique to the plate boundary A/B (Fig. 2.9b).

A different plate geometry develops when plate B subducts beneath plate A (Fig. 2.9c). P″ is being subducted beneath plate A immediately after its formation. Since plate A dislocates towards the SE with respect to the triple junction, the plate boundary A/B realigns towards the SE and results in a prolonged course (NW–SE) of the plate boundaries A/C and A/B. This geometry occurs in two places along the western coast of North America (Fig. 2.10). The East-Pacific Rise (R), the Central American subduction zone (T), and an equivalent of the San-Andreas transform fault (F) meet at the entrance of the Gulf of California in a complex triple junction. Further north, off the Canadian Pacific coast, this geometry is repeated by the North American, the Pacific, and the mostly subducted Juan de Fuca plates.

The rotational poles of three plate pairs whose boundaries meet in a triple junction have a common geometric relationship. If we know the orientation and the angular velocities of two rotational axes, the orientation and angular velocity of the third axis can be calculated using a vectorial equation. The angular velocities ω of the three pairs of plates have the following relation:

$$_A\omega_B +_B \omega_C +_C \omega_A = 0.$$

The rotational poles of all three plate pairs are located on a common great circle (Fig. 2.11).

2.3 Relative Plate Velocities—Past and Present

Relative plate movements from the geologic past can be calculated by two methods: (1) by using the magnetic stripes on both sides of a mid-ocean ridge and (2) by tracking the patterns and ages of volcanic chains on a given plate that moved over a hot spot (▶ Chap. 1). From the magnetic stripe pattern, the width of a stripe of oceanic crust formed within a given time span can be measured (Fig. 2.12). This can be used to calculate the average spreading rate of the ocean floor for the

given time span. However, it is necessary to know the age or respectively the time span of the magnetic stripes (Fig. 1.6). The age data can be deduced from a comparison of the stripe pattern with the dated magnetic time scale or by the determination of fossil remains of micro-organisms from sediments resting directly above the ocean floor. Stripes located at two plates which are connected at the mid-ocean ridge become narrower towards the common pole of rotation (Fig. 2.12) thus reflecting the above mentioned relation between the velocity of the plate movement and the distance to the common rotational pole of two plates.

The age of volcanoes formed above a hot spot may be determined using various age dating methods. The age difference and the distance between a given volcano and the presently active hot spot or the distance between two different volcanoes of the volcano chain may be used to calculate the average drifting rate of a plate for the given time span. For instance, the conspicuous kink in the volcanic chain formed by the Hawaiian hot spot (see Fig. 1.5) is 3300 km from Hawaii. The age difference between the volcanoes at the kink and those at Hawaii equals 43 m. y. This results in an average movement of 7.7 cm/yr of the Pacific Plate in relation to the hot spot.

2

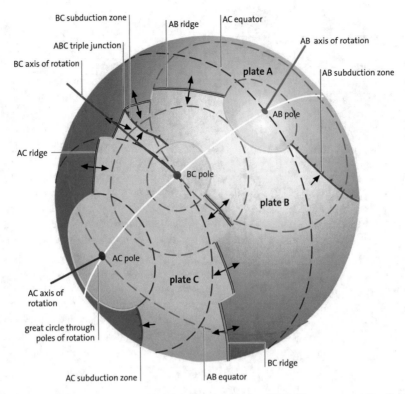

BC subduction zone

ABC triple junction

BC axis of rotation

AB ridge

AC equator

AB axis of rotation

AB subduction zone

plate A

AB pole

AC ridge

BC pole

plate B

AC pole

plate C

AC axis of
rotation

great circle through
poles of rotation

BC ridge

AC subduction zone

AB equator

■ **Fig. 2.11** Geometric relationship of three plates around a triple junction (Dewey, 1972). The common poles of rotation of the three pairs of plates are all located on one great circle (white line)

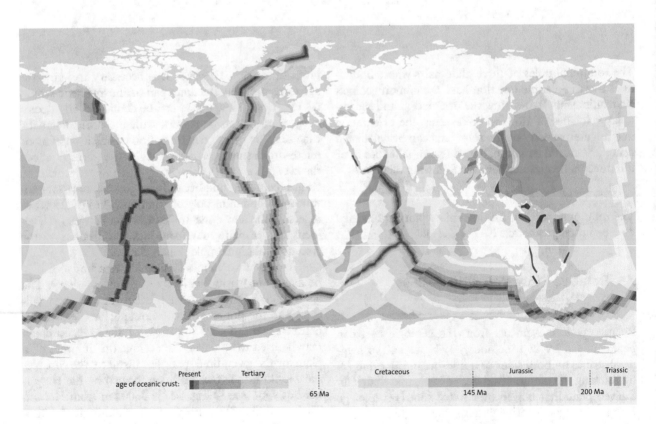

age of oceanic crust:

Present   Tertiary   Cretaceous   Jurassic   Triassic

65 Ma   145 Ma   200 Ma

■ **Fig. 2.12** Map showing distribution of ages of oceanic crust. The oldest oceanic crust is of Jurassic age and is located in the NW Pacific (ca. 185 Ma) and near the edges of the Central Atlantic (ca. 175 Ma). Small fragments of older oceanic crust are captured between continental blocks in the Mediterranean Sea.

Present velocities are calculated from young time intervals. The model NUVEL (Northwestern University VELocity model, established by the Northwestern University in Evanston, Illinois) uses average values of the last 3 million years as the best approximation for the present plate movements. In most cases, fairly good results are obtained. The Atlantic Ocean, currently the ocean with the slowest production of new crust, is spreading (measuring the total separation on both plates) at 2 cm/yr south of Iceland, and 4.5 cm/yr at the point of maximum spreading in the Southern Atlantic (LePichon, 1968; DeMets et al., 1990). Along the East-Pacific Rise, velocities as great as 15 cm/yr are achieved (▢ Fig. 1.2); these are actually the greatest spreading rates between two plates on Earth. Spreading rates in the Indian Ocean approach 7.5 cm/yr. Velocities at subduction zones and transform faults have been calculated using the plate movement pattern derived from spreading rates and hot spot activities. Subduction rates approaching nearly 10 cm/yr have been calculated in the western Pacific and movement along the San–Andreas transform fault achieves a total movement of nearly 6 cm/yr.

2.4 Direct Measurement of Plate Movements

Since the 1970s several attempts have been made to measure plate movements directly using Satellite-Laser-Radar (SLR) technology and Very-Long-Baseline-Interferometry (VLBI). SLR sends out radar pulses which are reflected by satellites and then detected. Distances can be measured with this method to an accuracy of centimeters. The VLBI method uses cosmic radio signals as the characteristic patterns of signals produced by quasars (very far away luminaries which constitute "quasi-stellar radio sources"—quasar is an acronym) that can be detected at several stations on Earth. From the difference of the arrival times of the same signals it is possible to determine the distance of the stations to an accuracy of millimeters if the measurements are made over time spans of several years. These techniques have been replaced during the last 10–15 years by measurements with the Global Positioning System (GPS).

Measurements using GPS have an accuracy of less than 1 mm. The current GPS is based on 24 satellites. In principle it is the measurement of waves and their Doppler effect as the satellite moves in relation to the gauging station at the surface. Each measurement occurs simultaneous with at least three satellites, more if possible. Exact synchronisation of the measurement cycles is essential. The GPS receivers measure with two frequencies of waves to enhance the accuracy. To min-

imize errors, the measurements of two or more simultaneously working receivers are merged (method of interferometry). The course of the satellites must be known with an accuracy of a decimeter and time measurement is achieved to within an accuracy of one part in a billion (during this time span light covers a distance of 30 cm). Moreover, the parameters of the Earth's rotation must be determined constantly with high precision. The location of the pole is known to within an error of less than 10 mm.

Following installation of the global reference system, it was possible after a period of only two years to determine plate movements to within an accuracy of millimeters (Reigber and Gendt, 1996). To determine the movement of a plate, at least three gauging stations must be installed on that plate. However, changes in the distance between two stations located on different plates can be determined directly. Comparing the relative movements between stations at Hawaii (Kokee Bay) and in the Bavarian Forest in Germany (near Wettzell) and with a number of gauging stations on other plates, an amazingly good accordance is achieved with values calculated using the NUVEL model. However, some significant deviations do exist (▢ Fig. 2.13a).

For instance, the deviations are significant when examining the values between Hawaii and stations at the western boundary of the North American Plate. This is a tectonically complex area along the boundary between the Pacific and North American plates where young compressive and tensional deformation becomes manifest in folding and basin formation. This generates individual movements of small blocks that do not coincide with the movements of the larger plate to which they belong (see ▶ Chap. 8).

Three points located on the Pacific Plate (Hawaii and Tahiti in the open Pacific Ocean, California west of the San Andreas fault) should not have any relative movement amongst each other following the strict definition of plates (NUVEL values: 0 mm; ▢ Fig. 2.13a). Nevertheless, Tahiti and California move away from Hawaii by an amount of several millimeters per year. As stated in ▶ Chap. 1, this suggests that the plates may be subject to a certain internal deformation related to earthquake activity. However, amounts of movement of this "intraplate tectonics" are at least an order of magnitude smaller than those at the plate boundaries. Such intraplate movements were also detected east of the Atlantic. The Bohemian massif, which encompasses the station at Wettzell, moves closer to Scandinavia by an amount of approximately 3 mm/yr although both regions belong to the Eurasian Plate. Convergence should be expected between Wettzell and southern Africa, because the Alpine–Mediterranean mountain range is still under compression (calculation with NUVEL indicates a convergence of 8 mm/yr); however,

2

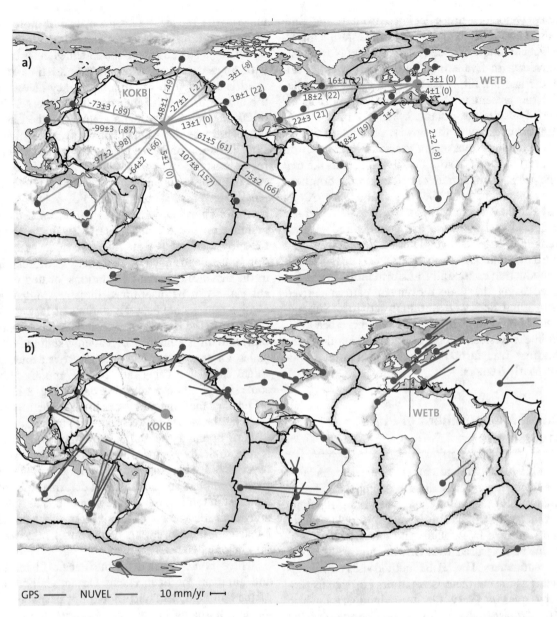

◻ Fig. 2.13   a Map showing changing distances as determined by GPS measurements (Reigber and Gendt, 1996). Reference points are Kokee Bay at Hawaii (KOKB) and Wettzell in the Bavarian Forest in Germany (WETB). The data (red numbers, in mm/yr) cover the time span from January 1993 to May 1995. Values calculated using the NUVEL model are given for comparison (blue numbers in brackets). Negative values indicate convergence between the gauging stations. **b** Comparison of absolute plate motion velocities calculated using the GPS technique and the NUVEL model (Reigber and Gendt, 1996). The length of the arrows is related to the velocity (after Spektrum der Wissenschaft 01/1996, p. 116/117)

GPS data show a slight drifting apart of the two points. This indicates extensional movements within the Mediterranean and/or the African continent.

To measure *absolute plate motions,* a stationary coordinate system is required that is centered in the Earth's center of gravity and defined by the geographic pole (z axis) and the spring point of the ecliptic (x axis). Directly compared absolute plate motions calculated by GPS and with NUVEL show a rather extensive agreement (◻ Fig. 2.13b). Arrows indicating the movement of plates demonstrate the NW and NE drift of the continents at both sides of the Atlantic as it was detailed in the example of the Tristan da Cunha hot spot (◻ Fig. 2.6). The fast divergence between Pacific and Nacza plates and the partly complex convergent movements along the subduction zones in the western Pacific are expressed as well.

2.5 Apparent Contradictions in the Plate Motion Pattern

During early stages of plate–tectonic studies, the boundary between North America and Asia (Eurasia) was considered to be problematic because the plate motion pattern is apparently contradictory and uncertain in eastern Siberia. We suggest that the Eurasian and the North American plates belong to one giant plate, Laurasia. Overall the boundary between the two plates in Northeast Asia is diffuse and difficult to define—only very small relative movements are known from mountainous areas in the region (◘ Fig. 2.17). However, these movements are more likely to represent intraplate tectonics and not a plate boundary. The northern Mid-Atlantic Ridge passes through the Arctic Ocean with spreading rates less than 1.5 cm/yr. In the area of the Verkhoyansk Mountains in eastern Siberia tensional as well as compressional movements occur. In this area, the location of the rotational pole of the Arctic spreading zone is assumed and it is a zone of transition from tension to compression. The tensional movement of approximately 1 cm/yr in the Arctic Ocean is compensated by a similar small convergent movement in central Japan and along the eastern coast of northern Japan (about 0.9 cm/yr). The Laurasian Plate (◘ Fig. 1.2) consists of a large plate that comprises most parts of Eurasia and Laurentia (= North America; the name is derived from the latinized word for the St. Lawrence River); the two continents are connected in the area of the Verkhoyansk mountains. Consequently, we propose the term Laurasian Plate because of the problem of unclear boundaries in Northeast Asia. The Laurasian Plate thus defined is subdivided into the Eurasian and North American plate components (◘ Fig. 2.17).

Fault–Plane–Solutions of Earthquakes

The orientation of planes of movement at plate boundaries can be deduced from earthquake data. In the case of an earthquake triggered at a fault plane, the two blocks move by creating an instantaneous offset up to several meters. This results in the generation of the two types of seismic body waves. Primary (P–) waves oscillate in the longitudinal direction of propagation. They are faster than secondary (S–) waves that oscillate transversally. If all of the seismic data from a given earthquake collected around the Earth are put into a diagram, the quadrants and the two separating planes can be determined with their spatial orientation (◘ Fig. 2.14). In the two quadrants that are in the direction of movement of each block, the first motion of the primary waves is away from the earthquake focus and an observer on the Earth's surface first receives a push; the wave starts with a compressive movement (compressive first motion shown as black quadrants in ◘ Fig. 2.14). First motion in the other two quadrants, shown in white, is in the opposite direction; it starts with a tension and is dilatational. Each of these motions is registered by the seismograph.

Using the principles discussed above, the quadrants and the two separating planes can be determined with their spatial orientation (◘ Fig. 2.14). One of these planes represents the slip plane generated by the earthquake; the other one is an aiding plane that has no use in nature. However, initially it is not possible to decide which one of these two planes was the slip plane. Commonly this can be

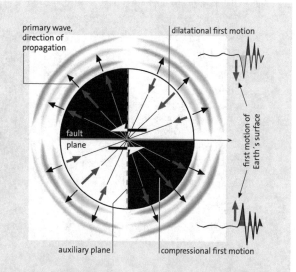

◘ **Fig. 2.14** Principle of fault-plane-solutions of an earthquake hypocenter. The fault plane of the earthquake (orthogonal to the paper plane) and an orthogonal virtual aiding plane define four quadrants. First motions of the primary waves oscillating in the propagation direction and expressed by vertical motions in the soil indicate the sense of movement of the blocks displaced during the earthquake. Seismograms subdivide two quadrants with compressive and two quadrants with dilatative first motion

deduced from geological observations if the approximate orientation of a fracture zone is known. On the other hand, careful analysis of seismic data generated by the aftershock activity following every large earthquake provides the opportunity to identify the slip plane because of

2

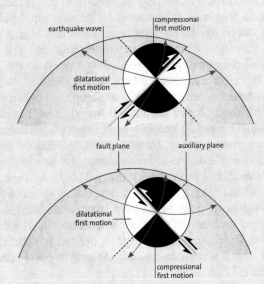

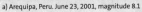

□ **Fig. 2.15** Ambiguity of fault-plane-solutions illustrated by an earthquake that produces an overthrust. Illustration is a schematic vertical section through the Earth

the shift of the seismic centers. If the slip plane is known, the sense of movement is easily detected (□ Fig. 2.15). Direction of movement in the slip plane is orthogonal to the aiding plane. A process similar to that used for analyzing P–waves can also be used for the analysis of S–waves that oscillate orthogonal to the direction of propagation. Using this method, which is called fault-plane-solution, the orientation of a slip plane and the sense of movement can be determined with high accuracy. Fault-plane-solutions allow for a reconstruction of plate boundaries and their movement patterns (□ Fig. 2.16). The analysis of earthquake first motion data impressively confirmed the concept of three different types of plate boundaries (Sykes, 1967).

□ **Fig. 2.16** Examples of fault-plane-solutions of recent earthquakes at the three different kinds of plate boundaries. Quadrants of the first motions of earthquake waves are shown in stereographic projections of the lower hemisphere. Earthquake **a** at a subduction zone (Peru), **b** at a mid-ocean ridge (Atlantic), and **c** at a transform fault (North Anatolian Fault). Black quadrants—compressive first motion. Red line—active fault plane. Earthquake data are from the internet catalogue of the National Earthquake Information Center (US Geological Survey)

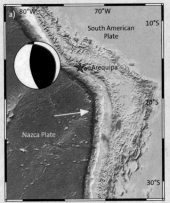

a) Arequipa, Peru. June 23, 2001, magnitude 8.1

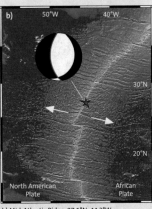

b) Mid-Atlantic Ridge, 27.1°N, 44.3°W. Dec. 13, 2001, magnitude 5.9

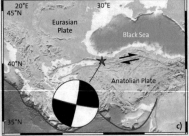

c) Izmit, Turkey. Aug. 17, 1999, magnitude 7.4

The African Plate has also been considered to be problematic. This plate is surrounded on three sides, west, south and east, by mid-ocean ridges (□ Fig. 1.5), and consequently has enlarged on these three sides for millions of years. When viewed from the perspective of the mid-ocean ridges, which were once thought to be stationary because they are fed by rising currents from the Earth's mantle, a space problem evolves. However, it is currently known from the technique of seismic tomography (see box) that convec-

tional currents are extremely complicated and that hot uprising and cool descending branches of currents are commonly oblique and sinuous. It is even possible that mid-ocean ridges are fed by horizontal currents that emit magma along a zone of weakness to the surface. Therefore, the geometry around the African Plate suggests that not all mid-ocean ridges are fed by vertically rising convectional currents. It is clear that the Mid-Atlantic and the northwestern Indian Ridge move away from each other. GPS data

(▣ Fig. 2.13) indicate that the Mid-Atlantic Ridge moves very slowly eastward since the eastward-directed component of the African Plate is larger than the westward-directed one of the South American Plate. Therefore, the Indian Ridge must migrate faster towards the east at several centimeters per year. This is confirmed by the seismic to-mography: A hot uprising zone in the upper mantle under the Mid-Atlantic Ridge is steeply inclined towards the west; under the Indian Ridge, a similar situation occurs albeit the westward inclination is at a shallower angle. This indicates a faster eastward migration of the ridge (▣ Fig. 2.18).

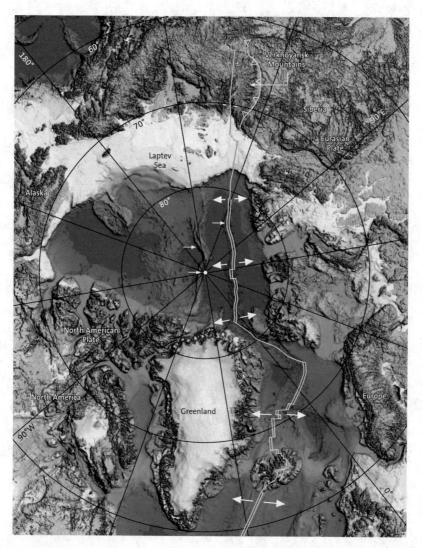

▣ **Fig. 2.17** Polar map view showing geometry of the Laurasian Plate as defined in the text. The plate boundary between the Eurasian and North American plates acts like a tear as the spreading center passes through the Arctic Ocean and opens into the Atlantic Ocean. Above the Gakkel Ridge, the two plates behave as one as they merge without a definable boundary across the Russian Arctic region into the Verkhoyansk Mountains of Siberia

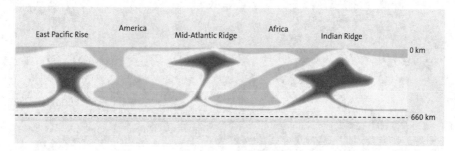

▣ **Fig. 2.18** Schematic E–W cross-section through the upper mantle to a depth of 660 km that shows mantle currents in the Pacific, Atlantic, and Indian oceans (Anderson and Dziewonski, 1984). Hot anomalies (red) indicate rising mantle currents that are expressed at the Earth's surface by mid-ocean ridges. Cool, descending mantle currents (blue) are characterized by high-velocity seismic waves. Because the Indian Ridge is drifting eastward from Africa, no subduction zone has developed on the east coast of Africa

Seismic Tomography

The method of seismic tomography analyzes vast amounts of seismic wave data that have passed through the three-dimensional space of the Earth's mantle. This data allows seismologists to determine whether a particular zone of the mantle is slightly warmer or cooler than its surroundings (Anderson and Dziewonski, 1984). Travel times measured in kilometers per second (km/s), are compared from waves that have different paths but propagate across similar areas within the mantle. The comparison allows researchers to determine whether the waves have the expected velocity in the area considered or whether they are accelerated or decelerated. If the rock is hotter, seismic waves decelerate; if it is cooler, they accelerate (◘ Fig. 2.19). Seismic tomography uses different types of waves in the analysis. The differences between the velocities vary and depend upon the type of wave. Cooler regions of the mantle can be correlated with descending currents, hotter ones with rising currents. In this manner, a three–dimensional picture of the currents in the Earth's mantle is generated.

This pattern generated by these currents is extremely complicated (◘ Fig. 2.20). However, in regions where long–lasting subduction has occurred such as the western parts of North and South America and the Alpine–Himalayan Mountain Belt, cold regions plunging obliquely into the lower mantle (◘ Fig. 2.21) to the core/mantle boundary can be detected (◘ Fig. 2.22). However in complex regions such as the western Pacific, patterns are relatively un clear. Although subduction zones can be detected seismically in the upper mantle only down to approximately 660 km depth, the cold, down plunging currents induce cold patterns into the

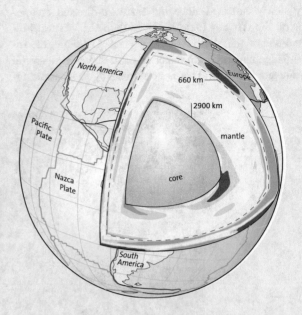

◘ **Fig. 2.20** Three-dimensional illustration showing the results of seismic tomography in the Earth's mantle (Dziewonski, ▶ https://igppweb.ucsd.edu/~shearer/mahi/SEDI/main/images/Tomo_earth.jpg)

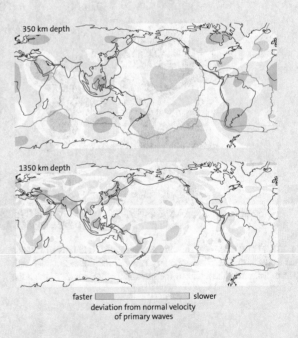

faster ▭▭▭▭ slower
deviation from normal velocity
of primary waves

◘ **Fig. 2.21** Maps showing deviations of velocities of the seismic primary waves (P–waves) from a normal value in horizontal cuts through the upper mantle. Upper map shows patterns at a depth of ca. 350 km (Anderson and Dziewonski, 1984); lower map shows the pattern in the lower mantle at a depth of ca. 1350 km (Grand et al., 1997). Orange areas indicate hotter, rising zones, blue areas cooler descending ones. Plate boundaries are shown for reference

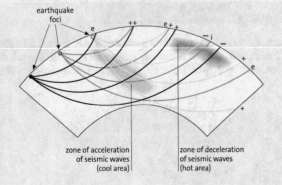

arrival time of earthquake waves: e = as expected;
i = indifferent (partly decelerated, partly accelerated);
− = decelerated; + = accelerated.

◘ **Fig. 2.19** Principle of seismic tomography. Temperature anomalies in the Earth's mantle cause increasing and decreasing travel times of seismic waves. From a large number of travel time measurements, warm and cool bodies can be segregated

lower mantle. The sub ducted plates are partly absorbed into the upper mantle, but more resistant parts with high density can penetrate to the lower mantle. However, penetration of the lower mantle occurs at far lower velocities; sinking rates of 1–1.5 cm/yr have been calculated beneath South America (Grand et al., 1997). The high resistance of the lower mantle leads to compression and deformation of the remains of the subducted plates (☐ Fig. 2.22).

Below the mid-ocean ridges, hot currents cannot be traced to great depths because the ridges are fed only by the upper part of the upper mantle. At 350 km below the East–Pacific Rise, the distribution of temperatures is irregular (☐ Fig. 2.21 top); at this depth, temperatures would be expected to be higher. Below the equatorial Mid-Atlantic Ridge and the eastern with seismic tomography because the relatively Indian Ridge tempertaures are unexpectedly thin, finger-shaped protuberances are difficult to low. Individual hot spots are difficult to detect record with this method.

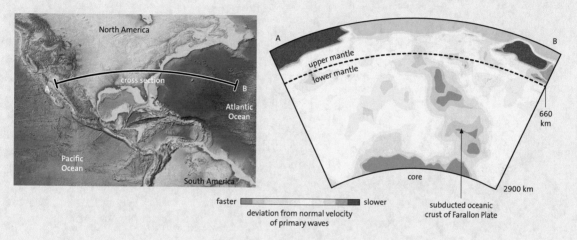

☐ **Fig. 2.22** Crosssection through the Earth's mantle beneath North America showing results from seismic tomography (Grand et al., 1997). The cool area (blue) obliquely plunging towards the east is interpreted as the remains of oceanic lithosphere of the Farallon Plate that was subducted beneath the North American Plate over a period of about 100 million years

Continental Graben Structures

Contents

© The Author(s), under exclusive license to Springer Nature Switzerland AG 2022
W. Frisch et al., *Plate Tectonics*,
Springer Textbooks in Earth Sciences, Geography and Environment,
https://doi.org/10.1007/978-3-030-88999-9_3

A continental graben structure or rift is a narrow, elongated, fault-bounded structure in the Earth's crust (◘ Fig. 3.1). Grabens consist of a central axial depression flanked by steep walls and elevated shoulders that plunge steeply into the rift axis and slope gradually towards the exterior (◘ Fig. 3.2). The most famous example is the East African rift system. Rift systems may be cut and apparently offset by transform faults; examples include the Upper Rhine Graben in Central Europe and its southern continuation in the Bresse and Rhône grabens (see below). Graben structures occur in regions where the crust and lithospheric mantle are extended and thinned (◘ Fig. 3.3). Broad regions of extension are typically expressed by numerous grabens and intervening higher horst blocks such as the Basin and Range Province in western North America. Graben systems also occur in oceanic crust along mid-ocean ridge systems and will be discussed in ▶ Chap. 5.

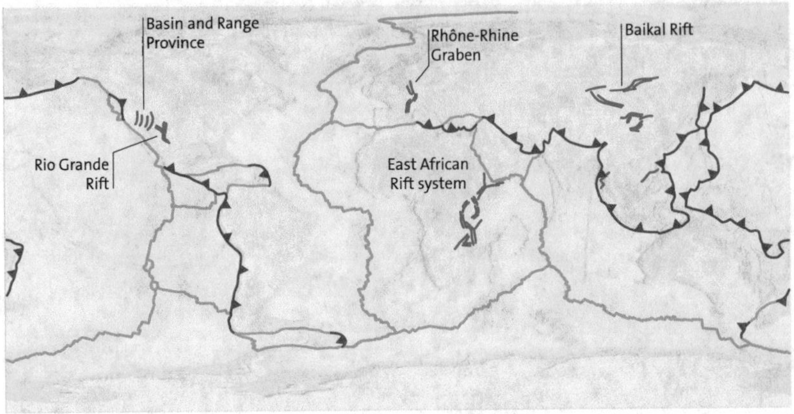

◘ **Fig. 3.1** Tectonic map of Earth showing large, young graben systems

The amount of extension across a graben varies considerably ranging between approximately 5 km across the Upper Rhine Graben to 50 km across the Rio Grande Rift in New Mexico. The brittle extension, generated by fracturing associated with earthquake activity in the upper crust, extends downward to a depth of approximately 15 km. At greater depths, ductile flow occurs without fracturing the rocks; rather, deformation takes place as solid-state plastic flow. Graben subsidence is accommodated along normal faults that dip towards the central graben axis at angles of 60–65° (◘ Fig. 3.2): the hanging wall, the block located above any point of the fault plane, moves downwards with re-

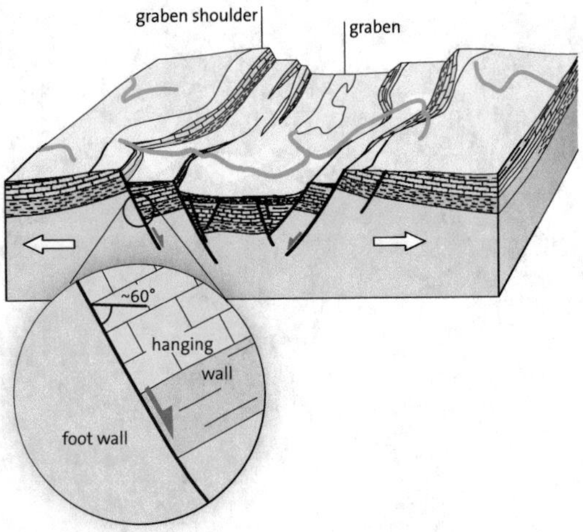

◘ **Fig. 3.2** Schematic block diagram of a graben

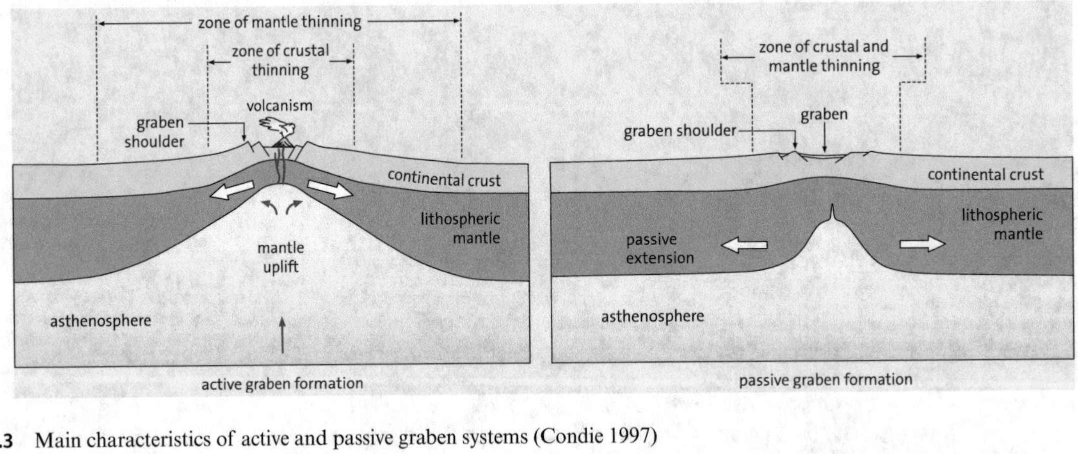

◘ **Fig. 3.3** Main characteristics of active and passive graben systems (Condie 1997)

spect to the foot wall and causes the subsidence of the graben. Normal faulting is linked to horizontal extension orthogonal to the graben axis.

In continental settings, as the lithosphere extends, the asthenosphere tends to rise (■ Fig. 3.3) and heat-flow rate increases; as a consequence, melting in the uppermost asthenosphere or overlying lithospheric mantle may occur. The melts penetrate the crust and feed volcanoes at the surface or form magma chambers at depth. Because the magmas are derived directly from the mantle, they are basaltic in composition, hence the close association of basaltic volcanism and graben rifting. However, when magmas are trapped at depth and accumulate in magma chambers, they potentially undergo additional processes that result in change of magma composition. Assimilation of adjacent continental crust and magmatic differentiation by removal of mafic minerals that have high melting points and sink to the bottom of the magma chamber produce intermediate to granitic melts. These various magmatic processes explain why many rift areas are associated with volcanism and plutonism of various compositions.

Active and Passive Graben Structures

Based on the relations between topographic expression and method of formation, Condie (1997) defined two classes of grabens, active and passive (■ Fig. 3.3). *Active grabens* are generated by upwelling of the asthenosphere, commonly over hotspots; the overlying mantle lithosphere and crust respond to this process and both are thinned as a result. The mantle lithosphere and lower crust deform plastically and the upper crust is faulted to form the graben structure; both are thinned and basaltic volcanism is generated. Extension of crust at an active graben structure is much wider in the deeper ductilely reacting part of the crust and the lithospheric mantle than in the brittle upper crust (■ Fig. 3.3; Thompson and Gibson 1994). The wide zone of the asthenospheric doming causes the bulge of the Earth's surface at active rifts to also be broad, commonly several hundred kilometers wide.

At *passive graben structures,* extensional forces are the primary cause. Initially, the extension is limited to the narrow zone of the rift, both in the deeper crust and in the lithospheric mantle (■ Fig. 3.3). This process can result in the complete tearing off of the lithospheric mantle which then leads to asthenospheric material rising to the base of the crust (Turcotte and Emerman 1983). The surficial bulge is thus restricted to the narrow graben zone and thermal uplift of the rift shoulders is reduced as is basaltic magmatism. However, extension of the lithosphere may also lead to a wider updoming of the asthenosphere and the lithosphere above. Thus a passive rift may change into an active rift system and the passive stage is no longer detectable. Although most present graben systems seem to be active rifts, it is assumed that both processes, updoming and crustal extension, act together. The primary reason for the graben formation is thus an academic chicken-and-egg question. Nevertheless, strong updoming of the asthenosphere causes strong magmatic activity and wide graben shoulders.

3.1 Symmetric and Asymmetric Crustal Extension

Crustal extension is believed to occur in two different modes symmetric (McKenzie 1978) and asymmetric (Wernicke 1981); models have been proposed for each. The *symmetric model* is based on many present graben systems (■ Fig. 3.4a). It assumes symmetric, brittle extension of the crust along normal faults in the upper 10–15 km, and ductile deformation at depth. Both the crust and lithosphere thin accordingly. Crustal thinning and brittle deformation cause the surface of the Earth to subside and generate the graben morphology.

If a 30 km-wide strip of 30 km-thick crust is stretched by 5 km, the resulting stretched crustal section has thickness of only 25.7 km. Such a situation is approximated in the southern Upper Rhine Graben. Assuming an original thickness of 100 km for the total lithosphere, the initial 30 km-wide tract is reduced in thickness to 85.7 km following stretching. However, the ascent of hot asthenosphere causes the lower part of the lithospheric mantle to be transformed into asthenosphere. The lithosphere–asthenosphere boundary is defined by thermal and state properties of approximately 1300 °C; it is not a material-based boundary. Therefore, lithospheric mantle can be transformed into asthenosphere by an increase in temperature and vice versa.

The original bulge of the surface, caused by a hot, relatively light bulge of asthenospheric mantle material, leads to erosion at the graben shoulders, a process that also results in a reduction of thickness of the crust. Thermal subsidence is developed after the heat source disappears and the mantle bulge cools and increases in density. The area of subsidence broadens because

3

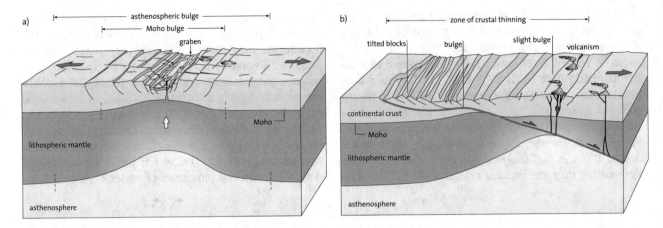

□ **Fig. 3.4   a** Symmetric and **b** asymmetric model for the evolution of a graben system. The asymmetric model also explains the early-stage evolution of metamorphic core complexes. The "Moho" (Mohorovicic discontinuity) is the boundary between the crust and mantle

the area of mantle uplift is generally two to three times wider than the graben structure. Therefore, old inactive graben structures may have morphologically unobtrusive shoulders.

The *asymmetric model* of graben formation was initially developed for the Basin and Range Province but also applies to some rifts associated with the formation of passive continental margins. Asymmetric grabens are characterized by a gently dipping master fault, termed a detachment fault, that cuts at low angles through the crust from one flank of the graben down to the base of the lithosphere (□ Fig. 3.4b). The overriding upper plate of the detachment is characterized by steeply inclined normal faults that form in response to the extreme amount of brittle crustal extension, in some cases greater than 200%.

In asymmetric grabens, the crust of the upper plate is extended and thinned at a different location from that of the lithospheric mantle, the lower plate of the detachment (□ Fig. 3.4b). This asymmetry gives rise to the following morphology: above the area of crustal thinning, the surface subsides because light crustal material is replaced by denser mantle material; above the area of lithospheric mantle thinning, the surface bulges because lithospheric mantle is replaced by slightly less dense, hotter asthenospheric material. In some cases, the bulge brings lower crustal material rapidly to the surface.

The asymmetric model has not been identified in a simple graben system (Roberts and Yielding 1994), but rather has been used to explain wide areas of extension such as the Basin and Range Province in the western USA. Such asymmetric areas of extension are connected to the uplift of metamorphic domes that will be discussed below.

In asymmetric rift zones, the zones of crustal and lithospheric-mantle thinning overlap but are off set whereas in symmetric rift zones they coincide. However, subsidence by crustal thinning is much greater because the difference in density between crust and mantle is much greater than that between lithospheric and asthenospheric mantle.

3.2 Sediments and Ore Deposits in Graben Structures

Typically sediments in graben structures are characterized by immature terrestrial deposits that are deposited by rivers that source the steep flank of the graben shoulder. Immature sediments are characterized by an abundance of mineral grains and rock fragments. Normally such minerals are readily weathered prior to reaching the site of deposition but because of the steep topography and short transport distance, they survive the sedimentary cycle in grabens. Many fluviatile sediments in graben structures are mostly composed of conglomerates rich in rock fragments and arkoses (sandstones that contain abundant feldspar) and have a relatively low percentage of quartz. Lacustrine deposits are rich in clays and, under arid or semi-arid conditions, saline sediments. Saline lakes, for instance, occur in the East African graben system.

Graben structures may also come under marine influence. Marine sediments in graben structures are mostly mudstone, marl (limy mud) or limestone. Strong evaporation in arid climates where partly or completely isolated basins fill with seawater leads to concentration of salt in the water followed by precipitation of salt. In the Upper Rhine Graben, marine ingressions are indicated by salt deposits. If a graben evolves into a narrow ocean like the Red Sea, saline deposits are typically preserved at the base of the marine sedimentary sequence.

Petroleum and natural gas are important deposits in some continental rift systems. Restricted water circulation in a narrow graben sea may lead to benthonic anoxic conditions where free oxygen is sequestered by the bottom fauna. Lack of oxygen in the lower part of the water column leads to oxygen-poor sediment which in turn prevents decomposition of organic matter. This generates an enrichment of organic material in the sediment and results in characteristic dark gray or black colors. Basin subsidence lowers organic-rich sediment into the so-called petroleum window, a temperature range between approximately 80 and 170 °C. Here petroleum forms by complicated reactions involving the organic matter. At temperatures over approximately 150 °C, gas deposits are formed. Oil shales of Messel that originated in a maar funnel (a volcanic penetration tube) within the subsiding Upper Rhine Graben provide a good example of sapropelic (organic-rich) sediments formed in an isolated basin. The incredible fossils preserved at Messel construe a unique deposit of global importance (UNESCO World Heritage Site) concerning the life of the Eocene.

3.3 Volcanism in Graben Structures

Magmatic rocks that form in graben structures are typically alkaline—they have an excess of alkalis (Na_2O, K_2O) compared to the content of silica (SiO_2) or alumina (Al_2O_3); alkaline rocks with deficiency in silica are also termed "undersaturated in silica". Alkaline magmas primarily develop from lithospheric mantle that undergoes a small amount (mostly less than 10%; Wilson 1989) of partial melting (see ▶ Chaps. 6, 7). However, tholeiitic magmas, which reflect a higher portion of partial melting (mostly more than 15%), are also common in graben systems. They accompany a rapid rate of extension of the lithosphere, especially where associated with hot spots. Rapid extension increases the rising and melting of hotter asthenospheric mantle rocks (▶ Chap. 6). Mid-ocean ridge tholeiitic basalts are formed in areas of rapid extension (▶ Chap. 5), another indication that tholeiitic basalts are more important in areas that undergo strong extension of the lithosphere accompanied by increased upwelling and melting of asthenospheric material.

In graben systems such as the East African Graben or the Rio Grande Rift, the alkalinity of melts increases outward from the graben axis towards the rift shoulder. This indicates that melt formation is highest below the graben axis where tholeiitic magmatism is favored. Other graben systems such as the Cenozoic Kenya Graben and the Permian Oslo Graben show a decrease of alkalinity with time that indicates increasing rates of extension and melt formation during the evolution of the graben system (Condie 1997). The East African graben system shows a shift from alkaline basalts and differentiates in the south at Tanzania and Kenya to tholeiitic basalts to the north in Ethiopia. This pattern parallels the increasing rate of extension from south to north (see below).

Volcanism in graben systems can be bimodal. In the northern Rio Grande Rift, tholeiitic basalts (basic, SiO_2 content of about 50 weight–%) occur beside rhyolites (acidic, SiO_2 content of about 70 weight–%). Intermediate rocks with SiO_2 contents in between are missing, however. This is not explainable through simple differentiation of an original basaltic magma. The East African Rift is dominated by alkali basalts (SiO_2 content less than 50%), phonolites (about 55% SiO_2, but very high content of alkalis: Na_2O along with K_2O about 12–14%), and trachytes (about 65% SiO_2, alkalis about 10–12%). Phonolites develop by differentiation from alkali basalts that are substantially undersaturated in silica; however, trachytes develop from less strongly undersaturated alkali basalts. Carbonatites also occur in rift systems. These are carbonate rocks that are derived directly from the Earth's mantle and are composed of calcite or dolomite with an accompanying extremely low content of SiO_2 of only a few percent.

The generation of basaltic magmas occurs in the mantle whereas acidic magmas are generated in continental crust or by mantle melts which are, in most cases, strongly influenced by continental crust. Acidic magmas occur in graben regions of greater crustal extension and higher, continuous magmatic activity. Igneous rocks include basalts of slightly alkaline or tholeiitic composition, and significant volumes of acidic volcanic rocks; absence of intermediate rocks indicates a clear bimodality (Barberi et al. 1982). Following intrusion into the continental crust, the primary basaltic mantle melts generate acidic crustal melts with their enormous heat. This explains the bimodal distribution of mag-matism. In contrast, graben systems that display little crustal extension are characterized by low magma production, interrupted volcanic activity, and strongly silica-undersaturated alkali basalts; intermediate and acidic rocks are rare.

3.4 The Upper Rhine Graben in Germany

Although the Upper Rhine Graben in Germany is not one of the largest nor most active graben systems, it is along with the East African graben system a type locality for the study of graben systems. The term "graben" (*German* ditch) was used by miners for blocks that were dropped down at faults (Pfannenstiel 1969) and was introduced into the geologic literature by Jordan (1803). Élie de Beaumont (1841) was the first geologist to describe the Rhine Graben. He understood that the facing Vosges and the Black Forest regions were broad, plateau-like uplifts separated from each other by the Upper Rhine River plain (◻ Fig. 3.5). He further noted that the Rhine plain was down-dropped and bounded by parallel faults that had dip directions towards each other. This is the classic geometry of a graben system.

3

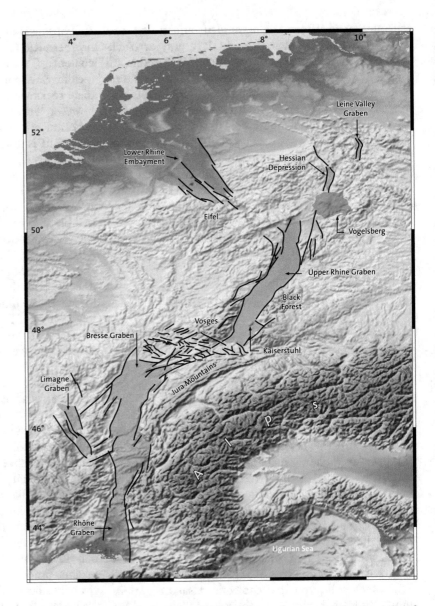

The acceptance of the term "graben" in scientific literature was solidified by the classic work "Das Antlitz der Erde" ("The Face of the Earth") by Eduard Suess (1885–1909).

The Upper Rhine Graben extends more than 300 km from Basel (Switzerland) to Frankfurt (Germany) and forms a part of a larger fracture system that runs from the mouth of the Rhône River to the North Sea (■ Fig. 3.5). The bordering faults have dip angles that range from 55 and 85° near the surface; however, a majority of the faults dip between 60 and 65°. The faults at the flanks are parallel and all faults dip towards the center of the graben and decrease in inclination at depth. The graben has a fairly constant width of approximately 36 km with a crustal extension of approximately 5 km (Illies 1974a). Crustal thinning is 6–7 km maximum and the continental crust in the southern part of the graben is thinned to 24 km

(■ Figs. 3.6 and 3.7). The graben parallels the axis of an elongated, stretched bulge that is mirrored in the graben shoulders on both sides, the Vosges to the west and the Black Forest to the east. Regionally, the graben shoulders are tilted 2–4° away from the graben.

The presently active earthquake foci occur mostly at a depth of less than 15 km. This indicates that brittle faults disappear at depth, and the crustal rocks below are deformed ductilely and not fractured. Ductile deformation of rocks rich in quartz (most of the rocks of the continental upper and middle crust are rich in quartz) initiates at temperatures of ca. 300 °C because quartz reacts from stress with plastic deformation at these temperatures. Seismic data indicate that the lower crust, dominated by rocks poor in quartz or without quartz, also reacts ductilely because of the higher temperatures; a pervasive horizontal lamination is interpreted to be the result of plastic flow (Illies 1974a).

■ **Fig. 3.6** Block diagram of the
Upper Rhine Graben. Note that
the upper crust is characterized
by normal faults whereas
the deeper crust is ductilely
extended (by plastic, fractureless
deformation). The Kaiserstuhl is
a Miocene volcano

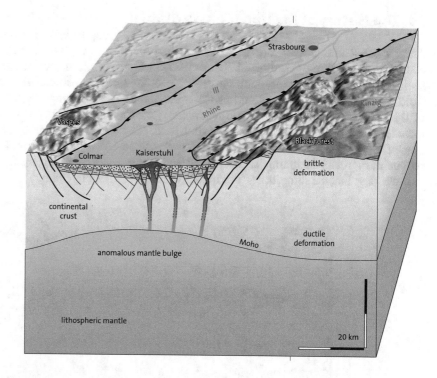

Seismic and gravity data indicate that in the Earth's
mantle directly below the base of the crust, an anomaly of rocks with relatively low density exists. Here, hot
and probably partly molten mantle rocks that rise because of the lower density, feed volcanism that is related to the graben formation. This suggests that a
mantle bulge is responsible for the bulge of the crust
(■ Fig. 3.6).

Formation of the graben, as indicated by initial normal faults, started in the Eocene at ca. 45 Ma. The first
sediments were deposited in the down-thrown graben
block. Extensional forces orthogonal to the graben axis
enabled the opening of the graben.

Today, the surface bulge extends more than 200 km
orthogonal to the graben axis. Uplift of the graben
shoulders varies regionally. More than 2 km of uplift
have been documented along the southern end. There,
the pre-Tertiary peneplain erosional surface (eroded
today) would be more than 2500 m above sea level
(■ Fig. 3.7; Illies 1974b). Total structural displacement across the graben varies from more than 5 km in
the south to 4 km in the north. The graben shoulders
are not significantly developed in the northern part, although the subsidence of the graben is generally greater
and the Tertiary sedimentary fill has a thickness of
more than 3 km. The result is a significant topographic
gradient parallel to the graben axis. As explained below, subsidence of the northern part occurred distinctly
later than that of the southern part.

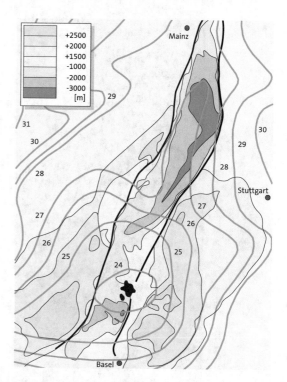

■ **Fig. 3.7** Map showing topographic and structural features of the
Upper Rhine Graben as well as the amounts of uplift of the graben
shoulder and subsidence in the inner part of the graben. Colored areas indicate the present level of the Early Tertiary erosional surface
relative to sea level (extrapolated into the air in the blue areas). Green
lines indicate the level of the crustal base (Moho) in kilometers below the sea level

3

3.5 The History of the Upper Rhine Graben

The Upper Rhine Graben has been filled with nearly 20,000 km³ of Tertiary sediments (Roll 1979). Most sedimentary rocks, both pre-and syn-rift, are eroded from the area of the graben shoulders. Along the edges of the graben, coarse-grained clastic sedimentary rocks include conglomerate and immature sandstone. The graben center is dominated by finer-grained clastic sedimentary rocks including siltstone and mudstone; non-clastics include limestone, dolomite, marl, and evaporites (salt rocks). Marine incursions generated saline to brackish conditions. During arid periods, evaporate deposits including halite and potash salts formed in the narrow, restricted seaway. Potash salt is particularly soluble in water and thus it is precipitated only at very high concentrations of evaporate minerals. Because of the low density and high mobility of evaporitic sediments, they rise as diapirs after being overlain by denser rocks. Some salt diapirs nearly reach the surface and are in contact with Quaternary sediments (Pflug 1982).

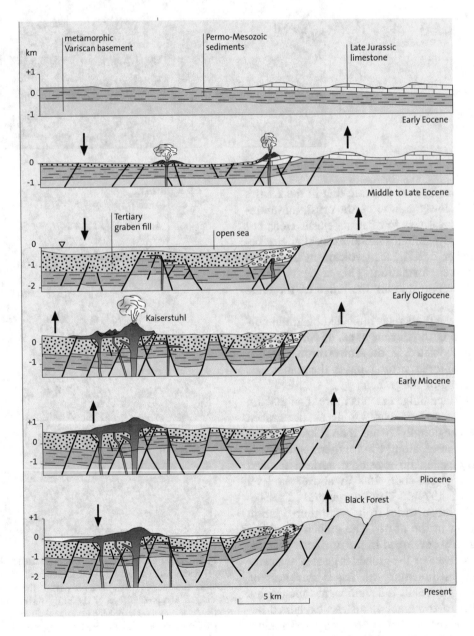

◨ **Fig. 3.8** A series of cross sections showing the evolution of the southern Upper Rhine Graben in the area of the Kaiserstuhl volcano (Schreiner 1977)

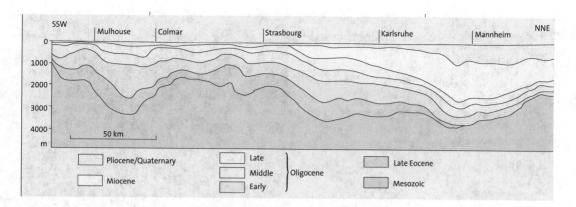

Fig. 3.9 Thickness of sediments in the Upper Rhine Graben in a profile parallel to the graben axis (Pflug 1982). Differences in the sedimentary thickness indicate that graben formation initiated in the southern area during the Early Tertiary, and that subsidence migrated to the north during the Late Tertiary

The Upper Rhine Graben in the Middle European Stress Field

Presently the extensional forces in the Upper Rhine Graben are oblique (SW–NE, with an azimuth of 050–060°) and not orthogonal to the graben axis. Consequently, the principal axis of compression is not parallel but oblique to the graben axis (NW–SE or 140–150°; Fig. 3.10). However, when the graben formed, the extensional forces acted orthogonal to the graben. An anticlockwise rotation of the stress field occurred during the evolution of the graben system during the Late Tertiary and caused a left-lateral component to develop that overprinted the normal faults. Therefore, as the Vosges and Black Forest diverged, the Vosges also moved parallel to the graben, southward in relation to the Black Forest (Fig. 3.10). The northern continuation of the Upper Rhine Graben trends along the Lower Rhine Embayment (Fig. 3.10). Neither a mantle bulge nor graben shoulders occur in this region. The graben system varies in width and opens towards the NW. Throughout its development, the exten-

Fig. 3.10 Directions of the present maximum horizontal stress (red arrows) in the Upper Rhine Graben and Lower Rhine Embayment (Blundell et al. 1992). The change in the orientation of the stress field is shown in the two schematic diagrams. The older stress field caused the formation of the Upper Rhine Graben, the younger one led to the formation of the Lower Rhine Embayment and to left-lateral movements in the Upper Rhine Graben. Volcanoes related to the graben formation are shown in green

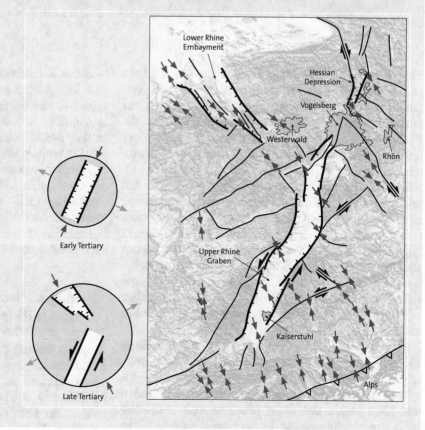

3

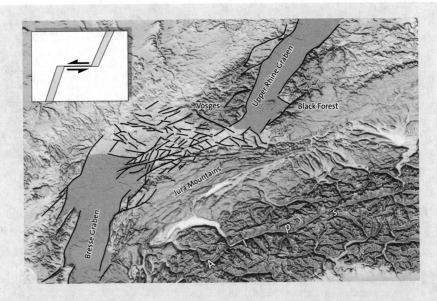

■ **Fig. 3.11**   Map showing the transition zone from the Upper Rhine Graben into the Bresse Graben. The transition is accomplished by a bundle of faults that in total mark the locus of a transform fault. A simplistic representation of the connection between the two graben structures is shown in the insert. The distance between the two graben axes remained unchanged through the course of time

sional stress has been orthogonal to the Lower Rhine Embayment since the Early Miocene. The missing mantle bulge indicates that it is a passive rift structure and that extensional forces are the primary cause for the graben development.

The southern continuation of the Upper Rhine Graben, the Bresse Graben occurs approximately 120 km to the west (■ Fig. 3.11). As in the Upper Rhine Graben, subsidence in the Bresse Graben ceased during the Early Miocene. Offset between the Upper Rhine Graben and Bresse Graben is only apparent; in fact, the distance between the two graben segments remained unchanged since their formation. The situation is that of a transform fault, although such a fault is not developed as one distinct fracture. Rather, the rift structure is transformed by a complex system of mostly W–E trending minor faults linking the Upper Rhine Graben and the Bresse Graben (■ Fig. 3.11). The individual faults in the transformation zone carry out left-lateral (sinistral) offset (see box in ■ Fig. 3.11).

The uplift of the shoulders initiated in the southern part of the graben during the Eocene (■ Fig. 3.8). Vertical displacement of more than 1000 m between the graben basin and its shoulders existed at the end of the Eocene as major stream systems were carved deeply into the rift shoulders and spewed coarse conglomerate onto the rift plain. Subsidence of the graben allowed marine ingressions from both the south and north. During parts of the Oligocene, a marine passage existed between the Alpine Molasse zone in the south through the Upper Rhine Graben into the enlarged North Sea Basin. Conglomerate input from graben shoulders decreased, indicating reduced tectonic activity and a lowering of relief. In the Late Oligocene, subsidence concluded in the southern part of the graben while further to the north fresh water lakes expanded as subsidence continued.

The last marine tansgression occurred in the Early Miocene, this time from the Lower Rhine Embayment, which was now established as a new branch of the rift system (■ Fig. 3.5). The southern half of the graben was filled with sediment by this time and completely lacks deposits of younger Tertiary age. Meanwhile, the Miocene Kaiserstuhl volcano was formed (■ Figs. 3.6 and 3.8). As the southern rift waned, subsidence continued to the north, a trend clearly documented by a longitudinal section through the graben (■ Fig. 3.9). In the Late Miocene and Early Pliocene, sedimentation was interrupted across the entire graben system signifying general regional uplift. Sediments of Late Pliocene age are dominated by fluviatile deposits, present only in the northern half of the graben. Quaternary sediments locally attain thicknesses of more than 200 m indicating continued tectonic activity coupled with present earthquake activity.

3.6 Magmatism and Heat Flow in the Upper Rhine Graben

Magmas feeding the volcanoes associated with the rift formation of the Upper Rhine Graben are strongly undersaturated in silica and originate from a depth of about 80–100 km, the base of the lithosphere. The magma has been modified at shallow depth by differentiation. Therefore, a broad variety of volcanic rocks evolved within the graben system. Volcanic rocks of the Upper Rhine Graben are of minor volume when compared to the voluminous volcanic rocks of the East African graben system (see below).

The most famous volcanoes of the graben system are the Kaiserstuhl and the Vogelsberg (■ Fig. 3.10). The Kaiserstuhl is constructed of lavas and tuffs that erupted during the Early and Middle Miocene. Its location corresponds to where the crust is thinnest and large faults exist (■ Fig. 3.7). The volcanic rocks are strongly alkaline and undersaturated in silica. This is reinforced by the presence of carbonatites which contain mostly silica-free carbonate minerals. Globally, carbonatites are very rare rocks and are almost exclusively related to graben structures. They are derived from mantle-sourced magmas that contain carbonic acid. The occurrence of carbonatites in the Kaiserstuhl is one of only a very few in Europe.

The Vogelsberg lies in the northern continuation of the Upper Rhine Graben system, the Hessian Basin, which was created as a temporary part of this rift system (■ Fig. 3.10). With an area of around 2500 km², the volcanic area resembles a large shield volcano, but recent drilling has shown that there are many stacked basaltic and trachytic layers from different production vents. The volcano was formed in the Lower and Middle Miocene 15–20 million years ago.

Because of the relatively low depth to the mantle and associated magmatic activity, graben systems are also zones of high heat flow. Below the Upper Rhine Graben, the temperature at the upper boundary of the mantle is at least 200 °C higher than beneath the graben shoulders. Surficial heat flow is about 50–80 milliwatt per square meter (mw/m²) outside the Upper Rhine Graben but increases to values between 80 and 120 mW/m² in the Upper Rhine Graben (Blundell et al. 1992). Locally values greater than 150 mW/m² are possible. Temperature at 1 km depth at these localities is 80 °C (Werner and Doebl 1974), whereas in areas with a normal geothermal gradient it would be 30 °C.

Such high thermal anomalies cannot be explained by heat conduction through the rocks alone because rocks are poor heat conductors. Rather, hot circulating water that migrates rapidly along fault zones is responsible. Cold surface water percolates to the depth, becomes heated, and ascends by convection. Such water cycles are expressed at thermal springs, common along the master faults of the Upper Rhine Graben, such as the health resort of Baden–Baden.

3.7 The Large East African Rift System

Along the greater East African rift system and the three-pointed graben star of the Afar Depression, where the East African rift system, the Red Sea, and the Gulf of Aden meet, different stages of continent break-up are represented (■ Fig. 3.12). The East African Rift, from which the Central African Rift between Lake Malawi and Lake Turkana branches off, is a presently active system with abundant volcanism (■ Fig. 3.13). The East African rift system has not matured enough to have formed a new plate to the east, although the crust is nearly severed at some places and the term "Somalian Plate" is used by some geologists (■ Fig. 3.14). At its northern end, the Afar Depression, which because of its triangular shape is also called the Afar Triangle, has partly generated new oceanic crust.

The region is characterized by two broad topographic uplifts, each more than 1000 km in diameter (■ Fig. 3.12), and each underlain by mantle diapirs with broad mushroom-shaped heads. The northern uplift includes Ethiopia and Yemen and has the three-pointed graben star at its center. The southern uplift area is in Uganda, Kenya and Tanzania and is marked by the intersection of the Kenya and Central African rifts. The faults of the rift systems are generally parallel to structures of the Precambrian basement, an observation that suggests that the graben structures follow old zones of weakness in the crust.

The East African Rift has evolved since the Late Oligocene or Early Miocene. It comprises the Ethiopian Graben and the Gregory or Kenya Rift (■ Fig. 3.12). The Gregory Rift Valley, eponymous for the geological term "rift", has shoulders that rise more than 3000 m above sea level and 1000 m above the inner part of the graben. Collective vertical displacement along the main graben faults much as 4 km. The graben area is cut by densely clustered faults that parallel the edges of the graben and define a horst-and-graben structure. A horst is a higher block between two down-dropped graben structures. The total graben has a width of 60–70 km but the width of the inner graben is 17–35 km. The base of the inner graben is covered by Pliocene and Quaternary volcanic rocks and sediments and the clear dominance of volcanic rocks contrasts sharply with the rarity of volcanics in the Upper Rhine Graben. Mt. Kilimanjaro, an active volcano, towers nearly 6000 m high and along with the older and partly eroded Mount Kenya, is located on the eastern graben shoulder.

Towards the inner part of the graben, small but steep erosional surfaces limit the transport of sediments, and consequently, deep depressions evolve that are commonly the sites of large lakes. The largest is the 650 km long Lake Tanganyika in the Central African graben system (■ Fig. 3.12); its depth is nearly 1500 m and its bottom lies 700 m below sea level. The lake is situated behind a 2000 m-high mountain range that forms a rain shadow from the trade winds. There-

3

□ **Fig. 3.12**   Map of the
principle element of the East
African graben system

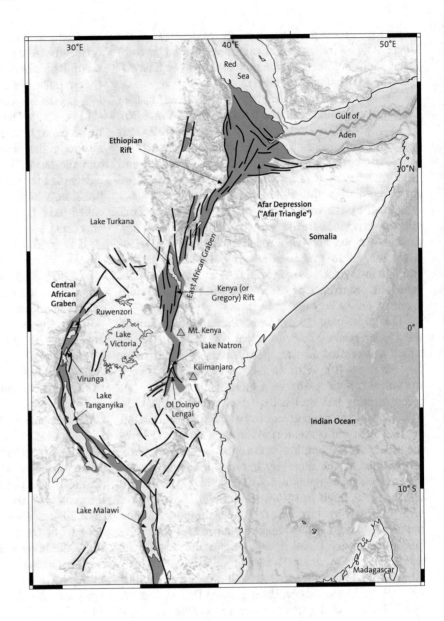

fore, the region is relatively dry with a total annual pre-cipitation less than 1000 mm, and the evaporation rate is high because of its position 3–9° south of the equa-tor. The sediments of the lake are mostly biogenic and chemical in origin with only minor input by rivers. The depth of the water body coupled with the low produc-tion of sediment causes the longevity of such lakes. Some of the smaller lakes in the East African graben system, such as Lake Natron in Tanzania, contain high concentrations of sodium hydroxide and with associ-ated high evaporation rates are dominated by evapo-rite deposits such as sodium carbonate, a by-product of volcanism. Chemical processes are strongly enhanced by bacterial activities (Kraml and Bull 2001).

Since the Miocene, the Gregory Rift has produced alkaline volcanic rocks, especially strongly alkaline basalts and phonolites. The volcanic rocks have a to-

tal volume of about 100,000 km³, half of which is ba-saltic. Interestingly the Gregory Rift is the site of the only active carbonatite volcano on Earth, the Ol Doinyo Lengai in northern Tanzania (□ Fig. 3.12). The Central African Rift has produced much less vol-canic rock material, although the 4500 m high vol-canic chain of the Virunga Mountains in the border region of Congo, Rwanda and Uganda belong to this graben system.

In the Ethiopian Rift, basalts have clearly dom-inated (ca. 300,000 km³) since the Eocene; these are dominantly slightly alkaline and range to tholeiitic in composition. The huge volumes of low-viscosity flood basalts have formed 2000 m-thick basaltic plateaus (see ▶ Chap. 6). Acidic differentiates in Ethiopia are mostly alkaline rhyolites to trachyandesites. Nonbasaltic rocks have a volume of about 50,000 km³.

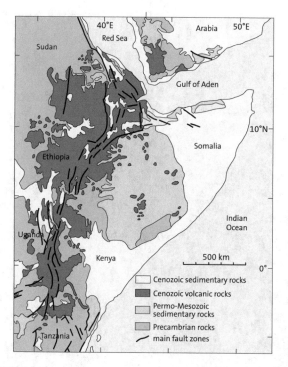

Fig. 3.13 Generalized geologic map of East Africa

the lithospheric mantle below the continental crust in a location that is enriched in elements such as alkalis that are incompatible with the mantle rock. In contrast, the tholeiitic magmas of the Ethiopian Rift are mainly derived from rising asthenospheric material that is depleted in these incompatible elements (Wilson 1989). The differences between the Gregory and Ethiopian rifts are also explained by the different extension rates: total crustal extension in Ethiopia approaches 60 km, but in contrast, much lower rates, 35–40 km in northern Kenya, 5–10 km in southern Kenya, and less than 5 km in northern Tanzania are present in the Gregory Rift. The graben system opened like a pair of scissors, which explains the increasing magma production and decreasing alkalinity (tholeiites originate from higher partial melting in the mantle rocks) from south to north.

The average extension rates in the East African graben system, 0.4–1 mm/year, are one order of magnitude less than those at a slow-moving constructive plate boundaries such as the nearby Red Sea or Gulf of Aden. Such a graben system is generally not considered to be a plate boundary but rather the result of intraplate tectonics. Activity in the East African graben system is waning, which is also expressed by its limited seismic activity. This is typical of extensional triple junctions, such as the Afar Triangle, where three arms meet—one arm is shut down after the other two have formed ocean crust between continental blocks.

The larger volumes of basalts in Ethiopia and their slightly alkaline to tholeiitic chemistry signify a higher percentage of partial melting in the mantle source compared to those in the Gregory Rift. The alkaline magmas of the Gregory Rift are assumed to originate from

Fig. 3.14 Block diagram of the East African graben system. The lower cross section through the Central African Rift and the Kenya Rift demonstrates the strong thinning of the lithospheric mantle that causes a negative gravity anomaly (Baker and Wohlenberg 1971). 1 Gal (galilei) = 1 cm/s² (unity of acceleration). 1 mGal = 10^{-3} Gal

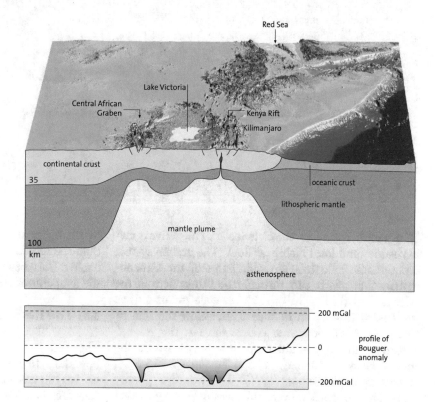

3

The Afar Depression

The Afar Depression is a lowlying triangular area (■ Fig. 3.15), at the center of a three-point graben star, where the East African graben system (Ethiopian Rift), the Red Sea, and the Gulf of Aden meet. Here, the transition from a continental graben to an initiating ocean basin can be observed. Underlying the depression, a mantle diapir rises and overlying continental crust is strongly thinned and fragmented. In fact, between the separated continental fragments, new oceanic crust has been generated, although it is uncertain whether continuous bands of ocean floor already exist or not. Because the region is above the sea level and thus accessible for direct observation, in the 1970s many aspects concerning the theory of plate tectonics were established here.

Numerous bundles of faults, visible on satellite images of this arid region, pervade the depression. Along these faults, some partly expressed as open-spaced cracks, basaltic lavas with a tholeiitic composition similar that of mid-ocean ridge basalts, periodically discharge. Narrow stripes of quasi-oceanic crust are produced at several spreading axes and have produced a complicated pattern of microplates (■ Fig. 3.15). The tectonic and volcanic activity is concentrated along the inner part of the graben system, an area characterized by both horizontal and vertical movements of blocks. Regional crustal thickness of 30–40 km in the upland of Ethiopia is thinned to less than 16 km in the northern Afar Depression (Makris et al. 1975). The principle of isostasy is illustrated in the Danakil Mountains where a block (microplate) greater than 20 km in thickness rises as a highland above the surrounding plain. The adjacent Danakil Depression is an area below sea level and marks a portion of the plate boundary (■ Fig. 3.15). This trend continues northward towards the central part of the southern Red Sea depression where crustal thickness is ca. 6 km, typical of normal oceanic crust.

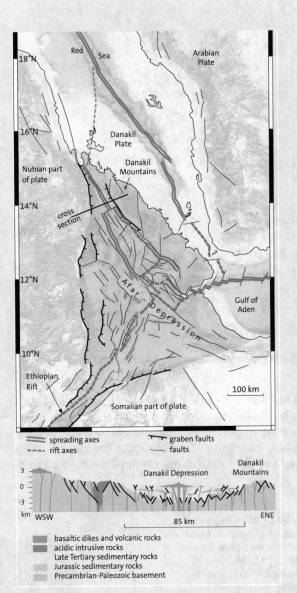

■ **Fig. 3.15** Map showing structural elements of the Afar depression or Afar Triangle. The region consists of a mosaic of blocks with thinned continental crust including the small "Danakil Plate". Basaltic rocks lie in the separation of the crust within the narrow spreading axes between the blocks. This relation is shown in the cross section through the Danakil depression

The entire region is characterized by negative gravity values and local high heat flow. The graben system is underlain by a 1500 km wide bulge of the asthenosphere that nearly cuts through the lithosphere in the Kenya Rift (■ Fig. 3.14); a 20 km-wide intrusion has protruded to a depth of only 3 km below the sole of the graben. The intrusion is detectable in the gravitational profile by a slight positive anomaly within the broad negative anomaly of the graben (■ Fig. 3.14; Baker and Wohlenberg 1971). The negative gravity anomaly mirrors a widespread mass deficiency caused by the bulge of the asthe-nosphere, which is less dense than the replaced lithospheric mantle. The mantle lithosphere has been transformed into asthenosphere by the temperature increase.

3.8   The Red Sea—From Rift to Drift

The Red Sea is a relatively recent constructive plate boundary that consists of a band of oceanic crust up to 100 km in width that was formed during the Late Tertiary by the separation of Arabia from Af-

◨ Fig. 3.16 Block diagram of the Red Sea region. Note the graben-in-graben structure, the high elevation of the graben shoulders, and the central graben fissure on oceanic crust in the middle of the Red Sea. The foreground cross-section that passes through the southern Afar Depression indicates that the continental crust is not severed there

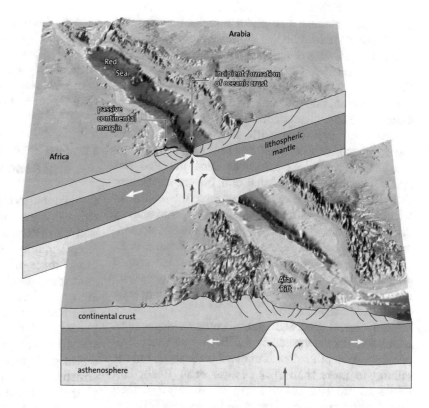

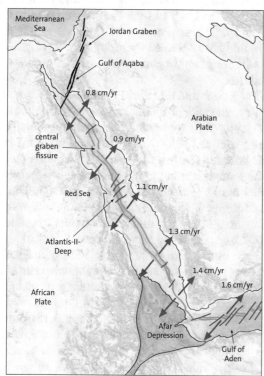

◨ Fig. 3.17 Spreading rates in the Red Sea and the Gulf of Aden (DeMets et al. 1990). The central graben fissure in the Red Sea is marked by a double green line. The Atlantis–II–Deep is one of several depressions that contain metal-rich oozes

rica (◨ Fig. 3.16). Here, the rift stage transformed into the drift stage and the Red Sea forms a nascent ocean. The oceanic band contains a central graben that marks the plate boundary and attains a depth of more than 2000 m. Because of the small amount of drift and the closeness of the continental margins, the central graben is not yet developed as a midocean ridge in the true morphological sense.

Plate divergence in the southern Red Sea is 1.4 cm per year and decreases towards the NW (◨ Fig. 3.17). In the Gulf of Aqaba it merges into the transform fault of the Jordan Graben (▶ Chap. 8). The Gulf of Aden represents a more advanced stage of ocean-basin formation with a mid-ocean ridge, a feature that continues eastward into the Indian Ocean. Here the spreading rate along the ridge increases to 7 cm per year (◨ Fig. 1.2). Spreading direction of ocean floor in both the Red Sea and Gulf of Aden is in the same, SW–NE, although the spreading axes of both oceans have different orientations.

The history of the Red Sea dates back to the middle Cenozoic. The Red Sea rift is cut through the Arabian–Nubian Shield, an area of Precambrian continental crust (▶ Chap. 12) that is mantled by Upper Cretaceous and Lower Tertiary sedimentary rocks. The formation of the graben initiated during the Oligocene and by Late Oligocene; 25 million years ago, violent volcanic activity with basaltic eruptions commenced

3

in the Afar Triangle, a southern extension of the Red Sea. During the Middle to Late Miocene, sea water intruded the graben system from the Mediterranean Sea; the graben-restricted sea had a blind end to the south. Restricted water exchange, high rates of evaporation in the arid region, and low inflow of freshwater generated more than 3 km of salt deposits.

Sea floor spreading with a rate of 1–2 cm/year started in the southern Red Sea in the Pliocene at approximately 5 Ma. At the same time the Red Sea opened to the Gulf of Aden and the Indian Ocean. Due to the total separation between the continental blocks of Africa and Arabia, the tectonic activity shifted from the edges of the graben to the new spreading zone in its center, the zone where new oceanic crust was generated along a narrow, newly formed central graben; this formed the present graben-in-graben structure (☐ Fig. 3.16). The outer graben shoulders are presently tectonically inactive.

The central Red Sea graben contains depressions with water depths greater than 2000 m. The water temperatures at depth increase to more than 60 °C and the salinity to more than 30% (Brewer et al. 1969). Hot brines are trapped in the depressions because of their high density: mud ooze formed on the seafloor is rich in iron, copper, lead, and zinc sulfides, ferric and manganese oxides, and calcium sulfates (anhydrite), gold, and silver. Total concentrations of non-ferrous metal in the oozes amounts to more than 10% by weight and are potentially of economic interest. The economic potential is presently offset by the very difficult and expensive mining procedures that would be required to extract the ore. However, if the ocean floor became obducted onto the continental margin in the geologic future, like the Oman ophiolite (▶ Chap. 5), mining might be economical.

The formation of these ore deposits is directly related to the opening of the Red Sea. The brines form an ideal environment in which to precipitate the ore content of hot water solutions that ascend from the depth. Salt solutions on the thinned continental crust along the sea floor next to the central graben were trapped in the depressions due to their high density and highly saline brines were formed. The basaltic lavas that extruded along the axis of the Red Sea heated the deep seawater and increased the solubility of the metals. Isotopic composition of the sulfur in the sulfides indicates that the total metal content is only partly derived from the basalts: the ratio of the isotopes sulphur-34 to sulphur-32 suggests that the ore metals are at least partly derived from the Precambrian crust of the adjacent continental blocks. Kuroko-type deposits that are present in local Precambrian rocks are present in the region and are rich in non-ferrous metal sulfides and gold (▶ Chap. 7).

The geologic setting and range of processes that formed the Red Sea were ideal for the generation of metalliferous ore deposits. Submarine basaltic volcanism, faulting, water temperatures, and presence of metal ions in crustal rocks all combined to produce the ores. The ore stock of one of these depressions, the "Atlantis–II–Deep" (☐ Fig. 3.17) is estimated to contain 3,200,000 t zinc, 800,000 t copper, 80,000 t lead, 4500 t silver and 45 t gold (Seibold and Berger 1982).

3.9 The Extensional Area of the Basin and Range Province

Continental crustal extension ranges between two styles, (1) areas with narrow zones of tectonic deformation characterized by a single, central graben and adjacent, uplifted graben shoulders, and (2) areas with broad zones of tectonic deformation characterized by numerous, parallel graben and horst blocks or metamorphic domes (see below). Examples of the first style have been described above and include the Upper Rhine Graben system and the greater East African rift system. When rifting continues over long periods of geologic time, such systems can evolve into constructive plate margins and young ocean basins—two plates are generated from a single precursor. Examples of the second style include the Basin and Range Province of western USA and the Pannonian Basin in Central Europe. These graben-horst systems attain great widths and have operated over long periods of geologic time without generating new oceanic crust or new plate boundaries.

The Basin and Range Province comprises two kinds of extensional faults, low-angle normal detachment faults with associated metamorphic core complexes (described below) that are responsible for most of the horizontal extension (over 200% in some places), and high-angle normal faults that create the numerous parallel basins (grabens) and ranges (horsts). The Basin and Range Province is extended in an east–west direction across 600 km and encompasses an area of 550,000 km^3 (similar to that of France). It is oldest to the south in southern Arizona (ca. 30 Ma) and youngest to the NW in Nevada and Oregon where it is still active today. As both a relief map and a geological map clearly show, the characteristics of the ranges contrast sharply with those of the basins (☐ Fig. 3.18). Ranges consist of elevated blocks that expose multiple rock types ranging in age from Precambrian to Cenozoic and display wide-ranging internal styles of geologic structure. Basins contain unconsolidated or loosely consolidated erosional debris derived from highlands and deposited in the depressions. Voluminous amounts of Late Cenozoic basalts were deposited within some basins. Elevation of the ranges and basins as well as local relief varies greatly across the province. For example, along the Utah–Nevada border in the center of the province peaks are almost 4000 m high and basin floors

Fig. 3.18 Topographic relief map and simplified geologic map of a portion of the Basin and Range Province in North America. The southern Snake Range along the Nevada–Utah border rises to 3982 m at Wheeler Peak, Nevada. Most basins are 1700 m above sea level across the central portion of the relief map. Ranges consist mostly of wide-ranging Paleozoic sedimentary rocks, some of which are metamorphosed across metamorphic domes. Faults, not shown for simplicity, bound the ranges

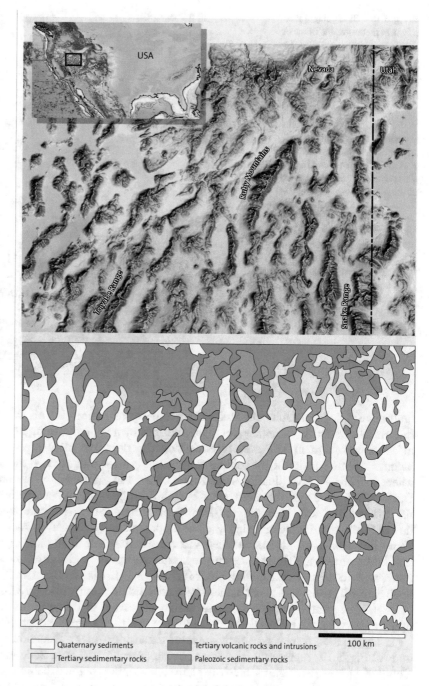

Quaternary sediments Tertiary volcanic rocks and intrusions 100 km
Tertiary sedimentary rocks Paleozoic sedimentary rocks

fall to 1700 m (relief of 2300 m), in south–central Arizona peaks are 3200 m and basin floors are 850 m (relief of 2350 m), in SW Arizona, the oldest portion of the province, ranges are 700 m and basins are 100 m (600 m relief), and in NW Nevada, the youngest portion of the province, ranges are at 2500 m and basins are at 1200 m (1300 m relief). Sickness of Cenozoic valley fill ranges from a few hundred meters, especially in younger basins, to several thousand meters.

Crustal thickness in the Basin and Range Province averages only 30 km, a low value when compared to the crustal thickness of the adjacent Colorado Plateau, nearly 50 km. Before the Cenozoic collapse of the Basin and Range Province, its crust was even thicker than that of the Colorado Plateau, as during the Mesozoic the Basin and Range Province was the site of the Sevier Mountains, a major thrust belt. The present Basin and Range Province exhibits a high geothermal gradient that is caused by anomalously light, hot, and partly molten asthenosphere.

3.10 The Development of Metamorphic Domes

Metamorphic domes, commonly referred to as metamorphic core complexes, develop in regions of asymmetric rifting along a sub-horizontal, large normal

3

◘ **Fig. 3.19**   Block diagram of
a typical metamorphic dome or
metamorphic core complex. The
dome is caused by thinning and
brittle faulting of the hanging
wall (upper plate), under which
the foot wall (lower plate) rises in
a dome-like fashion. The rocks
in the foot wall have risen from
the middle crust to the surface.
Mylonites are ductilely deformed
rocks that formed along the
shear zone but are now exposed
at the surface; they originally
formed at middle-crustal levels.
The mylonites (see photo,
with band of very fine-grained
ultramylonites) are overprinted
by brittle deformation during the
uplift and subsequent cooling.
By comparing this figure with
◘ Fig. 3.4b, the stages of
evolution of a metamorphic
dome can be visualized

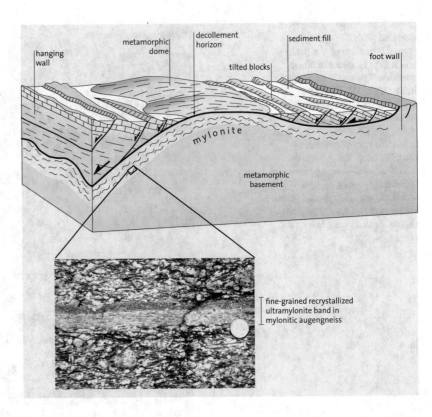

fine-grained recrystallized
ultramylonite band in
mylonitic augengneiss

fault zone called a detachment fault. The geometry of
the overall system is extremely complex and is shown
in simplified form in ◘ Fig. 3.4b. Note that the detachment fault buckles in the middle continental crust
where it merges into a ductile shear zone (Lister and
Davis 1989). At depth the shear zone breaks through
the lower crust and breaches the lithospheric mantle (◘ Fig. 3.4b). In the hanging wall or upper plate
above the shear zone, steeper inclined and curved normal faults attend the shear zone and separate the upper plate into domino-like gliding blocks (◘ Fig. 3.19).
Such curved faults and shear zones are called listric
faults (*Greek* shovel-shaped). Because of the concave
upwards curving, the blocks above the faults are tilted.

◘ Figure 3.4b displays an early stage of extension
and ◘ Fig. 3.19 represents late-stage development.
Comparing the two figures helps explain the origin
of the metamorphic core. As the extension shown in
◘ Fig. 3.4b expands, the upper plate thins; at the same
time the heat underlying the extension zone bows the
lower plate upward. The movement along the decollement horizon causes rocks with considerably higher
grade of metamorphism from the middle and lower
crust to abut rocks in the upper plate that consist of
non-metamorphic or lower-grade metamorphic rocks.
The shear zone bulges upwards at the place where the
hanging wall (upper plate) is thinned the most; eventually the shear zone tears through the upper plate and
the foot wall is exposed at the surface (◘ Fig. 3.19).
This type of exposure is called tectonic exhumation (as opposed to erosional exhumation). The ther-

mal rise of the footwall (lower plate) coupled with
the thinning and eventual rupturing of the brittle upper plate exposes the dome-like high-grade metamorphic rocks—the metamorphic core or dome. The metamorphic core has extended by ductile thinning while
the upper brittle plate fractured by faulting. No won-
der such complicated and unexpected juxtaposition of
contrasting rock types was not adequately explained
until 30 years ago!

Good examples of metamorphic domes are found
in the Basin and Range Province, e. g., the Ruby Mountains (◘ Fig. 3.18). In map view, the metamorphic core
or dome forms an irregular elongate—to kidney bean-
shaped body; dimensions can range to tens of kilometers. Depending on geometry and topographic relief,
the core zone may only expose the mylonite zone (explained below), or it may also expose the underlying
non-mylonized metamorphic rocks of the middle and
lower crust. Rocks that form the core can have diverse
geologic histories. They may be Precambrian basement
rocks or Phanerozoic rocks formed at or near the surface, buried during mountain building to lower-crustal
levels, and exhumed by the extensional processes discussed above. The metamorphic core is overlain by brittle rocks of the upper plate. Upper plate rocks can consist of sedimentary rocks and volcanic rocks, the latter
formed when material from the elevated asthenosphere
melts and rises to the surface along fault conduits.
Across much of west–central Arizona, these faulted
and tilted dark volcanic rocks form a distinctive moon-
like landscape.

The large shear zone between the hanging wall and foot wall is ductilely deformed below ca. 10 km depth causing the formation of mylonites. Mylonites are formed by intense ductile, fractureless deformation and consist of fine-grained, recrystallized metamorphic rocks. They overprint the middle and lower crustal rocks of the core complex. Mylonites thus represent the surficial expression of the decollement shear zone that originally formed 10–20 km below the surface. Rocks originally formed at dramatically different crustal levels adjoin along the shear zone. Another characteristic of metamorphic domes is the rapid uplift required for their formation, typically rates of several mm/year or kilometers per million years. However, the incredible uplift rates are not necessarily the sites of towering mountains. This is because the rapid uplift is accompanied by rapid horizontal extension—as much material is transported horizontally as is pushed up vertically. Therefore, rapid rates of erosion (impossible at this magnitude) are not required to expose the deep crustal rocks of the metamorphic core.

The crust of the Basin and Range Province has been extended up to two times its original width by the mechanism described above. However, the topography of the region is controlled by the later brittle faulting, the high-angle normal faults that flank the numerous ranges, features responsible for a much smaller fraction of crustal extension than the metamorphic domes.

3.11 A Brief History of the Basin and Range Province

The geologic history and geologic controls that acted upon the Basin and Range region are complex and beyond the scope of this book. However, a brief account is presented here. Much of the Basin and Range Province originally comprised the Early and Middle Paleozoic passive margin of western North America and was the site of unusually thick sedimentary rocks. In the Late Paleozoic and Mesozoic, the region was an active margin characterized by subduction, terrane accretion, and orogenesis. The culminating Cretaceous Sevier orogeny generated thick continental crust, a large thrust belt, and towering mountains. The hinterland bisected the present Basin and Range area, and the future Colorado Plateau and Rocky Mountains provinces lay near sea level to the east.

The initial tectonic collapse of the hinterland initiated in the Paleocene and Eocene as flat-slab subduction formed under the American Southwest. Subduction ended as the East Pacific Rise (the Pacific mid-ocean ridge) intercepted western North America in the Oligocene. As the eastward-plunging Farallon Plate was consumed along portions of the North American Plate, the westward moving Pacific Plate contacted the continent. Although North America has a westward component of motion, the Pacific Plate moved at a much greater rate to the NW. A "space problem" developed along the margin of North America as the Pacific Plate attempted to pull it westward. Consequently, the thick crust below the old Sevier hinterland was extended to fill the gap and the great crustal extension commenced.

The interception of the Pacific and North American plates can be tracked along the northward-moving Mendocino triple junction. In fact, the current location at Cape Mendocino, California, marks the approximate northward extension of the Basin and Range Province; its inception near the latitude of San Diego marks the inboard location of the oldest Basin and Range extension in southern Arizona. Since the Middle Miocene, ca. 15 Ma, much of the motion between the two plates has been taken up by transform faulting, especially along the famous San Andreas Fault, the current plate boundary. The San Andreas Fault and its precursors have continuously carved eastward into North America resulting in several hundred kilometers of North America being "captured" by the Pacific Plate.

Passive Continental Margins and Abyssal Plains

Contents

© The Author(s), under exclusive license to Springer Nature Switzerland AG 2022
W. Frisch et al., *Plate Tectonics*,
Springer Textbooks in Earth Sciences, Geography and Environment,
https://doi.org/10.1007/978-3-030-88999-9_4

4

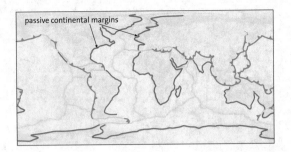

☐ **Fig. 4.1** Map showing global distribution of passive continental margins. The Atlantic and Indian Ocean are primarily bordered by passive continental margins

Nearly the entire Atlantic Ocean and large portions of the Indian Ocean are surrounded by passive continental margins (☐ Fig. 4.1). In contrast, the Pacific Ocean is mostly bordered by active continental margins. As mentioned in ▶ Chap. 1, passive continental margins are not plate boundaries because the continental plate is firmly attached to the adjacent oceanic plate. Nevertheless, some displacement occurs at faults along these margins which, because of the small movement involved, is considered to be intraplate tectonics. Passive continental margins develop when continental rift systems separate the continental block and generate juvenile oceanic crust between the separated parts. The oceanic crust is generated by a new spreading axis as is currently the case in the Red Sea (☐ Fig. 3.16). The initial oceanic crust is welded to the adjacent thinned continental crust and the intraplate margin is formed. Therefore, the oldest parts of oceanic crust in an Atlantic-style ocean are always in the immediate neighborhood of the passive continental margins (☐ Fig. 2.12).

Deep ocean basins cover about 40% of the Earth and are the most dominate type of crustal material. In spite of its relatively low crustal and lithospheric thickness, normal oceanic crust is mechanically stiff and relatively tectonically stable. As oceanic crust ages, it is covered by increasing amounts of sedimentary material. However, even on the oldest oceanic crust, the sedimentary thickness is relatively thin, only a few hundred meters thick on average. The thin sediment reflects a lack of nearby sediment source from the continents for much of the deep-sea basins. Composition and thickness of sediment are controlled by a variety of different factors, including sediment source and volume, proximity to an active or passive continental margin, climate, and ocean chemistry.

4.1 Continuous Subsidence of the Continental Margins

Passive continental margins are regions of thinned continental crust that are characterized by an array of seaward-dipping listric normal faults (☐ Fig. 4.2). Curva-

ture of the faults causes the crustal blocks to be tilted towards the continent. Resulting topography consists of asymmetric basins and ridges. Because the thinned continental margins mostly lie below the sea level, this topography is mirrored in the variable thickness and facies characteristics of the sediments (☐ Fig. 4.3). As the adjacent ocean expands, the highly elevated graben shoulders typical of young rifts (▶ Chap. 3) gradually lower as they drift out of the zone of mantle bulge. On both sides of the Red Sea, the graben shoulders are still prominent as high mountain ranges (☐ Fig. 3.16) and rivers draining towards the sea are accordingly short. Along older passive continental margins such as those around the Atlantic Ocean, the former graben shoulders are eliminated by erosion and subsidence and the broad continental margin drains towards the expanding ocean. If the drainage basin is large and sediment influx high, such margins can be the site of incredible volumes of clastic sediment. Such sediment

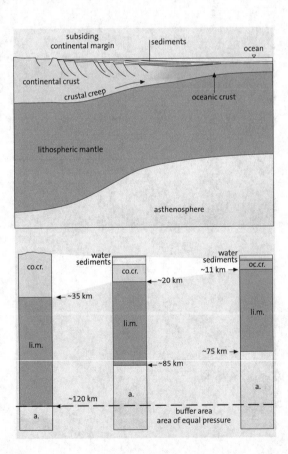

☐ **Fig. 4.2** Diagram of a passive continental margin. Above the thinned continental crust and the bordering oceanic crust a thick wedge of sediment is formed by the prograding shelf. Oceanward, crustal creep in the ductile framework of the continental crust (arrow) adds to its thinning. The rock columns indicate the equilibrium of buoyancy: in each case the same pressure prevails at a plane of balance in a depth of 120 km. For the calculation of the pressure the following average values of density were taken (in g/cm³): water 1, sediments 2.2, continental crust 2.8, oceanic crust 3.0, lithospheric mantle 3.3, asthenosphere 3.25

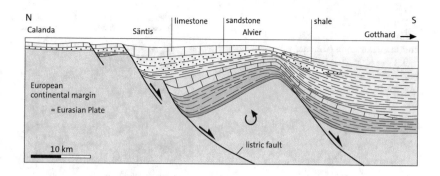

☐ Fig. 4.3 Formation of tilted blocks at the European passive continental margin in the Early Jurassic. Changes of sediment thickness are caused by the tilting of blocks above the listric (curved) faults. Reconstruction is from the Helvetic nappes in the Swiss Alps (Trümpy 1980)

accumulation can actually cover the original continental margin and cause the continental shelf to prograde seaward over time; the Gulf of Mexico passive margin along the southern USA is a prime example. Along the east Texas coast, the continental shelf has grown by 300 km during the Cenozoic.

Passive continental margins typically show a continuous subsidence during their evolution. As sediment covers and masks the old rift topography described above, a smooth, broad, gently seaward-dipping coastal plain-continental shelf is generated. Such shelf areas tend to be the broadest on Earth, as great as 500 km wide along portions of North and South America. The subsidence is orchestrated by several factors:

1. Extensive continental crust extension and thinning before continental separation, causes isostatic subsidence. The less-dense crust is replaced by denser mantle material (☐ Fig. 4.2). Down to depths of 10–12 km, normal faults accommodate subsidence whereas below this depth, plastic deformation thins and lowers the crust. The significant difference in thickness of thinned continental crust (15–20 km) and oceanic crust (6–8 km) drives the tendency to compensate this variation by ductile flow of the deeper crust towards the ocean ("crustal creep"; ☐ Fig. 4.2). This factor enhances crustal thinning and subsidence at a passive continental margin.

2. The newly formed oceanic crust and the adjacent continental margin cool as the ocean opens and expands. The facing passive continental margins drift away from the spreading center and are thus removed from the heat source of the rising mantle current. Both the oceanic lithosphere and the asthenospheric mantle cool as this process continues; the cooling increases density and generates subsidence.

3. Increasing sediment load is an additional factor. The processes described above generate accommodation space for sediment at continental passive margins, much of it below sea level. This condition attracts and preserves sediment; increasing sediment adds weight that generates additional subsidence as the underlying crust and mantle are depressed (the weight of continental glaciers also depresses the surface as seen at Hudson Bay and the Baltic Sea).

Once sediment loading initiates, it tends to be a self-generating process; as the edge of the continent is pushed down, more sediment enters the area and generates more subsidence.

4.2 The Sedimentary Trap at a Passive Continental Margin

The continuous subsidence of passive continental margins generates large sedimentary traps that lead to some of the largest accumulations of sediments on Earth. Both carbonate and clastic production can be high. Excess sediment progrades over the edge of the shelf and spills into deeper water additionally increasing the total volume and load of sediments (☐ Fig. 4.4). In low latitudes with warm water and low clastic influx, carbonate production is high and the accumulation of sediment on the shelf keeps up with the subsidence rate so that over a long time span, a thick series of shallow water carbonates develops. The balance between subsidence and sedimentation rate maintains the surface at the same elevation because carbonate sedimentation keeps up with subsidence, but stalls or ends if the shelf is above sea level. The controls of clastic sedimentation on passive margins are somewhat different. Water depth increases as sea level rises and sediment is trapped in near-shore settings; as sea level falls, sediment bypasses the shelf and progrades into deeper water. These processes are explained in more detail below.

During the initial stages of continental breakup, connections to an open ocean are commonly topographically restricted. This results in distinctive sedimentary deposits. In warm climates salt deposits form when evaporation rates are high (Red Sea, ▶ Chap. 3). Under both arid and humid conditions, deeper portions of a restricted sea lead to the deposition of black oozes, sediment rich in organic material. The circulation of oxygen-rich water in a narrow sea is reduced so sediment on the bottom has insufficient free oxygen to decompose the dead organic material from biogenic production. This results in the formation of sapropels, which during diagenesis consolidate to form

4

■ Fig. 4.4   A complete
sedimentary cycle at a passive
continental margin (Einsele
1992). A sequence is subdivided
into several tracts that are
controlled by transgression and
regression of the sea

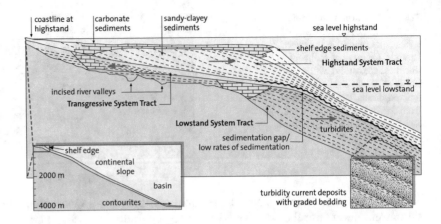

black shales rich in bitumen. These black shales commonly become source rocks for petroleum deposits and explain why passive margins are so important as major petroleum provinces (see below).

As passive margins mature, they become the site of widespread normal marine sedimentation. Because these sites are relatively tectonically inactive, they are excellent recorders of ancient sea level trends; they passively record the long-term rise and fall of sea level across the shelf. Sea level fluctuations of tens to more than 100 m, with periodicities of hundred of thousands to tens of millions of years, have been recorded in the geologic record. The reasons for the cyclicity, which is world-wide, seem to be caused by two major culprits: (1) plate tectonic processes and the resulting locations of continents, and (2) orbital parameters of the Earth. When these two factors have worked in concert, they have produced major glacial episodes throughout geo-logic history. But independently, they each produce cycles with their own diagnostic periodic signature.

The spreading rate at mid-ocean ridges controls the width of the ridge and, therefore, its volume. This in turn influences the total volume of the ocean basins. When ridges are relatively large, their excessive volume pushes water out of ocean basins and onto continents resulting in high sea level. Smaller ridges have the opposite effect. These cycles operate at long frequencies of tens of millions to hundreds of millions of years. The resulting sea level cycles are called first-, second-, and third-order cycles. A first-order cycle corresponds to the length of a so-called supercontinent cycle. Sea-level highs in the Ordovician and Cretaceous are examples and correspond to rapid sea-floor generation in successive super-cycles. Second- and third-order cycles correspond to the approximate length of the geologic periods and major divisions of the periods, respectively.

Tracts of Sequence Stratigraphy

A sequence is subdivided into system tracts (■ Fig. 4.4). Each tract represents a stage of sea level change. Although cycles have been defined differently by different workers, one of the most widely used methods begins a cycle during falling sea level. Low sea level commonly causes erosion on top of the underlying deposits resulting in a widespread unconformity. The resulting sequence boundary is marked by a gap in the stratigraphic record, which typically is characterized by regional angular discordance (the layers are non-parallel). The discordance develops during the time during which the regression and erosion occupy; the slight tilting of the bedrock originates through unbalanced regional subsidence. More oceanward, the sequence boundaries can be difficult to identify as the layers display little or no interruption of sedimentation and are concordant (parallel) with each other. The lower sequence boundary is an unconformity landward on higher portions of the shelf, and a correlative conformity seaward on lower portions of the shelf or across the shelf break (continental slope). The Lowstand System Tract (LST) develops on the sequence boundary during the time of low sea level. Because upper parts of the shelf are exposed to subaerial erosion, clastic sediments are transported beyond the shelf edge where they generate a sediment wedge beyond the shelf break well below sea level. River valleys incise into parts of the shelf as the time of unconformity development continues. Sediment bypasses the shelf, accumulates at the shelf edge, becomes over steepened, and feeds suspension currents that generate turbidity deposits. The resulting turbidites develop thick packages at the continental slope but thin distally into the adjacent ocean basin. In turbidity currents the sediment slides down with high velocities into great depths as a dispersed suspension. The layers of turbidites formed after the deposition of the suspended material have thicknesses that range from a few centimeters to one meter or more and have a graded bedding which is characterized by a decreasing grain size (normal grading). Coarser material is deposited rapidly as fines slowly settle out of suspension, commonly long after the turbidity current has waned (◻ Fig. 4.4).

As sea level rises, the sediment delivered from the land is trapped at the shoreline. As the transgression accelerates, sedimentation accumulates to form the Transgressive System Tract (TST). This results in a broad zone of landward-overlapping shelf sedimentation. The sediment thickness tends to be variable and is dependent on the rate at which sea level rises and creates the accommodation space versus rate of sediment supply. During rapid transgression, very little sediment may accumulate as accommodation outstrips sediment rates. During development of the transgressive system tract, little or no sediment reaches the shelf edge and the deeper water beyond (◻ Fig. 4.4) and what little does, is fine grained. These thin, fine-grained deposits are termed condensed sections and are powerful markers of times of rising sea level.

The Highstand System Tract (HST) encompasses the time of sea-level highstand—the time between transgression and regression. As sea-level rise diminishes, so does the formation of new accommodation space. Therefore, sediments are spread laterally by marine processes until all space on the shelf is filled. Towards the end of this tract, sediment spills off the shelf into deeper water. Ensuing regression drives the locus of sedimentation towards the the shelf edge (shelf edge sediments) as the shorelines progrades across the shelf, in some cases all the way to the edge.

Orbital parameters operate at much shorter frequencies and are called Milankovitch cycles after the scientist who first studied and defined them. The most important Milankovitch cycles operate at 100,000 and 400,000 years and reflect the excentricity of Earth's orbit and change in tilt of the axis. Not coincidentally, the Pleistocene global glaciation cycles operated at the same frequencies; but, of course, plate tectonic processes had moved the continents into ideal locations for generating glaciers.

The interplay of sea level fluctuations and subsidence causes cyclic sedimentation at passive continental margins. Global correlation of cycles falls within the realm of sequence stratigraphy. A sequence is defined as genetically related strata (accumulation) bounded by surfaces of erosion or non-deposition (unconformities). Most marine sequences initiate as deposits formed during transgression and conclude as deposits formed during subsequent regression. A transgression is a landward shift of the sea and can be caused by rising sea level, subsiding land, or both. A regression is a seaward retreat of the sea from the land and can be caused by falling sea level, rising land, or both.

4.3 Processes on Continental Margins

Sedimentation on passive continental margins is controlled by the balance between terrigenous sediments (delivered from the land) and biogenous sediments (typically carbonates on marine shelves—sediment produced in the ocean). Because clastic (*klastein*, *Greek* to break) sediment is derived from the land, high production is favored in areas of high relief, areas adjacent to large river systems, and regions subject to humid climates. Nearly all clastic-dominated shelves have their sediment delivered by rivers. In contrast, carbonate sediments form in areas where clastic sediment is sparse: areas of low relief, lack of rivers, commonly sub-humid climates. Carbonate production in the ocean is mainly from biogenic activity and is especially favored in shallow, warm, well-circulated waters. This is where lime-secreting organisms thrive, the area referred to as the "carbonate factory". Under some conditions, carbonates can form in cooler, deeper water, although many deep-water carbonates are derived from the shallow, adjacent shelf and transported into deeper water by turbidity and gravity processes.

Although ultimate clastic sediment supply to the oceans is from the land, distribution across the shelf and into deeper water is complex and strongly controlled by the system tracts discussed above. During marine transgression, sediment, especially sand, is trapped in shoreline environments, especially estuarine and barrier bar systems. During highstand, large delta systems cover the shelf with clastic sediment and fill available accommodation space. During lowstand, sediment bypasses the shelf and is transported into deeper water. An extreme example of this occurred during Pleistocene lowstand when dunes migrated from the western Sahara desert over to the edge of the dry continental shelf from where they were transported into the deep sea by turbidity currents (Sarnthein and Diester-Haass 1977).

Sedimentary processes can cause the erosion of portions of continental shelves. The erosional effect of turbidity currents is caused by the rapid movement of large volumes of sediment across the shelf edge and into deep water. This carves deep submarine canyons across the continental slope, typically near the mouths of large river systems that provide the appropriate delivery of sediments. An example is the submarine canyon at the mouth of the Hudson River on the east coast of North America (◻ Fig. 4.5). During the Pleistocene ice age, global sea level was more than 100 m

4

Fig. 4.5 The submarine canyon at the mouth of the Hudson River along the passive continental margin of North America

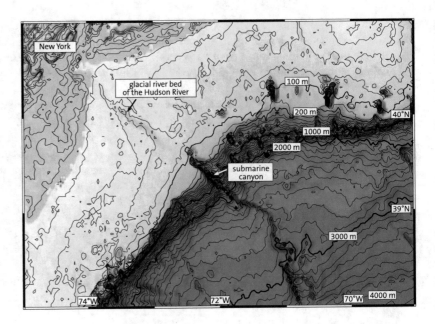

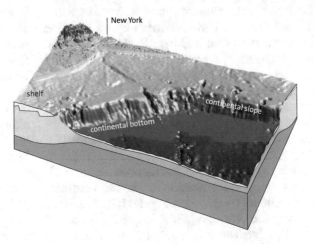

lower than today because sea water was locked up in the large glacial mass. Therefore, rivers could carry their sedimentary load closer to the shelf edge and provide material for the generation of turbidity currents.

Slumps and slump deposits are common along the edges of continental shelves. The combination of high water content in the pore space of unconsolidated sediments coupled with high sedimentation rates produces unstable sediment mass and generates large slumps. The sediment in these cases is not dispersed but rather moves as a fluid mass, even in areas with very low gradients. Massive slumps can occur on slopes that range from less than 1° to approximately 6°. Dimensions can be staggering; off Namibia, late Tertiary slumps occupy areas of several ten thousand square kilometers (Dingle 1980). The Storegga slides of Norway are the largest known slumps on Earth with a total volume of 5600 km³. A massive slope failure occurred about 7000 years ago and produced a tsunami wave which left a series of tsunamigenic sediments in the Norwegian fjords.

Turbidite sequences along the continental slope-rise form a link between continental slope and ocean basin by slowly prograding towards the basin. Some of this material is redistributed by slope-parallel, deep oceanic currents called contourites that follow the contours of the continents (■ Fig. 4.4, insert on left side; ■ Fig. 4.10). They are produced by the global oceanic circulation system and are partly driven by dense, gradually descending, salt-rich, and cold water currents in high latitudes. Their slope-parallel configuration is controlled by the Coriolis force, which also redirects them towards lower latitudes. Therefore, the contourites run in a counterclockwise direction in the northern hemisphere and clockwise direction in the southern hemisphere. Contourite currents achieve velocities of some decimeters per second.

The Gulf Stream generates contourites by its deep-sea countercurrent. It is an example of a current with strong evaporation in low latitudes that causes an increasing salt content and thus density of water. As the Gulf Stream approaches the Arctic region, the water cools and gains additional density that causes its descent. Following its descent to the ocean bottom, it flows along the North American coast back towards lower latitudes. Sediments transported by these contour currents are eventually deposited as contourites. They resemble distal, fine-grained turbidites.

Carbonate response to changing sea level is somewhat different from that of clastics. Shelf areas dominated by carbonate sedimentation are generally able to keep up with rates of subsidence or sea-level rise and are capable of producing thick sequences of limestone and dolomite. However, during sea level lowstand, carbonate production is limited to the outermost edge of the shelf (◻ Fig. 4.4). During transgression, the carbonate platform broadens and shifts landward. The highstand is characterized by the broadest belts of carbonate sedimentation across the shelf. Partly consolidated carbonate sediments may form steep scarps where large carbonate grains are transported beyond the continental slope; meter-to decimeter-sized blocks can slide down the slope.

Carbonate platforms in warm climates are characterized by fringing reefs along the shelf break with widespread lagoons behind. Shallow-water carbonate sediments are readily dolomitized. The original calcium carbonate is altered by ion exchange in pore fluids rich in magnesium into calcium–magnesium carbonate (dolomite). Recent examples of broad tropical carbonate platforms with fringing reefs and widespread lagoons include the Bahama Bank adjacent to Florida and the Great Barrier Reef on the northeastern coast of Australia; the latter has a length greater than 2000 km. Both areas are passive continental margins. A large fossil carbonate platform occurs in the Triassic of the Northern Calcareous Alps in the Eastern Alps. Reefs of the Dachstein limestone (Late Triassic, ca. 210 Ma) bordered a huge lagoon with a surface area of more than 100,000 km^2. The total thickness of the Middle and Late Triassic carbonate sequences reaches 3000 m, documenting the strong subsidence of this passive continental margin in the northwestern part of the Tethys Ocean.

4.4 Petroleum Deposits—The Economic Significance of Passive Continental Margins

Passive continental margins contain extraordinarily important petroleum deposits. The source rock forms especially during the rift stage when narrow oceanic basins are restricted and black shales rich in bitumen are

◻ **Fig. 4.6** Petroleum deposits of the Middle East were formed in sedimentary rocks at a passive continental margin later deformed by continental collision. Different types of petroleum traps developed in the Zagros mountain chains and on the Arabian platform

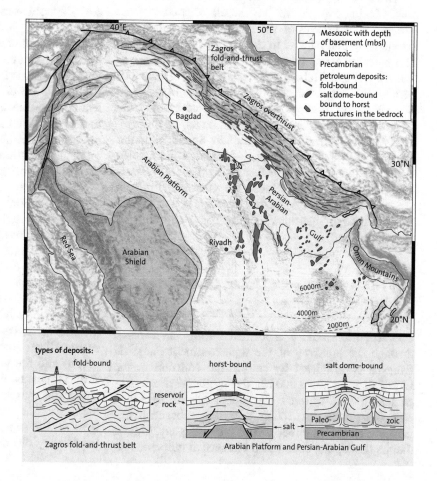

4

deposited. The oil begins to generate during the heating process caused by the superposition of younger layers (petroleum window, see ▶ Chap. 3). Oil and gas migrate after their formation into porous or fractured reservoir rocks; because of their low density, they have the tendency to ascend towards the surface. Only if they are captured in "traps", do they remain at depth and can be exploited. Faults and salt diapirs, typical of passive continental margins, are ideal petroleum and gas traps because they may trap and seal petroleum in porous layers. Oil and gas may be trapped in folded layers as well; here the low-density petroleum rises to the crests of anticlines and is trapped if the reservoir layer is overlain by an impermeable (mostly clayey) layer.

The rich petroleum deposits in the Middle East (Arabian Peninsula, Iran) formed on a passive continental margin that was subsequently folded during orogeny of the Zagros Mountains. Deposition of sediments and accumulation of organic material occurred at the passive continental margin of the Tethys (◘ Fig. 4.8). The petroleum was sourced from Mesozoic rocks, the reservoir rock is mainly fractured middle Tertiary Asmari Limestone. The petroleum was trapped by the successive events that led to the closure of the Tethys and the collision of Arabia with the Eurasian Plate.

Three of the petroleum traps above occur in this area (◘ Fig. 4.6). Many oil fields, located in the strongly folded layers of the Zagros mountain chains, are elongate and parallel to the NW–SE trending folds. The petroleum was trapped during folding in the anticlines. On the Arabian Peninsula and in the western part of the Persian–Arabian Gulf, the oil fields trend N–S. Folds occur above similarly oriented horst structures which formed along normal faults in the Precambrian basement of the Arabian Shield. Circular oil fields in the eastern Persian–Arabian Gulf formed above salt diapirs that were formed by the rise of Early Paleozoic salt deposits. Hence the greatest petroleum reserves on Earth are of various tectonic styles, though each of these originates in the same passive continental margin setting.

4.5 The Atlantic—An Ocean Opens in an Intricate Manner

The Atlantic Ocean was formed during the Jurassic through the break-up of the supercontinent Pangaea. Pangaea consolidated by continental collision in the Late Paleozoic about 300 Ma during the Variscan–Appalachian–Ural orogeny The supercontinent remained welded for more than 100 m. y. until its subsequent demise, which took another 100 m. y. to complete. Mantle diapirs (hot spots) punctuated Pangaea at numerous lo-

calities (◘ Fig. 6.5). Many of these evolved into graben structures, which like in the Afar Triangle, tend to generate three branches. The branches eventually joined to form continuous graben systems that yielded the break-up of the continent (◘ Fig. 4.7).

Graben systems that do not conjoin to another system failed because the extension concentrated on the continuous systems. The failed graben systems continued to subside long after abandonment and were subsequently filled with sediments from large rivers. Such failed graben systems are called aulacogens (*aulakogenesis, Greek* formation of a furrow). Before the Atlantic Ocean was formed, a number of triple-junction graben systems were formed and each had a failed arm that became an aulocogen. These are readily identifiable as each forms a bight into a continent and each has a major river draining into it. The mouths of the Niger and Amazon are two excellent examples (◘ Fig. 4.7). The early history of the Atlantic demonstrates that hot spots play an important role in graben formation and subsequent break-up of a continent; bulges of the asthenosphere are not long-stretched ridges, but primarily domes. Furthermore the Atlantic did not form in a consistent process at a continuous break line, but rather in several phases separated from each other.

4.6 Pangaea and Panthalassa

The supercontinent *Pangaea* (*Greek* the entire Earth), which was extant from about 300 to 175 Ma, was initially reconstructed by Alfred Wegener. Wegener used the continuity of mountain belts and other geologic features, as well as the fit of the continents, to make these reconstructions. Additional support for the reconstruction came from paleontologists and was based on the distribution of plant and animal fossils from the various parts of the supercontinent (◘ Fig. 1.1).

New reconstructions have enhanced significantly Wegener's initial fit of the continents (◘ Fig. 4.7). Modern reconstructions account for the thinned continental margins and use the 500 fathom depth (927 m) of the continental margins to yield the best fit. Remaining gaps and overlaps of the continental blocks need only minor local correction of this mean value and are due to variabilities in crustal thinning during break-up of Pangaea.

The break-up of Pangaea was complicated. The surrounding *Panthalassa* Ocean (*Greek* the entire ocean) and the westward extending branch of the Tethys Ocean contained numerous subduction zones along the margins of Pangaea (◘ Fig. 4.8; 260 Ma). Stresses in the subduction zones imposed tensional effects on the supercontinent, which acted on the thick, partly very old lithosphere of Pangaea that formed a

◨ Fig. 4.7   Restored continental fit across the Atlantic at the time of Pangaea (Bullard et al. 1965). Stretched continental margins are responsible for the apparent overlapping areas. Bulges over hot spots and subsequent graben structures with three branches played an important role during the break-up of a continent. Graben systems abandoned after the break-up of the ocean are called aulacogens

heat shield against the deeper mantle and resulted in the accumulation of heat below the lithosphere of Pangaea. Hot spots in the mantle caused bulges in the lithosphere that eventually burst at zones of weakness and subsequently, graben systems developed. Tensile stress in the lithosphere meant that graben systems could spread and the break-up of the supercontinent commenced. Note the important role that hot spots played in this process.

The *first phase* in the break-up of Pangaea occurred between North America and NW Africa in the area of

the present Central Atlantic. This break line continued westward between the two American continents, crossing the present Caribbean. Eastward it separated Europe and the Iberian Peninsula, which was still connected to the European mainland, from Africa including later microplates of the Adriatic region. Thus, the initial break-up of the supercontinent followed the shortest distance from the Tethys Ocean to the Pacific Ocean (◨ Fig. 4.8; 150 Ma). First indications of crustal extension along this fracture appeared in the Triassic approximately 200 Ma ago yet the initial break-up did not oc-

4

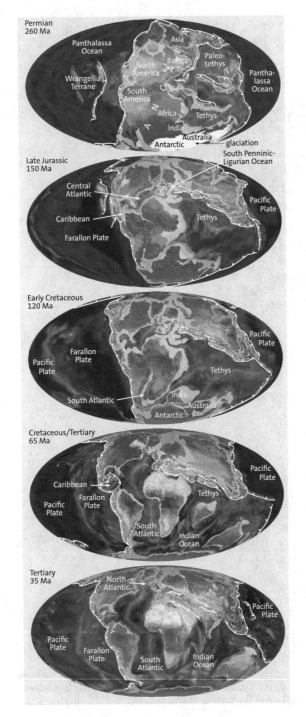

☐ **Fig. 4.8** Paleogeographic maps showing the break-up of Pangaea and the opening of the Atlantic Ocean (Blakey 2016)

cur until the Early/Middle Jurassic boundary at approximately 175 Ma. Initial direction of opening trended SW–NE. Oceanic crust was formed in the Caribbean, the central part of the Atlantic, and in the South Penninic–Ligurian Ocean which was later destroyed in the Alpine orogeny. Present ocean floor of Jurassic age exists only in the Central Atlantic and in the Gulf of Mexico (☐ Fig. 2.12). The Alps tectonically incorporated remnants of Jurassic ocean floor as ophiolites (▶ Chap. 5).

In the *second phase* of break-up, the South Atlantic Ocean between South America and Africa (☐ Fig. 4.8; 120 Ma) was formed. This break-up, however, did not start near the Central Atlantic, but rather initiated from the south and spread northward with oceanic crust first forming during the Early Cretaceous. This resulted in South America and Africa opening scissors-like with the southern parts of these continents drifting faster apart than more northward parts. The connection to the already existing Central Atlantic thus happened last. Because the breaking line had a very acute angle to the direction of opening, which was oriented approximately east–west, the zone of connection became exceedingly complicated. This complication generated numerous significant transform faults in the present equatorial Atlantic (☐ Fig. 2.2). The combination of closely spaced east–west-striking transform faults and only short north–south-striking ridge segments resulted in the WNW–ESE orientation of this ocean segment.

The *third phase* of break-up, the North Atlantic Ocean, developed several paths before the final location was established (☐ Fig. 4.9 top). An initial opening occurred between Newfoundland and Iberia during the Early Cretaceous, concurrent with the opening of the South Atlantic. The continuation of this fracture ran between the French Atlantic coast and the Spanish northern coast along the Pyrenees and continued farther into the North Penninic Ocean of the Alps. A 30° counterclockwise rotation of Iberia enabled the opening of the Bay of Biscay. The attempt to continue continental breakup here failed during the Late Cretaceous. Instead, rifting developed and spread northward between Greenland and the British Isles causing the separation of the Rockall Plateau, a submarine plateau with continental crust that today belongs to Europe (☐ Fig. 4.9; middle and lower). The rift between Rockall and the British Isles failed and jumped westward and formed a rift between Greenland and Canada; it too failed (see the map sequences on ☐ Fig. 4.9). Finally, as described below, during the Tertiary, the North Atlantic rift formed.

The final path of the North Atlantic Ocean was determined by hotspot activity associated with Iceland. Approximately 60 Ma, a hot spot originated at the location that would become Iceland as large volumes of basalts extruded. This resulted in the formation of a new spreading axis during the Early Eocene at 55 Ma that fulfilled the final successful continental separation between Greenland and Northern Europe along the Early Paleozoic Caledonian Mountains. These mountains constitute a zone of crustal weakness between the ancient consolidated shields of Greenland and Scandinavia and rifting exploited this weakness. North Atlantic separation reached only to northern Norway while the spreading axis in the Labrador Sea between Greenland

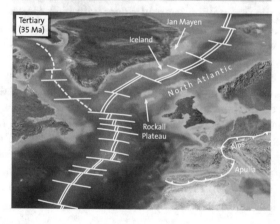

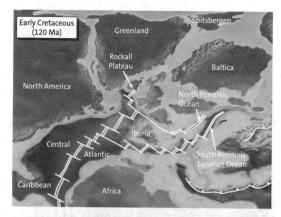

Fig. 4.9 Paleogeo-graphic maps showing the gradual opening of the North Atlantic Ocean. Following several failed attempts to continue rifting northward, the present spreading axis was eventually established (Blakey 2016)

Spitsbergen and the northeastern coast of Greenland and, at present, the Mid-Atlantic Ridge passes over the pole region towards the Siberian shelf (◘ Fig. 2.17).

As documented above, the spreading axis of the Atlantic, which currently appears uniform and continuous, was formed in many stages. Current spreading rates from less than 1 to about 4 cm/year indicate a rather uniform spreading axis. However, the history of ridge spreading and development documents a far more complex earlier history, a history replete with different segments of different ages with different histories, finally unified as a cohesive spreading center.

The oldest oceanic lithosphere in the Central Atlantic is Early to Middle Jurassic (~175 Ma). This old, cold dense oceanic lithosphere has a significantly higher average density than the asthenosphere below it and thus would be ripe for new subduction to develop. Therefore, the Atlantic Ocean has a limited endurance and the closure of this ocean and the collision of the continents at both sides will eventually occur; a new Pangaea would evolve and a Wilson cycle (▶ Chap. 11) would be ended.

4.7 The Large Abyssal Plains

The abyssal plains (◘ Fig. 4.10) are continuous to the flanks of the mid-ocean ridges and, although typically thought of as flat and featureless, they are morphologically complex and punctuated by numerous oceanic islands, ridges or elevated plateaus. The large plains occur at 4000–6500 m water depth, subside with increasing distance from the ridge, and are bordered at their margins by a passive continental margin or a subduction zone.

The age of the sediments that directly overlie the oceanic crust are only slightly younger than the oceanic crust itself and conclusively confirm the theory of seafloor spreading. Overall, basal sedimentary layers systematically increase in age from the ridge over the deep-sea plains to the continental margins and thus mirror the increasing age of the crust with increasing distance from the spreading axis. The age of the sediment, determined by fossils, is generally well known because of international research campaigns to explore the ocean floor by drilling; performed from research vessels, the global deep-sea drilling program (Deep Sea Drilling Project, DSDP, and its successors Ocean Drilling Program, ODP, Integrated Ocean Drilling Program, IODP, and International Ocean Discovery Program, IODP) has drilled oceans across the world.

The subsidence of the ocean floor that accompanies the increasing distance to the mid-ocean ridge is a consequence of cooling and thickness increase of the oceanic lithosphere with increasing age. Cooler lithosphere is denser and thus sinks deeper into the asthenosphere.

and North America remained active. Then near the end of the Eocene, approximately 35 Ma, the spreading axis between Greenland and Northern Europe was established east of Jan Mayen which was part of Greenland at that time (◘ Fig. 4.9 bottom; Braun and Marquardt 2001). Subsequently Jan Mayen was separated from Greenland as the Mid-Atlantic Ridge jumped westward and now runs between it and Greenland. The shift of the spreading axis emanates from the hot spot of Iceland that developed intense volcanic activity at that time. The North Atlantic then cleaved its way between

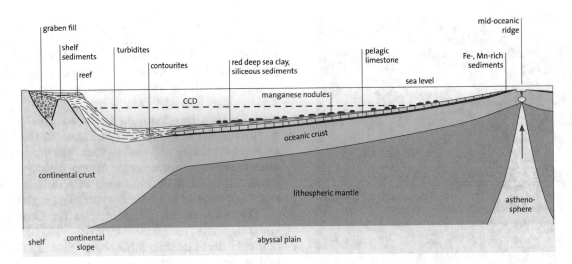

◼ Fig. 4.10 Cross section through a large ocean basin showing characteristic distribution of sediments related to the CCD (calcite compensation depth). Note that Walther's Law of Facies is well illustrated; see ocean floor under the letters CCD and note that deposits that formed above the CCD are overlain by deep-sea oozes that formed below the CCD and that these are overlain by terrigenous sediments

Moreover, the uppermost part of the asthenosphere is transferred into lithospheric mantle by cooling. Thus, old oceanic lithosphere is thicker, denser, and deeper under the surface of the ocean.

Subsidence of the ocean floor does not occur in a linear fashion, but is more rapid near the midocean ridge where it averages approximately 1000 m during the first 10 million years and slows with increasing distance to the ridge (e. g., 1000 m during the next 26 million years; (◼ Fig. 4.11). Therefore, the depth of the ocean floor increases primarily as a function of its age where the depth increases with the square root of its age. The subsidence follows approximately the equation

$$A = k \cdot \sqrt{t},$$

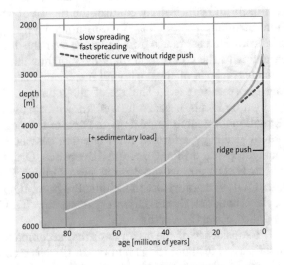

◼ Fig. 4.11 Subsidence curves of oceanic lithosphere. The water depth above the ocean floor is primarily controlled by its age; after 20 m. y., depth is independent of spreading rates at the mid-ocean ridge

where A is the subsidence in meters (emanating from the mid-ocean ridge), k is a constant with the approximate value of 320 and t the age of the ocean floor in million years (Parsons and Sclater 1977). Near the mid-ocean ridge, the subsidence varies with high and low spreading rates. The rate of subsidence is higher during the first million years at ridges with high spreading rates because the ocean floor drifts faster from the area of rising asthenospheric current, a process that additionally uplifts the ridge by ridge push.

As ocean floor increases in age, its depth deviates from the formula. In the Pacific Ocean, the formula holds to approximately 60 Ma and then subsidence decreases at a slower rate than predicted. The reasons for this phenomenon are not totally resolved, most probably it is a consequence of the gradual decrease of thickness growth of the oceanic lithosphere. After about 60 m. y. the thickness of the lithosphere reaches about 80 km and then grows very slowly (Nicolas 1995).

4.8 Sediments of the Abyssal Plains

Sediments of the abyssal plains are primarily very fine grained and are composed of terrigenous clays and organic materials. The former are derived from continents, either as distal suspended material or wind-blown loess, and the latter are derived from microorganisms that thrive in the oceanic environment. Microorganisms secrete either siliceous or carbonate tests that constantly rain down on the sea floor as the organisms die. Terrigenous material varies widely and ranges from volcanic ashes, to distal turbidite material, to wind-blown debris. Whether terrigenous or organic, this material is classified as pelagic (*pelagos, Greek* sea). In abyssal areas adjacent to continents, grain size slightly increases and reflects the continental origin—such sed-

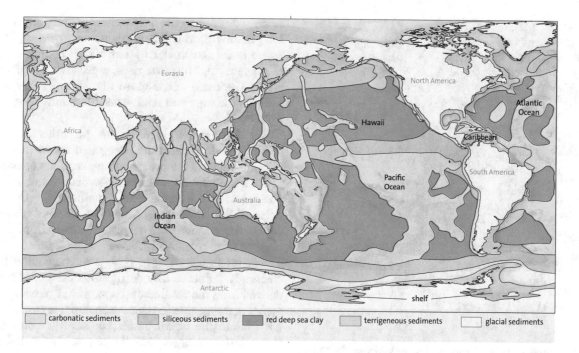

Fig. 4.12 Map showing present global distribution of sediments in the oceans (Berger 1974; Reading 1986)

iments are called hemipelagic (formed by oceanic processes but derived from the land). Because the abyssal depositional setting is a very passive one, they tend to reflect conditions such as climate, adjacent volcanic sources, or global wind patterns. Nearer continents, abyssal sediments reflect the continental conditions that are imposed on the adjacent deep sea environments and are referred to as terrigenous sediments (□ Fig. 4.12). The boundaries between terrigenous, hemipelagic, and pelagic materials are completely gradational.

The different sediment types represent specific sedimentary facies. The term facies denotes the whole character of a sediment and is related to a certain set of conditions that relate to the formation of the deposit. The lateral facies transitions on the ocean floor reflect the different conditions. As the plates migrate, the ocean floor enters zones of changing conditions and thus the facies change. This implies that the laterally changing facies should also occur in a vertical sequence with the youngest sediment on top (□ Fig. 4.10). This vertical and horizontal spatial relation of different types of facies is reflected in the principle called Walther's Law of Facies (Walther 1894).

The calcite compensation depth (CCD; □ Fig. 4.10) plays an important role in the sediment distribution of the deep sea. Below a certain deep level in the oceans, the solution of calcite exceeds the supply of calcite remnants of organisms supplied from above and dissolution of calcite occurs. Solution of calcite at these great depths occurs because the cold and deep oceanic water is rich in carbonic acid which causes the solution. Carbonic acid mainly comes from the decomposition of organic material. The supply of calcite mate-

rial comes from planktonic organisms that flourish in the uppermost layers of the water column. When the organisms die, their shells descend downwards through the water column. The planktonic layer consists of tests of foraminifera (single-celled animals), and the plates of coccolithophorids (single-celled algae).

Surficial oceanic water, especially in warmer tropical climates, are saturated to supersaturated with respect to calcium carbonate. However, in deeper waters at several kilometers depth, this condition changes and waters become undersaturated. Calcite solution occurs at the condition called the lysocline, although dissolution is delayed by an organic wrapping around the carbonate material. When this buffering effect is removed, calcite solution occurs. The level where complete dissolution takes place is the CCD.

The depth of the CCD varies. Generally it is deeper in equatorial areas than at high latitudes (□ Fig. 4.13). In the Atlantic the CCD is generally between 4500 and 5000 m; in the Pacific, it ranges from 4200 to 4500 m. It decreases to depths below 5000 m in all oceans near the equator whereas in subarctic areas it rises up to approximately 3000 m. In the geological past the CCD has fluctuated because of differing controls. For example, during the Cretaceous and Early Tertiary it was 1000–1500 m higher than today. Then, warmer climate and high volcanic CO_2 production led to a higher organic productivity and a lot of decomposition took place n the oceans and greater levels of CO_2 were dissolved in sea water; this favored the solution of calcium carbonate. Before the Late Jurassic the CCD was also higher because fewer carbonatic plankton populated the open oceans and, therefore, the deep water was undersaturated with respect to calcite.

4

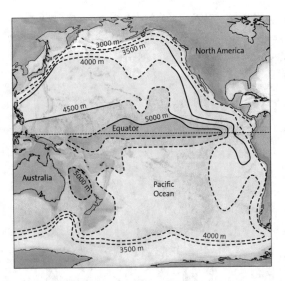

□ Fig. 4.13 Map showing depth to the CCD (calcite compensation depth) in the Pacific Ocean (Berger 1974)

The significance of this is that below the CCD, no carbonate sedimentation occurs. Thus the deeper portions of abyssal plains are dominated mainly by siliceous or clay-rich sediments. Siliceous sediments are composed of radiolarians (single-celled animals that secrete siliceous tests and live in equatorial areas) or diatoms (siliceous algae, thriving in high latitudes). The siliceous material consists of amorphous (non-crystalline) opal ($SiO_2 \cdot H_2O$) that is unstable and transformed with time into crystalline chalcedony and finally into quartz (SiO_2). The growth of sili-

ceous sediments in the deep sea occurs very slowly. The bio-geneous production is higher in upwelling zones where nutrient-rich deep water, which contains lot of phosphates and nitrates, rises, especially in equatorial areas. The trade winds move the warm surface water towards the west and cold water flows into shelf areas from depth.

In nutrient-poor areas with no other sediment source, red deep-sea clay is deposited below the CCD. It is red because the ocean bottom currents constantly provide oxygen-rich water that oxidize the ubiquitous fine particles of iron oxide to hematite ($Fe_2 + {}^3O_3$). Particles of clay originate by long-distance wind transport across the oceans from terrestrial areas, from volcanoes, or from cosmic dust.

Nearer the continents, pelagic sedimentation is increasingly influenced by terrigenous clastic sediments derived from the continents. Especially at passive continental margins, this terrigenous material is easily transported into the ocean basins because the continental slope directly merges into the abyssal plain. Subduction zone deep sea trenches or island arc systems provide an efficient barrier to such deep-sea transport from the land. Supply of terrigenous material is exceptionally high in front of large river mouths at passive continental margins (e. g., Ganges–Brahmaputra, Indus, Niger, Mississippi, Amazon) where huge sedimentary fans prograde into the deep sea. In subarctic regions of Antarctica or in the North Atlantic, moraine debris deposited by glacial ice is an important sediment source in adjacent deep sea basins (□ Fig. 4.12).

Manganese Nodules from the Deep Sea

The abyssal plains between 4000 and 5000 m water depth are littered with manganese nodules across broad areas (□ Figs. 4.10 and 4.14) and represent a considerable ore reservoir. The nodules are potato-shaped structures that are composed of crusts of iron and manganese hydroxides around a rock fragment; the nodules occur in fine-grained sedimentary layers and attain sizes of up to 20 cm. Important amounts of copper, cobalt, nickel and other metals also occur. These precipitates are derived from dissolved material in the sea water that in turn originates from the extraction of metal ions from oceanic crust at the mid-ocean ridges by circulating sea water (▶ Chap. 5). The nodules develop in areas with intensified bottom currents where sedimentation rates are extremely low. Areas of higher sedimentation rates mask the manganese nodules. The growth rate is in the order of several millimeters per million years and is extremely slow.

Manganese content in the nodules can be 30% whereas the contents of copper, nickel and cobalt ranges mostly around 1% each. On land, such concentrations would be considered very high-grade ores; however, mining of deep-sea nodules is currently not practical from an economic standpoint because of the great water depths involved and their great distance from land. Moreover, deep-sea mining would create significant damages to the fragile abyssal environment. The most concentrated fields of manganese nodules occurs in the eastern Pacific, north and south of the equator; however, they are present in virtually all abyssal plains of the three large oceans (□ Fig. 4.14). The global deep-sea reserves of manganese nodules is estimated to be 10 billion tons. Copper, nickel and cobalt represent the highest economic value. Their successful exploitation awaits new technologies.

□ **Fig. 4.14** Map showing occurrences of fields of manganese nodules on the abyssal plains (Berger 1974)

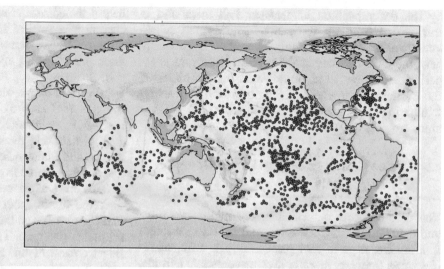

Sedimentation rates are low in the deep sea. The deep-water limestone rates are between 3 and 60 m per million years, those of siliceous sediments and red deep sea clay are between 1 and 10 m per million years (Berger 1974)—these are extremely slow sedimentation processes. Therefore, radiolarites and red deep sea clays encompass long time intervals. On the contrary, the terrigenous input in front of river mouths is very high and can exceed 1 km/m. y. even though such strong accumulations of sediments do not last over long time periods.

Deep-sea currents with velocities ranging to decimeters per second may winnow fine clayey or biogenous particles and deposit the fine material elsewhere. This process generates gaps in the sediment record, especially at swells on the ocean floor that are affected by higher current velocities. The bottom currents are controlled by surficial currents which themselves are controlled by the distribution of continents and climate. Moreover, they are influenced and deflected by the Coriolis force. Long-term ocean current patterns change through Earth history so that sedimentation in the deep sea is variable; the present distribution of deep sea sediments provides only a snap-shot of this complicated process.

The following discussion presents some of the factors that control present distribution of deep-sea sediments (□ Fig. 4.12): Calcareous sediments are restricted to areas above the CCD. In equatorial areas where the planktonic production is very high, the CCD is depressed by the rapid influx of calcite particles so calcite distribution is expanded. Siliceous sediments occur in zones of upwelling, especially along west-facing Pacific coasts from Mexico to Peru. Siliceous sediments rich in radio-laria (radiolarites) occur across the entire equatorial area below the CCD where carbonate sediments are absent. At high latitudes, siliceous sediments consisting of diatoms are widespread because the CCD is elevated and carbonate production is suppressed. In wide areas across the abyssal plains below the CCD, red deep-sea clay is deposited.

Clay mineralogy of deep-sea clays is variable (□ Fig. 4.15) and provides clues concerning their location of origin and condition of formation. Montmorillonite evolves as a product of weathering of volcanic ashes and is widely distributed in the Pacific where numerous volcanic islands and subduction related volcanoes occur. In the Atlantic, montomorillonite is rare as volcanoes are less prevalent. In contrast, illite evolves

□ **Fig. 4.15** Map showing distribution of clay minerals in the oceans (Berger 1974). M: montmorillonite, I: illite, K: kaolinite, M, I: dominance of montmorillonite over illite etc.

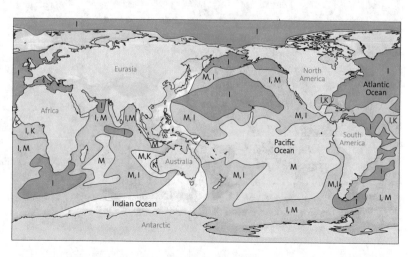

4

by the decomposition of feldspar and mica and is transported from the terrigenous sources and subsequently distributed by currents. Thus, it occurs not only near the land but also in the central ocean and is especially dominant in the middle and high latitudes of the northern hemisphere. Kaolinite, which evolves by weathering of feldspar in tropical areas, is concentrated along tropical coasts.

4.9 Facies Changes on the Large Oceanic Conveyor Belt

The oceanic crust is formed at the mid-ocean ridge and drifts away from the ridge, and subsides; through this process, a given spot passes through different facies zones that vary in water depth, geographic latitude, and distance to terrigenous source. The resulting stratigraphic record consists of characteristic sedimentary sequences that are successively built up as the conveyor

belt of the oceanic crust spreads laterally. The facies occurring in a vertical sequence, however, coexist at the same time also juxtaposed on segments of the ocean floor with different ages (Walther's Law of Facies, see above).

The mid-ocean ridges contain black and white smokers (► Chap. 5) that irregularly distribute thin, metal-rich sediments of extraordinary composition (◘ Fig. 4.10). The mostly dark brown sediments are rich in iron and manganese and contain a number of other metals which come from thermal water that leached the oceanic crust. However, the content of metals is mainly dissolved in sea water and is precipitated far from the mid-ocean ridges in the manganese nodules of the abyssal plains (see box). The metalliferous sediments at the ridges consist of fine spherules of iron and manganese hydroxide that are precipitated from the sea water, and iron-rich clay minerals (Fe–smectite). Additional, various sulfides and sulfates also occur. Such deposits are especially common along the East Pacific Rise.

The Bengal Deep Sea Fan

The Bengal deep sea fan occurs adjacent to the mouth of the Ganges–Brahmaputra river system and represents the largest deep sea fan on Earth. The Ganges and Brahmaputra carry enormous amounts of sediment that is eroded from the rapidly rising mountain chain of the Himalaya and is transported to the Indian Ocean. Much of it, especially the finer sediment fraction, is carried into the deep, open ocean. A smaller but significant fan occurs also on the west side of the Indian sub-continent adjacent to the Indus River which drains the Karakorum and a smaller part of the Himalaya. It has only one third of the volume of the Bengal fan.

The Bengal fan has an elongated shape in the N–S direction that is parallel to the elongation of the Gulf of Bengal between India and Southeast Asia (◘ Fig. 4.16). The fan extends from the rivermouth estuaries nearly 3000 km southward to the southern tip of Sumatra. In its northern part the sediments achieve a thickness of more than 10 km. Its volume is several times greater than the volume of the Himalayas lying above sea level. Each year 2 billion tons of sediment are deposited, or roughly a cubic kilometer, and yield an average rate of erosion of approximately 1 mm per year across the area of erosion of the Himalaya and Transhimalaya. The fan, which initiated at the beginning of the Eocene about 55 Ma, documents in detail the history of uplift and erosion of the Himalaya and the Tibetan Plateau. Major uplift started in the Eocene approximately 40 Ma and is mirrored by the coarsening and rapid accumulation of sediments in the Bengal fan. In the Early Miocene (20 Ma), parts of the Tibetan Plateau rose to great heights and the Himalaya was a high mountain range. Sedimentation rates were

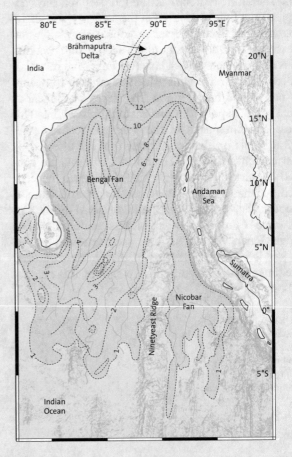

◘ **Fig. 4.16** Outline map of the Bengal fan and its canyon systems with sedimentary thicknesses shown in kilometers (Curray and Moore 1974). The Nicobar Fan is presently cut off from the sedimentary supply from the Ganges–Brahmaputra delta by the Ninetyeast Ridge

in particular high during the Quaternary as glacial erosion in the Himalaya delivered enormous amounts of sediment. The resulting turbidite sediments are exceptionally thick.

Turbidity currents have carved numerous submarine canyons across the fan surface into abyssal depths (◘ Fig. 4.16). The surface of the fan and the position of the submarine canyons constantly evolve as feeder channels change their position. Currently, most of the sediment material is funneled SSW and feeds the main Bengal Fan. However, at the Ninetyeast Ridge, an inactive remnant of a volcanic ridge formed above a hot spot, the fan is subdivided. The smaller part of the fan east of the ridge is called the Nicobar Fan and is currently cut off from the sedimentary supply of the Ganges–Brahmaputra delta. Since in the Early Quaternary, this part of the fan is isolated by the complex geometry of the subduction zone in front of the Andaman and Nicobar Islands.

As the ocean conveyor belt drifts away from the ridge area, commonly fine calcareous oozes are deposited on top of the metal-rich sediments or, where these are absent, directly on the basalts of the oceanic crust. Along smooth, rapidly spreading ridges like the East Pacific Rise, calcareous sediments accumulate in a uniform layer. On ridges with slow spreading and strong tectonic structuring like the Mid-Atlantic Ridge, the sediments fill in depressions until sedimentation covers irregularities to form continuous layers. Depending on the supply, the calcareous oozes can also contain minor amounts of siliceous organisms and clay.

If the ocean floor subsides below the CCD the limestones are overlain by either siliceous sediments or red deep-sea clay (◘ Fig. 4.10). In regions with high production of siliceous organisms, their remnants dominate and radiolarites or diatomites grade upwards from carbonates. Siliceous and clayey layers may be interbedded and in regions with low production of siliceous organisms, red deep-sea clay is formed. Nearer continental areas, deep-sea sediments are overlain by terrigenous clastic sediments. If the ocean floor arrives at a subduction zone, sediments in the deep-sea trench end the sedimentary sequence as it is subducted.

The above-described sequence of siliceous above carbonatic sediments does not apply to all fossil examples of ocean floor sequences. In the Alpine–Mediterranean region, ocean crust was formed in the Middle and Late Jurassic and remnants of this crust are preserved in numerous ophiolites exposed in the Alpine mountain chains. Here, ocean floor basalts are overlain by radiolarites and succeeded by limestones. The change from siliceous to carbonatic sedimentation in the Late Jurassic coincides with an explosion of distri-

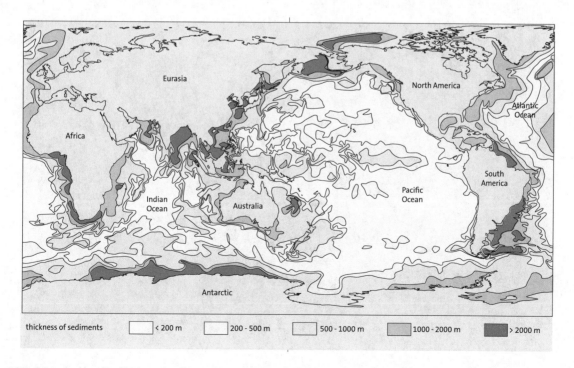

◘ **Fig. 4.17** Map showing distribution of sediment thicknesses in the global ocean basins and adjacent shelf areas (World Data Center for Marine Geology and Geophysics 2003)

4

bution of calcareous plankton. At this time, foraminifera and coccolithophorids proliferated in all oceans of the world and lowered the CCD.

Generally, the sediment pile on the ocean-floor conveyor belt is thickest where the oceanic crust is the oldest—the distribution of sediment thicknesses parallels the systematic increase in age and distance from the mid-ocean ridge (◘ Fig. 4.17). Therefore, the thickest sedimemt occurs adjacent to continental margins, especially the passive margins of large oceans. The basal sediments directly overlying the basalt yield similar ages to the age of the crust. However, in large areas of the Pacific, distal from the East Pacific Rise, sediments are very thin, especially in those areas where the red deep-sea clay with extremely low sedimentation rates is formed. Such patterns of sediment thickness and variation are related to the fundamental differences between the two ocean basins. The Atlantic is mostly bordered by passive continental margins. Sediments thus can be transported mostly unhampered into the deep-sea plains. In the Pacific, sedimentation paths are restricted by the relief of numerous island arc systems and deep-sea trenches.

Mid-ocean Ridges

Contents

© The Author(s), under exclusive license to Springer Nature Switzerland AG 2022
W. Frisch et al., *Plate Tectonics*,
Springer Textbooks in Earth Sciences, Geography and Environment,
https://doi.org/10.1007/978-3-030-88999-9_5

5

Mid-ocean ridges are the oceanic counterparts of continental graben structures. Both are zones of extension although mid-ocean ridges have substantially higher spreading rates and also mark plate boundaries where new oceanic crust and lithosphere are formed. The presence of normal faulting and development of a central graben document that as new crust is formed, it is still under extension as it drifts and spreads.

Mid-ocean ridges are present in all large oceans and form a system of submarine mountains over 60,000 km long, the longest linear elevations on Earth. More than 60% of the magmatic rocks of the Earth, approximately 20 km³/year, are generated at these ridges (Schmincke 2004). The name "midocean ridge" is used for constructive plate boundaries in oceanic areas whether or not the ridge is precisely in the middle of an ocean. The central location of a mid-ocean ridge is only maintained in the absence of subduction near ocean margins as subduction would destroy the symmetry. This is currently the case in the Pacific. In the Pacific Ocean, subduction velocities along the eastern margins are somewhat lower than those along the western edges (◘ Fig. 1.2); however, the ridge has shifted eastward, hense the name East Pacific Rise, because in the Mesozoic and Early Cenozoic, subduction was more rapid along the American continents than at the western edge of the Pacific. Ridges always migrate towards the direction of most rapid subduction.

In contrast to the East Pacific Rise, the Mid-Atlantic Ridge occupies a central position because the Atlantic is bordered mostly by passive continental margins. Therefore, spreading of the ocean floor was symmetric, and maintained the same velocity in both directions. This is also supported by the nearly equal width of magnetic anomaly patterns on both sides of the ridge (◘ Fig. 2.12). The Indian Ridge is centered between Australia and Antarctica because it is also bordered by passive continental margins. Further to the west, the Indian Ridge remains medial even though complex subduction patterns are present; however, symmetry is the result of several coincident factors including the subduction at the Sunda Arc and junction with another ridge from the SW.

In the equatorial Pacific, new lithosphere is formed and the plates diverge with spreading rates of up to 15 cm/year. The rates are substantially lower in the Atlantic with only 4 cm/year in its southern part and less than 1 cm/year in the Arctic region. The fastest spreading rates in the Indian Ocean are approximately 7 cm/year. Total extension in the Upper Rhine Graben is about 5 km over a time period of almost 50 m. y., this is an average of 0.01 cm/year. During the same interval, the North Atlantic south of Iceland, a relatively slow-spreading ridge with a rate of 2 cm/year, diverged 1000 km.

Oceanic ridges are cut by numerous transform faults, a feature relatively rare in continental graben systems. The individual segments of the rift axis are not displaced against each other by the faults but are maintained in their original position relative to each other (► Chap. 8). The individual segments are up to several tens of kilometers long; in comparison, the Upper Rhine Graben is never segmented over a distance of 300 km.

Earthquakes along mid-ocean ridges occur along both the ridge and associated transform faults and are coincident with each as demonstrated by earthquake distribution maps (◘ Fig. 1.8). All epicenters are shallow because of the thinness of the solid crust. Earthquakes occur in swarms and are of low magnitude and mostly related to rising magma. In most cases, fault-plane solutions (see ► Chap. 2) indicate normal faulting associated with the extensional regime of the mid-ocean ridges and strike-slip faulting along the transform faults.

5.1 Topography of the Ridges

Mid-ocean ridges are 1000–4000 km wide and are characterized by an uneven surface (◘ Fig. 5.1). On average the ridge crests have a water depth of about 2500 m which makes them significantly higher than the adjacent deep ocean basins. Ridges with a high spreading rate have an average water depth of about 2700 m (e. g., Pacific) and those with a slow spreading rate are about 2300 m deep (e. g., Atlantic). The greater depth of ridges with faster spreading rates results from the rapid drift of the plates and a lessening of the effect of vertical push-up pressure of the rising asthenosphere. Ridges coincident with hot spots, such as Iceland, can emerge above sea level due to the high production rate of magmatic rocks.

Ridge shape is related to the rising currents of hot, less dense asthenospheric mantle material. Basaltic melts rise to form a large magma chamber at a few kilometers depth and are approximately 15% less dense than the mantle peridotite from which they were extracted. This melt is 6% less dense than its solid equivalent, basalt or gabbro, in the ocean crust. Therefore, the less dense ridge has higher topography than the ocean basins with their relatively thick, cold, and dense lithosphere. However, a mass surplus exists at the ridge as indicated by a slight positive gravity anomaly compared that of the ocean basins (◘ Fig. 5.1). This surplus is related to the effect of the push-up pressure caused by the rising asthenospheric current that lifts up the ridge. The push-up pressure is subsequently re-directed into the horizontal drift movement and is an im-

Fig. 5.1 Topographic profile across the central Atlantic between New York and Dakar (Heezen and Hollister 1971). The gravity profile indicates a broad positive anomaly above the mid-ocean ridge (Talwani et al. 1965) that is generated by ridge push. 1 Gal (galilei) = 1 cm/s² (unity of acceleration). 1 mGal = 10^{-3} Gal

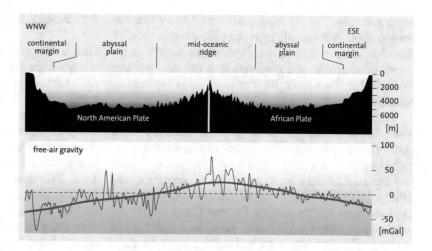

portant driving mechanism for the drift of the plates (▶ Chap. 1).

A central graben structure, which evolves by the extension of its flanks, is mainly developed at ridges with slow spreading rates. These ridges are under greater extensional forces because the supply of crustal material is low. Graben structures at ridge axes are narrower and shallower than their continental counterparts. The drift movement is mainly compensated by the rising magma and only to a lesser degree by extension. The subsidence of the graben compared to its shoulders rarely reaches more than several hundreds of meters. On the other hand, the total vertical displacement at the edges of continental graben structures is typically several kilometers.

5.2 Generation of Oceanic Lithosphere

Oceanic lithosphere develops when the peridotite of the asthenophere is split into a molten part of basaltic composition and into a solid peridotitic residual rock. The rocks of the oceanic crust are generated from the melts; the residual rocks are depleted of the melted material to form the peridotites of the lithospheric mantle. Normal oceanic crust has a thickness of about 6 km and the base of older oceanic lithophere is approximately 70–80 km thick. However, at the mid-ocean ridge, the asthenosphere nearly reaches the surface. Oceanic lithosphere consists of different layers; this layered structure is rather uniform because of its uniform process of formation. This contrasts significantly with the structure of continental crust that has a complex, long-lasting, and non-uniform history.

Asthenosphere is composed of a particular peridotite, lherzolite, named after Lac (lake) de Lherz in the French Pyrenees (■ Fig. 5.2). It is mostly composed of olivine, a magnesium–silicate undersaturated in silica, and with approximately 10% of the magnesium replaced by iron. Additional major components include

two different pyroxenes: en-statite and diopside. En-statite is an orthopyroxene (crystallizing after the orthorhombic crystal system), diopside is a clinopyroxene (crystallizing after the monoclinic system). Enstatite is very similar to olivine but has more silica (saturated in silica). Diopside and other clinopyroxenes also contain calcium and aluminum.

Lherzolite from the asthenosphere rises below the mid-ocean ridge (■ Fig. 5.2). At a depth of approximately 75 km, partial melting occurs due to pressure-release; melting points are generally depressed at

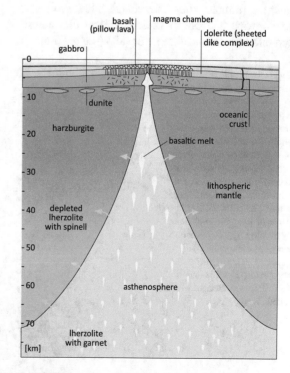

Fig. 5.2 Schematic profile across the plate boundary at a mid-ocean ridge. The basaltic melt is derived from the lherzolite of the rising asthenosphere and ocean crust crystallizes. Peridotitic residual rocks, from which the basaltic melt had been extracted, are depleted lherzolite and harzburgite. They form the lower and upper part of the lithospheric mantle, repsectively

lower pressure. The resulting partial melt has a basaltic composition that roughly resembles the composition of diopside; after solidification, basalts are mainly composed of diopside and plagioclase, a calcium–sodium feldspar. Finally a lherzolite remains that is depleted in basaltic melt.

Peridotite from which about 5% of basaltic melt was extracted, still contains diopside but now rises faster than during the initial stages of the melting process because of the lower densitiy of the enclosed molton blobs. Continuing pressure-release generates additional basaltic melt until diopside completely disappears from the peridotite. During this process, the original lherzolitic residual rock becomes a harzburgite (named after the village Bad Harzburg in Germany), a peridotite mainly composed of olivine and enstatite (◘ Figs. 5.2 and 5.3). Therefore, partial melting follows the formula

$$lherzolite = harzburgite + basalt,$$

where the basaltic melt immediately beneath the mid-ocean ridge comprises about 20% of the volume of the original lherzolite. The molton basaltic blobs merge to larger bodies to eventually construct a large magma chamber at shallow depth to form oceanic crust, which in spite of its layered structure, is almost completely of basaltic composition. Oceanic basalts are so-called tholeiites (named after the village Tholey, Germany) and contrast sharply in their chemical characteristics from alkaline or calcalkaline basalts (see ▶ Chap. 7).

5.3 **Rocks of the Oceanic Crust**

Only a very thin sedimentary cover is present in the vicinity of the mid-ocean ridge; sedimemnt accumulates as the crust ages over millions of years. The uppermost basaltic layer is composed of pillow lavas (◘ Fig. 5.4). As a basaltic melt with a temperature of about 1200 °C extrudes onto a submarine floor, blobs of lava are formed that are chilled by cold seawater and frozen into spherical or elongated structures that resemble pillows. The spherical shape develops because a sphere has the smallest surface area per given volume. Small surface area reduces heat emission. A similar effect occurs when water is dropped onto a hot plate. However, water drops become elliptical because of their own weight. The same happens with lava blobs on the ocean floor. The outer rim is rapidly chilled to form igneous glass while the inner part remains liquid and slowly solidifies to crystalline rock.

Pillow lavas form layers greater than 1 km in thickness and are crosscut by fine-grained crystalline basaltic intrusive sheets that either form vertical dikes or horizontal sills. Because the dikes and sills do not contact sea water, they cool more slowly and crystallize completely and, therefore, do not contain igneous glass. These fine-grained basaltic intrusive rocks are called dolerites and resemble the inner parts of basaltic pillows. The steeply inclined dikes represent feeder channels that formed as basaltic magma extruded from the magma chamber to the surface. If the overlying layer of

◘ **Fig. 5.3**   Profile through the oceanic crust at the mid-ocean ridge. The magma chamber, fed by melt rising from the asthenosphere, provides the melt for the dolerite dikes and pillow lavas above; laterally the melt solidifies to gabbro. These all contain the same chemical composition. Plagiogranites at the roof and peridotites at the bottom of the magma chamber may develop by differentiation (change of composition) of the melt. The insert shows the development of a dike–in–dike structure associated with a sheeted-dike complex

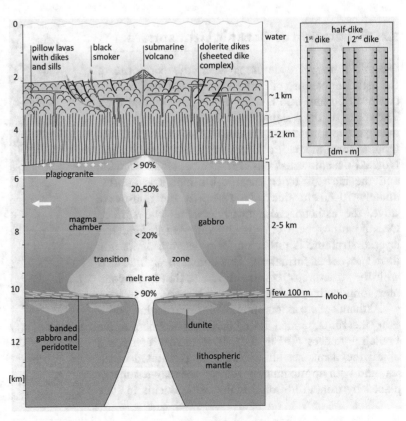

pillow basalts is thin, the roof may be uplifted by the internal pressure of the rising magma. This may cause a dike to bend and form a horizontal sill. At depth the number of dikes increases until a zone is reached that is entirely made up of dolerite dikes (■ Fig. 5.3).

This dike complex, also called a sheeted dike complex since it resembles vertical sheet-like layering, underlies the pillow basalt layer. It has a thickness of 1–2 km and develops during repeated intrusions of basaltic melt into fractures. One single intrusion event forms a dike of approximately 1 m width. Following eruption, melt remains in the fracture and solidifies. The edges of the dike are finer grained than the center because the edges cool faster and crystals have less time to grow. New dikes may intrude into slightly older ones that have not yet solidified completely because such dikes are easy to split. These structures are called dike-in-dike structure; the split half-dikes are fine-grained on one side and coarser grained on the other—the original center of the dike (■ Fig. 5.3).

Gabbros, the plutonic equivalent of basalts and dolerites, lie beneath the sheeted dike complex (■ Fig. 5.3). Gabbros are coarse-grained rocks with crystals several millimeters to a few centimeters in size and, like dolerites, mostly consist of diopside and plagioclase. These plutonic rocks represent the solidified magma chamber that was fed by mantle melts as described above. The magma chamber solidifies at its edges and as it is continuously filled from below, the solidified walls are spread apart. The magma is not entirely liquid but rather consists of a mixture of crystals and melt, the so-called crystal mush. At its base the chamber has a width of up to 20 km and is tapered towards the top (Nicolas 1995). The highest proportion of melt occurs at the base, due to the high temperatures, and at the top, due to pressure relief. Gabbro crystallizes out in the lateral transition zone (■ Fig. 5.3). The resulting gabbro layer has a thickness that ranges from less than 2 to approximately 5 km. The thickness is dependent on spreading rate of the ridge, thin in the case of slow and thick in case of rapid (see ■ Fig. 5.9). A thin gabbro layer with slow spreading rates is explained by low magma production. Overall, the thickness of oceanic crust thus varies between approximately 4 and 8 km.

Pillow Lavas

The shapes of pillow lavas are rather variable. Many details were uncovered during diving cruises of submersibles like "Alvin" that investigated and photographed the mid-ocean ridge of the northern Atlantic in the FAMOUS project ("French–Amerian Mid-Ocean Undersea Study") (Ballard and Moore 1977).

Normally, basaltic pillows are spherical objects that range from several decimeters to more than one meter in diameter. In vertical cross section, pillows have an upward-vaulted convex shape and a horizontally elongated body that is caused by flowage outward and downward due to the weight of the pillows. The bottom is typically characterized by concave–up indentations that result from the deformation of the warm, plastically deformable lava that drapes over the cold and solidified pillows below (■ Figs. 5.4a and 5.5). The skin of the pillows consists of basaltic glass. Glass is a product of sudden cooling by cold sea water. This process occurs so fast that crystalliza-

■ **Fig. 5.4** Photographs of pillow lavas. **a** Pillows with convex curvature at the top and concave borders against older underlying pillows (Costa Rica). The indentations along the bottom side develop when the new, internally still liquid and thus easily deformable pillows are deposited on top of the already solidified pillows below. **b** Pillow lavas viewed towards the curved upper surfaces of the pillows (Elba, Italy). **c** Internal shelves that indicate a multiphase leakage of the pillows when the lava was still liquid in their interior (see ■ Fig. 5.6). Because the shelves are formed horizontally, the outcrop indicates tectonic tilting by nearly 90° (Elba). **d** Variolas (spherules of up to 1 cm diameter with radially arranged actinomorphic minerals) as products of devitrification in the pillow rinds (Elba)

5

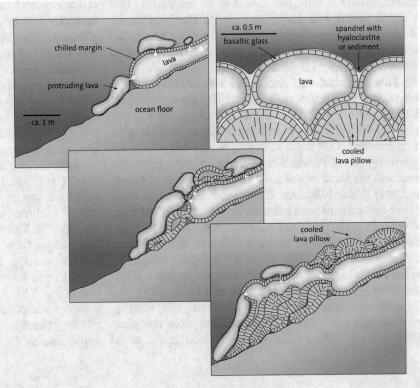

Fig. 5.5 Evolution of pillow basalts (upper right) and complex formations of lava tubes and pillows at mid-ocean ridges (Ballard and Moore 1977)

tion of single crystals is not possible. Glass, therefore, is *amorphous* (*Greek* without shape) and the atoms are not arranged in a crystal lattice but disorderly distributed as in a pane of window glass. Towards the centre of the pillow the melt crystallizes and crystals increase in size. Until final solidification, the pillow is deformable and the glassy skin tears open to produce furrows and cracks; further protrusions of lava may occur if the feeder channels of the pillow are partially fluid during subsequent events. If lava supply is continuous, instead of pillows, lava tubes several meters in length can develop, especially on slopes (**Fig. 5.5**). The tubes are blocked by obstacles such as large protruding pillows. The down-flow end of tubes commonly burst, lava squeezes out, and new pillows or tubes develop. This process can produce complicated shapes of lava tubes and interconnected pillows.

To study some pillow lavas, it is not necessary to use a submersible. If the ocean floor is obducted onto continental crust (see below), they can be studied on outcrop. For example, at an extraordinary outcrop along the coast of the island of Elba (Italy) it is possible to walk on Jurassic ocean floor and study the different pillow structures (**Fig. 5.4b**). Pillows that are split by fractures, develop shelves in their interior (**Fig. 5.4c**); these shelves are also found in recent pillows at the Mid-Atlantic Ridge. Shelves within pillows develop when a pillow tears open and part of the melt leaks out but is not fed by a supply of fresh lava (**Fig. 5.6**). The lava level recedes slightly and solidifies to form a glassy surface or "shelf" within the interior of the pillow. This process may be repeated to

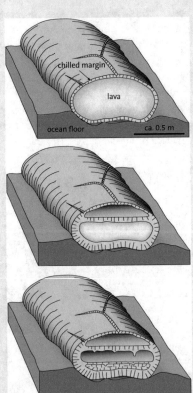

Fig. 5.6 Development of horizontal shelves within a pillow by leakage (Ballard and Moore 1977)

form up to 6 or 7 shelves in a single pillow. Cavities that form between shelves may be subsequently filled by sedimentary material if the cavities are opened shortly after their formation or they may be filled with minerals that precipitate from aqueous liquids; some may remain hollow. Because the shelves are always formed in a horizontal position, they indicate in outcrops whether the pillow have been tilted tectonically. The shelf outcrops at Elba indicate that the pillows have been later tilted nearly 90° (◻ Fig. 5.4c).

Pillow lava outcrops occur worldwide and in rocks of most ages. In North America, pillow lavas occur in the 1.7 Ga Vishnu Schist in Grand Canyon. Younger examples include abundant outcrops of Jurassic and Cretaceous pillow lavas in California, Nevada, and Oregon.

Glass splinters, hyaloclastites (*hya–los, Greek* glass; *klastein, Greek* to break), commonly form in the spandrels between the pillows (◻ Fig. 5.5). Glass splinters

burst off from the rind of the pillow because of the rapid temperature reduction and resulting stress. Basaltic lava becomes solid at 1000 °C, and a very short increment of time elapses as the outer glassy rim cools to the surrounding sea water temperature.

Because igneous glass is unstable over long periods of geological time (millions of years), it devitrifies (*vitrum, Latin* glass). Initially, small needle crystals are formed in a rose The around a nucleus. These form spherical objects that range from millimeters to ca. 1 cm in diameter that are called variolas (*variola, medieval Latin* smallpox). These are common in the rinds of the pillow lavas of Elba (Abb. 5.4d). Over time the entire glass can crystallize to form a fine-grained mixture of basaltic minerals rich in the green mineral chlorite.

Mostly gabbros have an isotropic (directionless) structure that is typical of most plutonic rocks. Locally, the tops of the gabbros contain plagiogranites. Plagiogranites are mixtures of plagioclase, hornblende and quartz. They develop either as the latest crystallization product from the original basaltic melt that generated the gabbros or by sea water penetration in form of overheated steam that penetrates through fissures down into the roof area of the magma chamber. The water decreases the melting temperature of the gabbros, which normally are completely solidified at about 1000 °C. However, in the presence of water, partial melting is possible at temperatures as low as 750 °C. Such a melt produces the plagiogranites. The water is partly incorporated into hornblende, a mineral that resembles clinopyroxene but also contains water in form of hydroxyl ions that are bound in the crystal lattice.

At the base of the gabbroic layer, banded and sheared gabbros and peridotites occur (◻ Fig. 5.3).

These are generally considered to be cumulates, crystalline bodies that develop by the sedimentation of precipitated crystals that drop through the melt because of their high density and are deposited at the bottom of the magma chamber. Cumulates exhibit a layering or banding like sediments. However, banded rocks at the base of the gabbros may also develop as shear structures (Nicolas 1995). Shear structures form as the glutinous crystal mush is transported by magma currents; the elongate, column-shaped crystals align with their long axis parallel to the direction of current flow thereby squeezing out residual melt because of the tight fit of the crystal columns. Horizontal flow in the mantle below generates an additional shear movement between the mantle rocks and the gabbro. This process is responsible for a commonly observed schistosity in the basal gabbro (with plagioclase) or peridotite (without plagioclase).

Seismic Layers

Seismic investigations have been a crucial factor in deciphering the structure of the oceanic crust. During the 1960s, three crustal layers were established based on different seismic wave velocities. These layers, in descending order were, (1) sediments, (2) basalts and dolerite dikes, and (3) gabbros. Advanced methods in seismic analysis now provide a more detailed picture that demonstrates that seismic velocities, although irregular, increase continuously with depth (◻ Fig. 5.7). Low velocities in primary seismic waves (P-waves) of approximately 2 km per second (km/s) characterize young and unconsolidated sediments; these form continuous and thicker layers only at some distance from the ridge.

Layer 2 ranges to 2.5 km in thickness and comprises pillow lavas and the sheeted dike complex. The top of the layer 2 is a topographically rough surface that is covered and leveled by overlying sediments. Seismic wave velocities increase substantially downward in this layer. The uppermost basalts, which near the ridge axis are characterized by pores and fractures and consequently by abundant circulation of seawater, have P–wave velocities of 3.5–4 km/s. As the age of layer 2 increases, pores and fissures are closed at depth by mineral precipitation. This leads to higher seismic wave velocities with increasing distance from the ridge. The P-wave velocities increase to

5

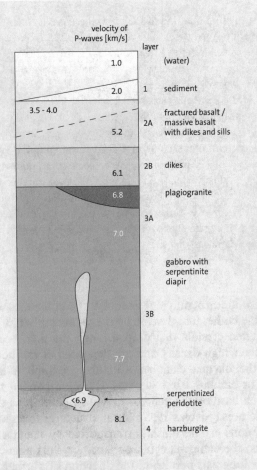

velocity of
P-waves [km/s]

layer	
1.0	(water)
2.0	1	sediment
3.5 - 4.0	2A	fractured basalt / massive basalt with dikes and sills
5.2		
6.1	2B	dikes
6.8		plagiogranite
7.0	3A	
		gabbro with serpentinite diapir
7.7	3B	
<6.9		serpentinized peridotite
8.1	4	harzburgite

◘ Fig. 5.7 Velocities of seismic primary waves and the deduced layered structure of the oceanic crust

more than 6 km/s in the sheeted dike complex because the dike intrusions become more massive and lack fractures. Velocities of P-waves increase to 7–7.7 km/s in the gabbro layer (layer 3) and accelerate with increasing depth, a factor attributed to differences in the rock composition. In the uppermost part of the gabbroic layer, plagiogranites rich in plagioclase and quartz generate lower velocities. The banded rocks in the lower parts are rich in pyroxene and generate higher velocities. Varying amounts of diapiric material from serpentinite rising from the mantle may complicate the seismic behavior of the layer.

More recent research has identified a layer 4 that corresponds to the rocks of the uppermost mantle. They are clearly defined by P-wave velocities of 8.10–8.15 km/s. Near the ridge axis the velocities of the mantle peridotites are slightly lower because of the higher temperature and increased portions of melted material. Serpentinized peridotites cause distinctly lower velocities of seismic waves.

Normally the layered structure of the ocean floor is not exposed. However, along the Vema Fracture Zone an exposed profile was investigated that confirmed the layered structure (see box below). On land a number of sections through ocean crust are exposed. These exposures occur at locations where ocean floor was obducted and thrust onto continental crust during compressional processes and orogeneses. Such accessible remnants of oceanic crust, which may be associated with rocks of the uppermost mantle, are called ophiolites and are important clues in the reconstruction of former oceans that are now completely subducted (see below).

5.4 Basalts of Mid-ocean Ridges

Basalts formed at mid-ocean ridges (MORB—mid-ocean-ridge basalts) are tholeiites and differ from basalts formed in other geotectonic environments, such as above hot spots or above subduction zones. Tholeiites can not be diagnosed from either hand specimen or microscopic petrographic investigation based on the mineralogical composition of the rock. Rather, chemical analysis is required for distinguishing the different basalts. A number of specific elements, especially those which occur in low concentrations or only as trace elements, play an important role.

In order to discriminate basaltic rocks, so-called immobile elements are used. Among these are titanium, zirconium, phosphorus, niobium, yttrium and the rare earth elements. The relatively low ion radii and high charge numbers of these elements restrict transportation in aqueous fluids. During processes of weathering or metamorphism, they remain immobile and thus characterize the unchanged primary magmatic composition of the rocks. On the other hand, so-called incompatible elements are preferentially incorporated into basaltic melts that are derived from partial melting of the mantle peridotite. They are incompatible with the peridotite of the Earth's mantle, because they do not fit into the crystal lattices of the minerals of the peridotite (in particular olivine and pyroxene). In spite of their partial mobility (e. g., potassium, rubidium, strontium) they can be used for basalt discrimination if the rock is not altered.

The concentration of incompatible elements is a function of the composition of the primary rock of the mantle and its degree of melting. Alkaline basalts that occur in intraplate environments above normal hot spots or at graben structures indicate a low degree of partial melting (a few percent) of the primary peridotite (most probably from the lithospheric mantle). Incompatible elements such as the alkalis, therefore, are substantially more enriched in the melt compared to tholeiitic basalts which are characteristic for highly productive hot spots and in particular for mid-

□ **Fig. 5.8** Ternary diagrams that discriminate basalts at mid-ocean ridges, basalts at volcanic arcs above subduction zones, and basalts at hot spots (Pearce and Cann 1973; Meschede 1986). For these ternary diagrams, the contents of the respective elements have to be converted to a common sum of 100%. To place the fields adequately, some element contents have to be enlarged or reduced by a certain factor. The field of basalts from mid-ocean ridges transitional to intraplate basalts is indicated by an asterisk

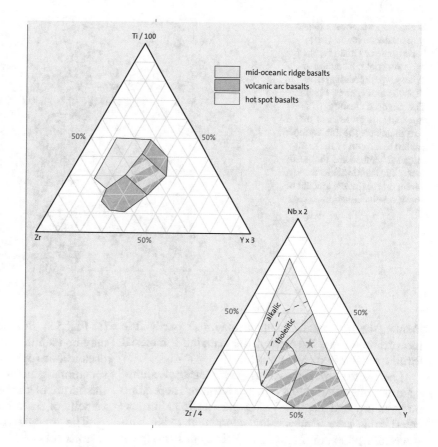

ocean ridges. Tholeiitic basalts derive from melts with melting portions of about 15–25%. The concentration of incompatible elements is diluted. Moreover, they come from the asthenosphere which was the source of MORB for billions of years and thus continuously emitted incompatible elements. Therefore, the source of MORB is already depleted in these elements when melting starts and is thus termed "depleted mantle".

A number of diagrams have been constructed using the concentrations of immobile and incompatible elements of different types of basalts. Basalts produced at different plate tectonic environments are distributed into different fields (□ Fig. 5.8). With these diagrams, it is possible to assign basaltic rocks to a certain plate tectonic environment. Such discrimination is also possible using the rare earth elements and the isotope relations of some elements (strontium, neodymium). Additional information comes from other associated magmatic or sedimentary rocks.

If hot spots are located in the area of a midocean ridge, basalts of intermediate chemical composition between MORB and intraplate basalt may occur. Such transitional basalts are slightly enriched in incompatible elements and thus referred to as enriched MORB (E-type MORB).

5.5 Fast and Slow Spreading Ridges and Rocks of the Lithospheric Mantle

The lithospheric mantle below the oceanic crust also displays a layered structure, though somewhat gradational (□ Fig. 5.2). In the uppermost part, dunites and lenses of chromite occur. Dunites (named after the Dun Mountains in New Zealand, which in turn are named after the dun colour of the dunites) are nearly exclusively composed of olivine. Chromite is a chrome-spinel (an iron–chrome oxide) that occurs in the form of ore deposits (see box below). The dunites form relatively small bodies, usually meters to kilometers across. They are surrounded by harzburgite that merges downward into lherzolite. As already mentioned above, the asthenosphere is composed of a lherzolithic peridotite. In the lithosphere, a basaltic component is removed from this lherzolitic peridotite by partial melting, first producing modified ("depleted") lherzolites and later harzburgites as residual rocks which remain in the lithospheric mantle (□ Fig. 5.2). Zonation of the lithopheric mantle mirrors the increasing depletion by basaltic partial melting from a few percent up to 20% where the lherzolite looses its clinopyroxene and thus changes into harzburgite. In the uppermost zone of the

5

■ **Fig. 5.9** Mid-ocean ridges with moderate to fast (left; harzburgite or Oman type) and slow (right; lherzolite or Trinity type) spreading rates (Boudier and Nicolas 1985). The diagnostic features of slow spreading are presence of the central graben, the thin layer of gabbro as a result of the small magma chamber, and the mostly lherzolitic mantle below the Moho due to the low amount of basaltic melt extraction

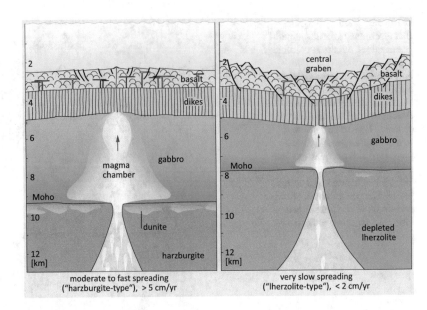

mantle, the hot basaltic magma may also partly fuse the orthopyroxene (enstatite); the remaining material forms irregular bodies of dunite.

Lherzolites contain various percentages of alumina-rich minerals that vary in composition depending on pressure and depth: > 75 km garnet, 7530 km spinel (in this case a magnesium–aluminum oxide with some iron), < 30 km plagioclase. This means that peridotites of the deeper lithosphere are spinel lherzolites and those of the asthenosphere below the typically 70–80 km-thick oceanic lithosphere are garnet lherzolites (■ Fig. 5.2).

The above structure of the oceanic lithospheric mantle only pertains to the lithosphere that developed at mid-ocean ridges with spreading rates of at least 2 cm/year. Most mid-ocean ridges belong to this category throughout Earth history. Slow-spreading ridges (0.5–2 cm/year) as currently present in the Red Sea and the Arctic Atlantic are characterized by lower production of melt as reflected by smaller magma chambers and subsequently a thinner layer of gabbro (1–2 km thickness). Partial melting of mantle peridotite at these ridges is between 10 and 15% whereas at fast-spreading ridges it is approximately 20%. Therefore, lherzolite, not harzburgite forms in the upper part of the lithospheric mantle; the lherzolite is depleted in basaltic melt and differs from the original, asthenospheric lherzolite with the occurrence of plagioclase and a lower content of diopside.

Oceanic lithosphere therefore consists of two types: (1) slow-spreading lherzolite type or Trinity type named after an ophiolite complex in California, and (2) medium-to fast-spreading harzburgite type or Oman type named after an ophiolite complex in Oman

(■ Fig. 5.9; Boudier and Nicolas 1985). These types may be identified in mountains where the remnants of ocean floor (ophiolites) are incorporated. The rock associations generate important conclusions concerning the nature of the mid-ocean ridge and in particular the velocity of spreading.

The topography of the mid-ocean ridges and numerous other factors are controlled by the interaction of magmatism and tectonics. Fast-spreading ridges, with spreading velocities of several centimeters per year like in the Pacific, are dominated by magmatism that approximately compensates the drift of the plates. Therefore, tectonic extension plays a minor role and a central graben structure is missing (■ Fig. 5.9). A high rate of magma production rapidly fills up the valleys and equalizes the relief along the extensional zone of the ridge. At slow-spreading ridges like Atlantic, extensional tectonics is much more dominant. Extension is only gradually compensated by magmatism and commonly is equalized by tectonic processes along with normal faulting. Here, a deep central graben structure with steep flanks develops. According to the low magma production the magma chamber is small and the gabbo layer is thin (■ Fig. 5.9).

5.6 Segmentation of Ridges by Faults

Mid-ocean ridges are segmented by transform faults that are orthogonal to the axis. Individual sections typically have lengths of 30–100 km. The sections of the ridge appear to be displaced at the faults; however, the distances between the spreading axes do not change. The faults are mostly characterized by topo-

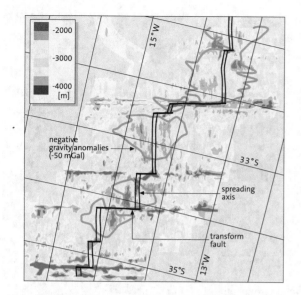

Fig. 5.10 Segmentation of the mid-ocean ridge in the southern Atlantic by transform faults (Neumann and Forsyth 1993). The segments along the spreading axis are bulged by magma chambers and at the ends they are marked by deeply incised valleys along the transform faults where the chambers are interrupted. The magma chambers cause negative gravity anomalies (circled with a green line)

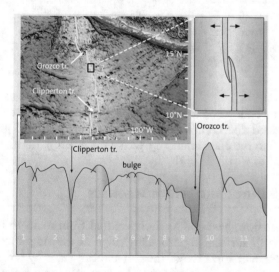

Fig. 5.11 Segmentation of the East Pacific Rise by transform faults as shown in map view and by a cross section parallel to the ridge axis (Macdonald et al. 1988). The segments are bulged by magma chambers at a shallow depth; overlapping of the chambers occurs at the ends of the segments. tr: transform fault

graphic furrows several hundred meters deep; some in the Atlantic approach 3 km in depth. The incisions are particularly distinctive at large faults. In contrast, the ridge segments form bulges parallel to the spreading axis (Fig. 5.10). The bulges are caused by the magma chambers in the crust. They are indicated by negative gravity anomalies because

the magma is lighter than the adjacent solidified crust.

Gaps between the magma chambers coincide with the ends of the ridge segments at the faults. Below the rapidly spreading East Pacific Rise the chambers are approximately 1–2 km below the surface. Magma production is particularly high and the magma chambers extend in the longitudinal direction of the ridge, extend past the transform faults, and cause a latitudinal overlap of the spreading axes over a distance of up to 25 km (Fig. 5.11). Magma chambers at the Mid-Atlantic Ridge are smaller. The tips of the segments are free of such overlapping chambers and the transform valleys are commonly deeply incised into the ridge (see box below).

5.7 Graben Formation in the Atlantic

Although all mid-ocean ridges have a zone around the ridge axis 20–40 km wide that is penetrated by ridge-parallel faults, a central rift valley is only developed where spreading rates are slow such as on the Mid-Atlantic Ridge. The central rift valley, locally 1 km deep, has a variable width between 2 and 10 km. The low spreading rate of 2 cm/year at the ridge in the North Atlantic south of Iceland means that rocks at the graben shoulder are younger than 1 Ma. The volcanically active zone in the middle of the graben has a width of only 1 km. Sediments in the graben attain thicknesses of a few meters where they are transported by currents to protected positions. The normal sedimentation rate of deep sea sediments is too low (millimeters to few centimeters per thousand years) to explain thick accumulations of sediments on top of the very young oceanic crust without transportation by currents.

The central graben of the Mid-Atlantic Ridge at 36° 50′ northern latitude has been studied in detail by the submersible "Alvin" (Ballard and Moore 1977); they used numerous photographs to document the variety of pillow lavas. Spreading of the ocean floor is slightly asymmetric, 0.7 cm/year towards the west and 1.3 cm/year towards the east. The total graben structure has a width greater than 20 km and a central rift valley between 2 and 3 km. The western flank of the central rift is steep and its straight shape clearly indicates a fault; relief is 500 m. In the graben center, 200–300 m-high volcanic edifices align and mark the boundary between the North American and African plates. Production of lavas is episodic, and after 10,000 years of activity, the volcanoes become extinct and new fissures open to feed new volcanic edifices.

An Oceanic Crustal Profile in the Atlantic Ocean

The profile through the oceanic crust and the uppermost mantle is not easily accessed. The remains of oceanic lithosphere are found within ophiolite complexes in large outcrops on land where they have been thrust onto continental crust or trapped between crustal blocks during collision; most have been widely dispersed or intensely deformed. Several ophiolite complexes have been only slightly deformed and provide excellent profiles through parts of the lithosphere; examples are the Semail complex of the Arabian Peninsula (see below) and the Troodos complex in Cyprus.

A section through the oceanic crust has also been discovered on the ocean floor where a large transform fault zones cuts across the mid-ocean ridge. This occurs along the W–E striking Vema Transform Fault in the equatorial Atlantic between 10° and 11° northern latitude. The exposure occurs in a deeply incised valley with relief of 3000 m that was formed by fault zone activity. The French submersible "Nautile" discovered an entire profile through ocean crust including rocks of the upper mantle. The profile is gently dipping but internally only slightly deformed (◘ Fig. 5.12).

The section at the Vema Transform Fault begins at the basal peridotites that are mostly altered into serpentinite by oceanic water, and is succeeded by a layer of gab-

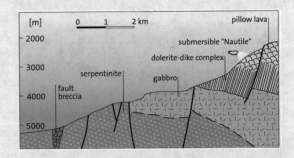

◘ **Fig. 5.12**   Cross section through the oceanic crust at the Vema Fault in the equatorial Atlantic (Auzende et al. 1989; Bonatti 1994). The exploration of this profile was performed with the French deep sea submersible "Nautile". At one of the faults the rocks have been fractured and breccias have formed

bro approximately 1–1.5 km thick. Next in the succession is a sheeted dike complex 1 km thick that consists of the solidified feeder channels to the overlying pillow basalts. The entire abnormal crustal thickness of only 3 km was either tectonically thinned or the production of basaltic melts was low. Because a relatively cool uppermost mantle can be assumed in the equatorial Atlantic and the spreading rate is rather low, the conclusion of low melt production is certainly correct.

Numerous fractures separate the basaltic crust into blocks that are abuted to each other. Open vertical fissures, a few centimeters to 4 m wide, are particularly abundant along the floor of the graben; parallel orientation to the graben indicates extension orthogonal to the graben axis. Normal faults that dip approximately 60° are most common along the walls of the graben. They are associated with tectonic breccias, rocks composed of angular rock fragments that develop along the faults. Measurements of magnetic directions fixed in the basaltic rocks vary and no section longer than 250 m is uniformly magnetized. This pattern reflects the intense faulting and rotation of blocks in the area of the graben.

The magnetic stripe patterns, the discovery of which played an important role in the history of plate tectonics (see ▶ Chap. 1), result from the recording of magnetic polarity in rocks of the oceanic crust. However, coincidences between magnetic polarities (normal or reverse magnetization of the respective stripes) measured remotely with shipboard magnetometers and polarities directly measured in basalts obtained from the seafloor are very weak; therefore, the pillow lavas cannot be responsible for the magnetic stripe pattern. In fact the pillow lavas form in irregular flows and are frequently

disturbed. Therefore, it is generally assumed that the sheeted dike complexes, which are strongly aligned parallel to the plate boundary, create the linear pattern.

5.8   Black and White Smokers

Heat flow, heat conductively emitted from rocks to the Earth's surface, is particularly high at the ridges axes. Here heat flow exceeds the global mean value by several times and it is estimated to be approximately a quarter of the heat that is emitted from the interior of the Earth. However, following an intense gathering of heat-flow measurements, it turned out that the conductive emission of heat at the mid-ocean ridges is variable and locally unexpectedly low.

Tubular chimney-like towers several meters high with temperatures greater than 350 °C were first discovered in 1977 along the East Pacific Rise at the Gulf of California and Galápagos Islands. Since then, such features have been observed along ridge axes in each of the three large oceans. Thermal water, rich in metal and salt, mixes with cold seawater whose temperature is only slightly above freezing. In the ensuing reaction cooling causes the dissolved load to precipitate out of

◻ Fig. 5.13 Submarine photograph of a white smoker at a water depth of approx. 1500 m. The smoker is in the Kemp Caldera, a volcanic collapse structure at the transition from the South Sandwich Island arc to the spreading zone of the East Scotia Ridge in the South Atlantic. The photo was taken by a diving robot during the PS119 expedition with the research vessel Polarstern and was kindly provided by Marum Bremen

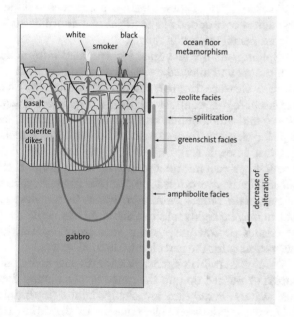

◻ Fig. 5.14 Effect of seawater circulation in young hot oceanic crust. Hydrothermal alterations cause ocean floor metamorphism, hot aqueous solutions leak at the midocean ridge as black and white smokers

the thermal water into dark and light mineral particles that form a turbulent suspension. The scene resembles smoke being emitted from a chimney and therefore these structures are called "black smokers" or "white smokers'" (◻ Fig. 5.13).

The smokers result from the exchange between sea water and hot rock deep below the ocean-floor surface. The hot mixture moves upwards through fractures eventually escaping to the surface loaded with the dissolved minerals (◻ Fig. 5.14). This hydrothermal circulation system cycles an amount of water

equal to the content of all oceans of the world (about 1,350,000,000 km³) every one million years (Herzig and Hannington 2000). A considerable amount of heat from the interior of the Earth is transported to the surface by convective flow in the uppermost kilometers of the crust. This locally results in regional cooling and is expressed in low values of (conductive) heat flow.

Black smokers form at temperatures about 350 °C and are expelled with velocities of up to 5 m per second. Water remains in the liquid phase at 374 °C, the critical temperature of water, if the pressure is greater than 22 MPa; these conditions exist at a water depth of slightly more than 2000 m, typical of most places on mid-ocean ridges.

Black smokers are colored by particles of iron, copper and zinc sulfides (pyrite, pyrrhotite, chalcopyrite, sphalerite) that are precipitated from the hot solutions. The sulfides are dissolved from the basalts, dolerites and gabbros of the oceanic crust where they occur in trace amounts. They are easily dissolved and concentrated in thermal waters. Thermal waters also dissolve other rock material and precipitate the dissolved particles at different locations where lower pressure and temperature exist. Near the surface, they stream through a zone of fractured rocks (breccias) that are a result of hydrofracturing from rising, overpressured water. Fractures formed through this process are filled with mineral precipitates. Therefore, mineral precipitation not only occurs where the solutions go into the seawater, but also below the ocean floor. Besides sulfides, anhydrite (calcium sulfate) and quartz are precipitated in a network of veins, and gold may also accumulate. The basaltic host rock can be silicified (infiltrated by silica) and chloritized. The water-containing mineral chlorite is formed by degradation of basaltic glass and pyroxene.

Smokers form in amazingly short time. Chimneys of several meters height and some decimeters in diameter grow within a couple of months or a few years. Sometimes edifices evolve into blocks as high as houses that eventually collapse into piles of rubble later penetrated by the rising solutions. The mineral precipitation combines the fragments of the smokers into a solid breccia.

Black smokers accumulate considerable amounts of copper and gold. Concentrations may reach economic values with copper contents higher than 20 weight-percent. Such ore deposits, now exposed in ophiolites, have been mined at Cyprus for thousands of years.

White smokers form at cooler temperatures than the black smokers; their temperatures are between 300 and 100 °C. Their expelling velocities are several decimeters per second. White smokers may evolve from black smokers and precipitate sulfates like barite and anhydrite as well as sphalerite and silica. In summary white smokers contain more light colored particles and less dark sulfides hence, the light color of the "cloud of smoke". Gold from the black smokers may be remobilized and further concentrated by white smokers.

The hydrothermal fields of smokers contain exotic extremophile biocoenoses. Endemic faunas are commonly characterized by exceptional growth rates and have a particular adaptation to the environmental conditions. Among the more advanced animals, worms, prawns and gigantic mussels live in symbiosis with sulfur bacteria. Prawns without eyes were found at the Mid-Atlantic Ridge. At the East Pacific Rise, tube worms are covered by a mantle of bacteria that probably have an isolating function in order to live at temperatures up to 80 °C (Campbell et al. 2001). The more than 10 cm-long tube worms tolerate temperature differences of up to 60 °C between their upper and lower end. Bacteria have been found in hot water of 110 °C.

Smokers play an important role related to the temperature, the chemical composition, and the pattern of currents of the oceans. With increasing distance from the plate boundary, the hydrothermal activity in the crust rapidly decreases. One reason is that the crust becomes cooler and the thermal gradient, which drives the convectional currents, decreases. Also, fractures and joints, where water circulation occurs, are closed by mineral precipitation and tectonic activity decreases so few new fractures are formed; these are closed shortly after formation by mineral precipitation. For these reasons, hydrothermal activity is restricted to a narrow zone near the plate boundary.

5.9 Ocean Floor Metamorphism

Ocean water that is emanated from the smokers percolates through fractures and joints of the ocean crust down to depths of several kilometers. During this process, water is heated to temperatures well above the critical temperature and causes substantial modifications to magmatic rocks by material exchange ("ocean floor metamorphism"). Circulating overheated water contacts the top of the magma chamber or the upper zone of the already solidified gabbro. Where hot mantle peridotite is accessible to the circulating waters along prominent faults, it reacts very sensitively.

This process modifies the rocks in the oceanic crust with mineralogical changes that correspond to those that occur in mountain building processes related to regional metamorphism. However, ocean floor metamorphism occurs under static conditions and is not connected to the deformation of rocks. Metamorphism occurs through the process of OH$^-$ (hydroxyl) ions binding to minerals and thus transforming the "dry" magmatic minerals into "wet" metamorphic ones.

At shallow depths with temperatures up to 200 °C, zeolites, a mineral group similar to feldspar, are formed by hydrothermal activity. They contain adsorptive water molecules that gradually dissapear with increasing temperature. This zone is called the zeolite facies zone (◩ Fig. 5.14). The deeper basalts and dolerites are transformed to greenstone (metamorphism in greenschist facies). Rock glass and pyroxene are mostly altered to chlorite and actinolite (a low-temperature amphibole), and plagioclase is partly altered to epidote. All three of the new minerals bind hydroxyl ions in their crystal lattice; their green color is responsible for the rock name. Plagioclase, a calcium–sodium feldspar, is modified by Ca^{++}-Na$^+$ exchange in water to become a nearly pure sodium feldspar (albite); the sodium comes from seawater. This type of alteration is called spilitization and the affected rocks are spilites. In the gabbro, where temperatures may be over 500 °C, pyroxene is changed into hornblende, another amphibole, and the rocks then become an amphibolite (amphibolite facies). Each of these metamorphic zones and mineral suites corresponds to regional metamorphic rocks in continental mountain ranges.

Splinters of oceanic crust incorporated into mountain ranges and affected by regional metamorphism acquire a cleavage through tectonic deformation. Such greenschists (in contrast to the non-schistose greenstone) and amphibolites can not be distinguished, except for cleavage, from ocean floor metamorphic rocks. In most cases, it is not possible to document an older ocean floor metamorphic event in ocean floor rocks subsequently overprinted by regional metamorphism in mountain ranges. Moreover, ocean floor metamorphism may be highly selective based on distinctive characteristics of fracture systems that create contrasting permeability for the percolating thermal waters.

Along transform faults, peridotites are commonly thrust or pushed up into the range of percolating thermal waters. Peridotites are changed into serpentinite by contact with water. The minerals of peridotite, olivine and pyroxene, are "dry"—no water is chemically bound to the minerals. Both olivine and orthopyroxene readily react with water and change into serpentine, a mineral group with water bound in the chemical structure. However, the material properties of serpentinites are fundamentally different from the ridgid peridotites. Serpentinites are easily deformable at the prevailing temperatures in the oceanic crust and they have a substantially lower density (ca. 2.7 g/cm^3) than peridotites (ca. 3.3 g/cm^3) or the gabbroic and basaltic rocks of the crust (ca. 3.0 g/cm^3). Therefore, serpentinites tend to rise as diapirs to the surface where, during active phases of faulting, they may be mixed with sediments. Such tectonic mixtures of serpentinite and cal-careous-shaly sediments are called ophicalcite.

5.10 Chromite Deposits

Chromite deposits are formed at mid-ocean ridges. Present deposits outcrop in ophiolites, remnants of ocean floor thrust onto continental crust. Chromite deposits are common in southern Europe, Turkey and Oman and some are of economic interest. The chromite origin is from harzburgite in the uppermost mantle where typically a few tenths of percent chrome oxide occurs. Enrichment may form deposits that contain several millions of tons of chromite, a mineral in the spinel group: $FeCr_2O_4$; some of the iron may be replaced by magnesium. The ore occurs in irregular, finely distributed lenses of nearly pure chromite or in a mixture of chromite and olivine surrounded by dunite, which also contains chromite. The dunite body is embedded in schistose harzburgite; the schistose fabric formed immediately after crystallization during a hot stage associated with current motion ("schistose peridotite"; ▢ Fig. 5.15). Dunite bodies and chromite lenses originally are found slightly below the Moho within the harzburgites (▢ Fig. 5.2).

According to one hypothesis, the formation of chromite lenses results from the continued melting of harzburgite as basaltic magmas stream through and leave behind dunite and enriched chromite as residuum;

olivine and chromite are those components from the original peridotite that have the highest melting points (Cann 1981). A more favored hypothesis assumes that olivine and chromite enrichments form in basaltic dikes in the uppermost mantle (▢ Fig. 5.15; Lago et al. 1982). These dikes are feeder dikes for the magma chamber below the mid-ocean ridge. Olivine and chromite are the first to crystallize from the melt when it cools under 1200 °C. Normally, when gabbro continues to crystallize, these minerals are resorbed. However, small chambers form in the feeder dikes that are caused by irregularities in the dikes (▢ Fig. 5.15, circular insert); olivine and chromite crystals that form along the cooler walls of the chambers, may adhere to the walls and clump together, or sink to the bottom of the chamber because of their high density. The resulting crystal mush filters molton streams that flow through them and strain out additional crystals, thereby causing further accumulation of olivine and chromite crystals. Chromite lenses are irregularly distributed and highly variable in size which makes prospecting and mining of deposits difficult.

5.11 Ophiolites

Ophiolites are the remnants of ocean crust and sub-adjacent mantle lithosphere. The name comes from the Greek language (*ophis,* the serpent) because the serpentinites (from *serpens, Latin* serpent) frequently associ-

▢ **Fig. 5.15**   Evolution of chromite deposits within dunite bodies in the mantle below the magma chamber at a mid-ocean ridge (Nicolas 1995). Olivine and chromite are the first mineral precipitates from the rising basaltic melts. They can accumulate in widened chambers where they filter out the ensuing streaming minerals and cause the concentration of olivine and chromite

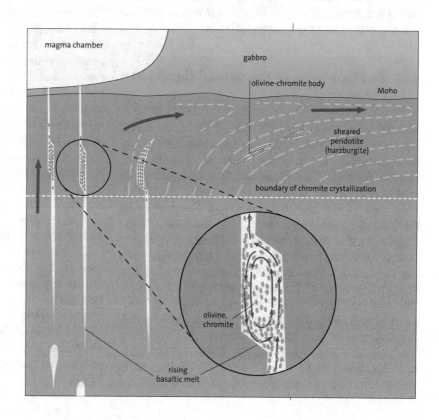

5

ated with ophiolites have a scaly surface that resembles the skin of a serpent. Since ophiolites are thrust onto continental crust by tectonic processes or trapped between crustal blocks, it is possible to study ocean floor material now on land.

There are two broad processes that can displace ophiolites into continental crust. (1) Parts of the ocean floor can be thrust onto a continental margin by compressional tectonics; an example is the Semail Nappe in Oman (Nicolas 1995). Large overthrust units are called "nappes" and thrusting of ocean floor and related material onto a continental margin is called "obduction" (*obducere, Latin* to pull over or to cover) in contrast to "subduction" (*subducere, Latin* to lead away downwards). Obduction especially occurs where young, relatively hot, and thus less dense ocean floor that is not yet able to subduct, is involved in compressional movements. (2) Ophiolite remnants are trapped in collisional orogens that evolve by the collision of continents or island arcs with other continental blocks. Following an initial subduction, ophiolite remnants are commonly dispersed and deformed and are usually characterized by high-pressure metamorphism. They mark the suture zone between the colliding crustal blocks; such a suture zone lined by ophiolites is called an "ophiolitic suture". Ophiolites indicate that ocean crust has been subducted at a given location where continental collision and mountain building occurred. The intervening ocean floor is nearly completely subducted into the depth of the mantle—the sparse accessible remnants represent only a small fraction of the original ocean floor. Yet, they are critical evidence in the reconstruction of plate tectonic processes of the geological past.

5.12 The Ophiolite of the Semail Nappe in Oman

The ophiolite complex of the Semail Nappe in Oman has been investigated by many researchers because of its importance in the development of plate tectonic theory (Nicolas 1995). It is one of the largest ophiolite complexes on Earth at 500 km long, 50–100 km wide, and 15 km thick (◘ Fig. 5.16a). The history of the Semail Nappe began as part of the great Neotethys Ocean that during the Early Cretaceous, separated Arabia from south–central Asia by some 1200 km. As the Neotethys narrowed and compressed ca. 100 Ma, the NNE portion of the oceanic plate was thrust over the SSW part. The thrust developed in a zone of crustal weakness along the mid-ocean ridge, a plate boundary and zone of thin, weak lithosphere (◘ Fig. 5.16b). The ophiolite nappe was initially thrust several hundred kilometers onto oceanic crust that belonged to the other side of the spreading axis. It rapidly migrated

southward and the entire complex was obducted onto northeastern Arabia at 80 Ma. The calculated thrusting velocity was approximately 3 cm/year.

When the Semail Nappe was thrust over the continental margin, the forces of friction increased, the depressed continent was compressed, and because of its lower density, it raised upwards. These frictional forces slowed the obduction of the ophiolite onto the continental margin and obduction ceased after the nappe was transported 100–200 km.

Driven by the global plate drift pattern, the convergence between Arabia and Eurasia (Iran) continued after the ophiolite obduction and the Neotethys Ocean was subducted by a new subduction zone, the Makran subduction zone. Here, the remaining oceanic crust is being subducted beneath the continental crust of Iran. This process will lead to the final collision and mountain building event between the Arabian and Eurasian plates in approximately 2 million years (◘ Fig. 5.16). The Oman ophiolites will likely be overthrust by the Eurasian Plate and subsequently deformed, dispersed and metamorphosed. Only an ophiolitic suture will remain of the once great Neotethys.

5.13 Alpine–Mediterranean Ophiolites

Ophiolites of the Alpine–Mediterranean region originated primarily from the borders of the Penninic–Ligurian Ocean, which during the Middle and Late Jurassic represented the northeastern continuation of the central Atlantic Ocean (◘ Fig. 4.8). These ophiolite assemblages, associated with coeval oceanic sedimentary rocks, were incorporated into the Alpine orogen much later, after they had drifted far from the mid-ocean ridge. During this time, the oceanic lithosphere had cooled down and become denser. Due to its high density, the lithosphere uncoupled from the continental margin and became subducted. Subduction is documented by high-pressure metamorphism. Subsequently, the ophiolites were scraped off and sandwiched between the colliding continental margins, where they mark the suture zone between blocks derived from Africa and from Europe. However, some ophiolites are not metamorphosed and were presumably obducted such as those on the island of Elba.

The Alpine–Mediterranean ophiolites evolved in a completely different manner from those in Oman. The Oman ophiolites were derived from a midocean location within their ocean of origin and were obducted across long distances. The Alpine–Mediterranean ophiolites formed at the margin of the ocean and most were subducted before being underplated and uplifted by the continental margin during subsequent continental collision.

Metamorphic Sole

During obduction, heat from the base of the overthrusting ophiolite is transferred downwards into the units below the thrust surface. The heat induces metamorphism and is accompanied by deformation caused by the thrusting process. This style of metamorphic zone below ophiolite nappes is called a "metamorphic sole". Metamorphic soles are characterized by metamorphism that is most intense at the tectonic boundary and rapidly decreases with depth because the generation of heat persisted for only a short time span.

In Oman, the heat in the sole of the Semail Nappe induced volcanism because the base of the nappe, immediately after its formation, was approximately 1000 °C. Basalts of the subducted ocean floor were heated to 900 °C, metamorphosed into amphibolites, and partly melted along with overlying sedimentary rocks (◻ Fig. 5.16c). On the upper plate, the melts created andesitic volcanism that has the characteristics of island–arc volcanism (Boudier et al. 1988). During continued thrusting of the ophiolite nappe over the ocean floor, the base rapidly cooled so progressively lower temperature metamorphism developed in the underlying basalts (e. g., greenschists at the temperature range between 500 and 300 °C).

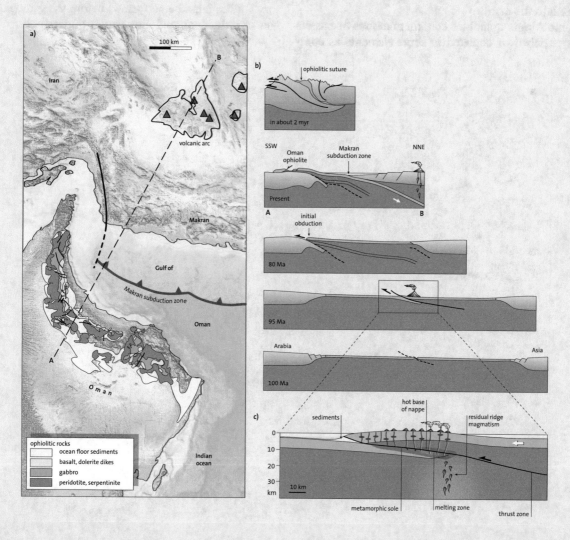

◻ **Fig. 5.16** Evolution of the ophiolitic Semail Nappe in Oman (Nicolas 1995). **a** Map view of the Semail Nappe, the recent subduction zone (Makran) and the related volcanic arc. **b** In the initial stage the north–northeastern limb of the ocean was thrust over the south–southeastern limb along the mid-ocenic ridge (100 and 95 Ma). Obduction of the ophiolites onto the Arabian continental margin began at 80 Ma. Presently, a volcanic arc exists in Iran that is generated by the newly formed Makran subduction zone. In the near geological future, an ophiolitic suture will form at the collision zone of the continental margins. **c** In the initial stage of thrusting, the hot base of the overthrusting plate induced an inverted metamorphic profile ("metamorphic sole") in the lower plate, as well as partial melting of lower-plate basalts and sediments to feed andesitic volcanism on the upper plate

5

Many ophiolites of the Alpine–Mediterranean region lack gabbros and sheeted dike complexes. Basalts and deep-water, ocean-floor sediments are associated with serpentinite bodies that represent altered mantle peridotites (see above). This configuration juxtaposes the uppermost units of the ophiolite profile with rocks that originated in the mantle. This peculiar phenomenon originated along faults and large fracture zones on the ocean floor where peridotites were metamorphosed to serpentinites that flowed upwards in diapiric structures and came into contact with the basalt layer. When later ophiolite bodies are scraped off from subducting lithosphere, they mainly contain basalts and serpentinites from the upper oceanic crust, whereas the deeper layers such as the dolerite dikes and gabbros continue to sink into the mantle.

Some Alpine ophiolites contain andesites or basalts either enriched or depleted in some elements as compared to normal ocean floor basalt (see ▶ Chap. 7). These ophiolites probably formed in marginal ocean basins above subduction zones; a modern example is the Sea of Japan. Commonly obducted ophiolites were interpreted as having formed in marginal oceanic basins. Such settings are difficult to subduct because they are relatively hot as they are warmed by continuous subduction magmatism and therefore buoyant. As shown in Oman, this generalization does not always prove correct because there the obduction was triggered by the initial overthrust at the hot mid-ocean ridge. The andesitic volcanic rocks in the Semail Nappe were formed during the overthrust (◻ Fig. 5.16c) and are not related to a marginal basin. The points illustrated above demonstrate that careful field and laboratory studies must be performed before the geologic origins of individual ophiolites can be understood.

Hot Spots

Contents

© The Author(s), under exclusive license to Springer Nature Switzerland AG 2022
W. Frisch et al., *Plate Tectonics*,
Springer Textbooks in Earth Sciences, Geography and Environment,
https://doi.org/10.1007/978-3-030-88999-9_6

Although the vast majority of volcanoes on Earth are related to plate boundaries–mid-ocean ridges and subduction zones, approximately 5% are classified as "hot spots". Hot spots are the product of mantle diapirs (*diapeírein, Greek* to perforate) or plumes (because of their shape) that rise through the mantle as finger-shaped hot currents and penetrate the crust. In spite of their relatively low numbers, they play an important role in the convection system of the Earth's mantle and are responsible for about 5–10% of the melts and energy emitted by the Earth. At present, approximately fifty hot spots have been identified in both continents and oceans (◨ Fig. 6.1). Although most are located in the interior of plates ("intraplate volcanism", e. g., Hawaii), some hot spots are either coincident with mid-ocean ridges (e. g., Iceland) or in close proximity (e. g., Azores, Tristan da Cunha). Mantle plumes and the resulting hot spots are responsible for the formation of large volcanic complexes; these include volcanic chains up to several thousand kilometers long and huge flood extrusions, commonly called large igneous provinces or LIP, that consist of basaltic lavas. Plumes are produced in the lowermost mantle adjacent to the core within the so-called D″ layer (see below; ◨ Fig. 6.2), at a depth of about 2900 km. The hot source in the mantle is generally considered to be fixed in its position over long periods of time. The existence of hot spots and their significance as fixed points was first established by J. Tuzo Wilson, one of the doyens of plate tectonics (Wilson 1963).

The boundary between the metallic core and the siliceous mantle is not only a material boundary but also a thermal one where temperatures change drastically across a short distance. Temperature decreases from about 3000 °C by several hundred degrees within the lowermost 200 km of the mantle. Plumes are apparently triggered by hotter areas at the base of the mantle by heat transfer from hot spots in the liquid outer core. However, a plume is only generated if the temperature anomaly is large enough and exceeds a limiting value. In spite of its high temperature, the rising mantle rock is not molten because of the high pressure.

The plume ascends because its material is hotter than the surrounding mantle and thus has slightly lower density and viscosity; the mantle generally is able to flow ductilely in spite of its solid state. The plume has a cylindric conduit approximately 150 km diameter. Experiments indicate that it develops a bulbous head that moves upward through the mantle (◨ Fig. 6.3). Processes associated with the rising head of the plume are complex; the head is slowed down by the viscosity of the surrounding mantle (in particular the lower mantle exerts a strong resistance); hot but less viscous material below the head causes expansion of the head so that its slower upward motion is compensated by the faster but smaller stalk below. As the head expands and rises, surrounding mantle material is also incorporated into it.

Melting and magma generation occur as the plume rises near the base of the lithosphere at a depth of 100–150 km. Here pressure is reduced and the temperature of the plume is 250 °C higher than that of the surrounding asthenosphere; this leads to the formation of significant amounts of magma by partial melting of the peridotite in the plume (White and McKenzie 1995). The content of water in the mantle rock is also of importance as water reduces the melting point. Portions of the upper mantle are "wet" or "damp" and accordingly larger amounts of melt are produced. The wet mantle plays a role in Hawaii as well as in the Azores

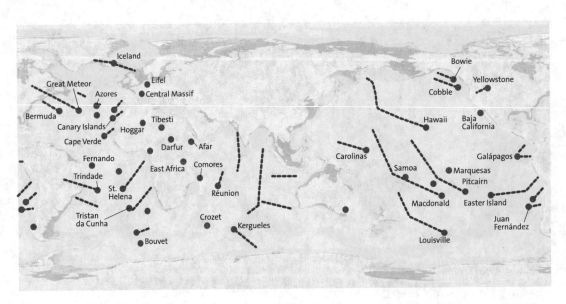

◨ **Fig. 6.1**   Map showing global distribution of hot spots. Volcano chains formed by the hot spots (dashed lines) indicate the plate movement over the hot spots

■ **Fig. 6.2** Cross section through the earth as viewed from the South Pole. Oceanic and continental lithosphere are shown exaggerated because they would be less than 1 mm thick in a realistic representation. Heavy parts of subducting plates sink to the base of the mantle. The liquid outer core of the earth generates convection currents that interact with the earth´s mantle. Mid-ocean ridges have shallow mantle sources. Modified after Condie (1997), Schubert et al. (2001), Tackley (2008), Torsvik et al. (2014). LLSVP: large low shear velocity province, under Africa ("Tuzo ") and under the Pacific ("Jason"); PGZ: plume generation zones of mantle diapirs. The dashed line indicates bulging of the geoid surface

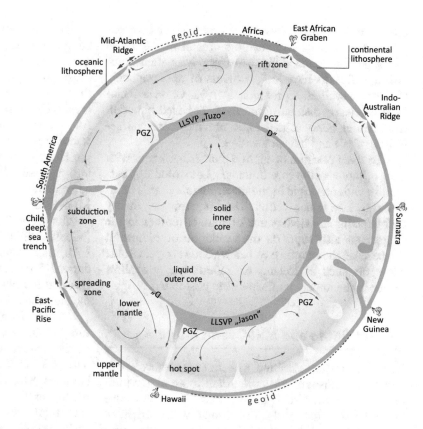

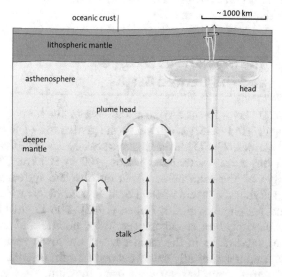

■ **Fig. 6.3** Model for the evolution of a mantle plume based on experiments with glucose syrup (Griffiths and Campbell 1990)

(see below). Water can be detected because it is bound as hydroxyl ions in basaltic minerals that evolve from the melts.

When the plume reaches the bottom of the rigid lithosphere, a broad mushroom-like head is formed that may achieve a diameter of 1000 km (■ Fig. 6.3). This head generates surface uplift because the push of the plume and the thermal expansion of the heated area causes bulging of the lithosphere; elevations greater than 1000 m can be attained. The magma generated in the head penetrates through the lithospheric mantle and crust. The final result is the extrusion of enormous volumes of basalt. Smaller plumes such as one under the French Central Massif do not have a distinctive head.

The consequences of hot spot magmatism can be extreme. The large volcanic edifices at Hawaii form the highest single mountain on Earth. They can also form huge flood basalt fields where magmatic extrusions occur within amazingly short periods of time, commonly within one million years. Such an extrusion occurred in the Deccan Trap basalt in India. Here the generation of huge volumes of basalt was related to the break-up of a portion of the large continent Gondwana. Huge, extensive intrusions and basalt flows can lead to thickening of ocean crust over wide areas such as the Ontong–Java Plateau in the western Pacific and other similar oceanic plateaus (■ Fig. 6.14).

Because of the enormous loss of magma material, more than 2 million cubic kilometers in some cases, the head of the plume cools rapidly and the magmatic activity subsides. However, the active life of some hot spots may persist more than 100 million years, commonly through pulsating events of volcanic activity.

Where hot spots underlie continents, the magma penetrates and is influenced and contaminated by continental crust. The basaltic melts differentiate and their heat may cause large volumes of crustal rock to melt. This process generates acidic magmatic rocks. Therefore, most continental hot spots coevally produce basaltic mantle melts and rhyolitic crustal melts. These two groups of rocks differ considerably in their composition. Basalts have SiO_2 contents of approximately 50% and rhyolites approximately 70%. Close associations of clearly distinctive magmas are called bimodal. Continental bimodal magmatic provinces are characteristic of rift and hot spot settings.

The fixed position of mantle plumes correlates with their origin near the core boundary but would not harmonize with an origin in the upper mantle that is dominated by strong convection. The "superplume" theory that explains the dominance of mantle plumes in the Cretaceous also supports the formation of plumes at the base of the mantle (see below). However, for some hot spots such as Iceland or Yellowstone, a shallow source in the upper mantle has been postulated.

6.1 Hot Spots and Mid-Ocean Ridges

Magmatic rocks associated with hot spots have a different mechanism of formation than those of the mid-ocean ridges even though both basaltic melts evolve in the uppermost mantle. The former, which originate within plates and are called intraplate magmatites, have characteristic chemical signatures that distinguish them from basalts of mid-ocean ridges (◘ Fig. 5.8).

Mid-ocean ridge basalts (MORB) are tholeiites. Tholeiites evolve through relatively high percentages of partial melting of peridotite (mostly 15–25%). Hot spots with high magma production generate flood basalts and island chains like Hawaii and also develop tholeiites, albeit of slightly different composition from that of MORB. Hot spots associated with low volumes of magmatic activity produce alkaline basalts. Compared to MORB, hot spot basalts contain distinctly higher contents of elements that are incompatible with mantle rocks including potassium, rubidium, phosphorus, titanium and the light rare earth elements. This mirrors the low degree of partial melting of alkaline basalts in the mantle (mostly below 10%) as well as a mantle source that is different from the MORB source. The MORB source is solely the asthenosphere. In contrast, the intraplate basalts of the plumes contain components of the lower mantle, the asthenosphere, and the lithospheric mantle. In intraplate tholeiites, the asthenospheric portion dominates, in alkaline basalts the lithospheric mantle dominates (Wilson 1989). If hot spots are located on a midocean ridge, their chemical composition indicates a particularly high asthenospheric portion. Alkaline basalts generated from a fading hot spot reflect the waning melt formation and are rather coupled to a lithospheric source.

Isotope chemistry is a powerful tool for differentiating the source of magmas. All isotopes of a given element have the same number of protons but vary because of their different numbers of neutrons and therefore, have different atomic masses. The mantle reservoirs of hot spots are richer in incompatible elements. As confirmed by strontium isotopes, these different reservoirs have not been substantially mixed during the last 2 billion years with the exception of the mixing processes that occur in areas where hot spots are coincident with mid-ocean ridges. Presently forming MORB have $^{87}Sr/^{86}Sr$ isotope ratios of 0.7027 whereas basalts from hot spots have ratios of about 0.7040 (Nicolas 1995). Because ^{87}Sr evolves from the decay of ^{87}Rb, there must be a deep mantle source for basalts derived from hot spots. The deep mantle has Rb/Sr ratios that are higher than those generated in the Rb-depleted asthenosphere, the source of MORB.

Helium isotopes also have a characteristic signature in basalts generated from hot spots; the high ratio of ^3He to ^4He is considered to be a signature of melts derived from the lower mantle. The primordial mantle contains ^3He whereas ^4He evolved during the Earth's history by the radioactive decay of uranium and thorium (alpha decay). Because the upper mantle degasses easier, it contains less ^3He and thus lower $^3He/^4He$ values than the lower mantle.

6.2 The Mysterious D″ Layer and the Dented Earth

The D″ layer forms a zone at the base of the mantle. It is suggested that it is critical in the generation of mantle plumes. The D″ layer is generally 200–250 km thick but has extremes that range from 100 to 500 km. The origin of the term comes from an obsolete system of classifying the layers of the Earth devised by New Zealand geophysicist Keith Bullen (1942). The D shell in his classification stood for the lower mantle and was later subdivided in the main part of the lower mantle (D′) and a seismically newly defined bottom layer (D″); this term alone has survived from the Bullen system. An internet site by A. Alden sarcastically suggests we call it "hell".

Current geophysical thought suggests that the D″ layer, because of its irregular and heterogeneous characteristics, represents the ultimate fate of subducted plates (Vogel 1994). Below the long lasting, active subduction zones around the Pacific Ocean, seismic waves travel relatively rapidly in the D″ layer, indicating the presence of slightly cooler and denser material. This characteristic suggests to geophysicists that the D″ layer consists of still relatively cool remnants of

Fig. 6.4 Above: Bulges of the earth's core, which are related to zones of reduced shear wave velocity (LLSVP, large low shear-wave velocity provinces) at a depth of about 2800 km (Torsvik et al 2010). Large igneous provinces of the Mesozoic and Cenozoic and present hot spots are mainly arranged along the outer edge and above the LLSVP zones ("Jason" and "Tuzo"). It is assumed that the LLSVP zones remained stable in the geological past. Below: Bulges of the geoid surface according to the gravity model EIGEN-6C of the GFZ Potsdam (▶ http://icgem.gfz-potsdam.de/). Bulging range to about 80 m depression to about 100 m. The deviations of the geoid surface from the rotational ellipsoid indicate, that the gravity minima roughly coincide with the LLSVP zones

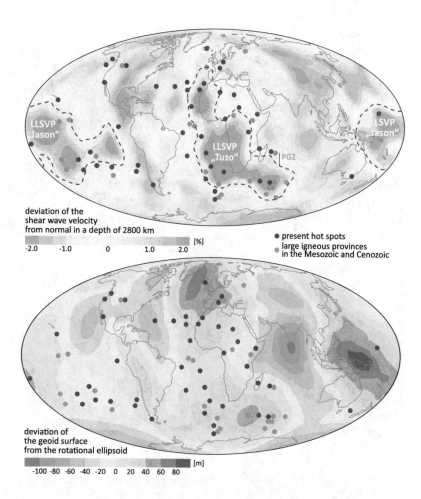

deviation of the shear wave velocity from normal in a depth of 2800 km

-2.0   -1.0   0   1.0   2.0   [%]

● present hot spots
● large igneous provinces in the Mesozoic and Cenozoic

deviation of the geoid surface from the rotational ellipsoid

-100 -80 -60 -40 -20 0 20 40 60 80   [m]

oceanic crust that was subducted into the lower mantle where it sank to the bottom because of its high density (■ Fig. 6.2). It is assumed that these rocks contain stishovite, a high-pressure form of quartz that has a density of 4.34 g/cm³, nearly twice that of normal quartz at 2.65 g/cm³. The decay of widespread perovskite ($MgSiO_3$) in the lower mantle could form stishovite (SiO_2) and magnesiowustite (MgO) under the high pressure in the D″- layer (almost 140 gigapascals at approx. 3000 °C). The weight of this dense material from subduction zones causes the core to be indented several kilometers (■ Fig. 6.4). The presence of cooler subducted material at the core boundary likely induces downward-directed convection currents in the liquid outer core.

Beneath the Central Pacific and the western half of the African Plate, however, lower velocities of seismic waves indicate that the basal mantle layer is slightly hotter, less viscous, and therefore, easier to move. These opposing zones are called LLSVP (large low shear-wave velocity provinces) (■ Figs. 6.2 and 6.4). Burke (2011) named them after two pioneers of plate tectonics, J. Tuzo Wilson and W. Jason Morgan, and named them Tuzo and Jason. It has been shown that these zones with a bipolar distribution of mantle diapirs and subduction zones on the Earth are related to each other (■ Fig. 6.2) The decreased shear wave velocity is suggested to be a result of increased temperature, which is probably a result of upward convection currents which transport the heat from the liquid outer core to the LLSVP zones.

Hot Spots of Pangaea

In theory, there should be a connection between the LLSVP zones over which there are bulges of the geoid and the concentration of hot spots. Geoid bulges form because large areas of thick continental lithosphere have poor heat conductivity and have an insulating effect on rising heat from the sub–lithospheric mantle (Anderson 1982). Mantle temperatures rise and the region forms a thermally uplifted bulge. Resulting bulges with 50 m of relief can form within 100 m. y. They tend to be centered near the equator where their configuration is most stable

Fig. 6.5 Bulge of the present geoid in the area of Africa–North Atlantic with related hot spots, superimposed onto the continental pattern at the end of the Jurassic (Anderson 1982; map after Blakey 2016)

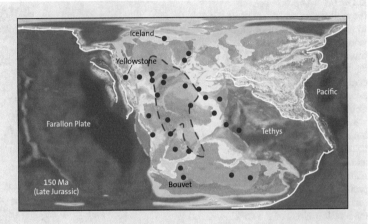

within the rotating Earth. Therefore, if large continental masses are concentrated near the equator, hot spots and resulting bulges will also concentrate near the equator. This causes the axis of Earth to adjust by true polar wandering.

An example of concentrated hot spots and large continental masses near the equator occurred during the break-up of Pangaea. Many hot spots are located beneath the African–Atlantic bulge at present, which is caused by the LLSVP "Tuzo" (▪ Fig. 6.4) and as reconstructed for the continental masses of Pangaea during the Jurassic (▪ Fig. 6.5). The concentration of so many hot spots in this configuration suggests that the present, currently active hot spots are old and initially formed during the time of Pangaea 300–175 Ma. Because mantle diapirs require approximately 100 m. y. to develop and ascend, most of the hot spots would have formed towards the end of Pangaea's existence. The hot spot bulges and the subduction zones at the margins of Pangaea (▪ Fig. 4.8) produced extensional stress in the supercontinent. The plume locations formed corners for subsequent rifting—the sites of three-pointed graben stars (Ch. 3). Rift zones propagated by extensional forces, followed weak zones in the crust, and eventually connected the hot spot corners—the fate of Pangaea was sealed (▪ Fig. 4.7). The crust of the supercontinent was fatally cracked and post-Pangaea plate patterns would rift it apart. The history of the Atlantic is closely connected to the activity of hot spots and corroborates the above picture. Individual parts of rifted continents drifted away from the zone of ascending mantle currents towards zones of descending mantle currents. New subduction zones formed at old passive margins and continents eventually rejoined, sometimes in approximately the same positions that they were torn apart previously. In the past 60 Ma, new collisions have occurred to form the Alpine–Mediterranean mountain range at the southern margin of Europe and Asia—a starting signal for the formation of a new Pangaea?

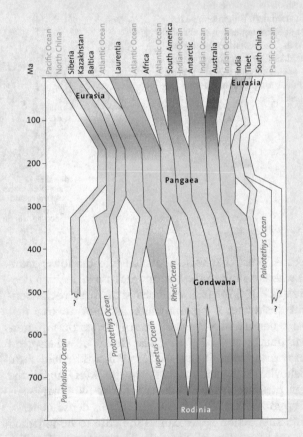

Fig. 6.6 Graphic illustration of major landmasses before, during, and after Pangaea. Pangaea lasted from ca. 300–175 Ma. The supercontinent Rodinia preceded it in the Late Precambrian. Colors used to show general affinities of continents following major rifting events

A second geoid bulge, which is caused by the LLSVP "Jason", occurs in the Pacific, directly opposite to the African–Atlantic bulge (▪ Fig. 6.2), where a conspicuous concentration of hot spots is located. The present Pacific bulge is the broader and higher one of the two. The length of the Pacific bulge corresponds to half of the circumference of the Earth along the equator (■ Fig. 6.4) and thus defines the present location of the pole of the Earth because the axis of rotation is vertically oriented to it. *If* the African–Atlantic geoid bulge, which is stretched in NW–SE direction and only slightly smaller, were the stronger bulge, the axis of rotation of the Earth would be oriented vertically to it through polar wandering—the equator would align with Greenland and Western Europe according to the opinion of some scientists (Chase 1979).

Supercontinents probably form at intervals of several hundreds of millions of years and then break apart. Pangaea was formed by several continent–continent collisions ca. 300 Ma and gradually broke up 130 m. y. later (■ Fig. 6.6). Pangaea formed when Gondwana (named after the kingdom of the Gond in Central India, i.e., the present Madhya Pradesh), an accumulation of the present southern continents as well as Arabia, India, and Southwest Europe, was welded together at the end of the Precambrian 550 Ma and later combined with North America, Baltica, and Asia (■ Fig. 6.7). Rodinia (after *rodina, Russian* fatherland), a supercontinent similar to Pangaea that comprised all large land masses, amalgamated ca. 1000 Ma and disintegrated ca. 750 Ma. The spatial distribution of the land masses in Rodinia was fundamentally different from that of Pangaea (■ Fig. 6.7).

An even older supercontinent, Panotia, may have existed ca. 1400 Ma. The history of supercontinents suggests that they form and break-up in cycles of several hundreds of millions of years duration. As suggested above, the processes of supercontinent formation and destruction are orchestrated by events within the mantle including the activity of hot spots.

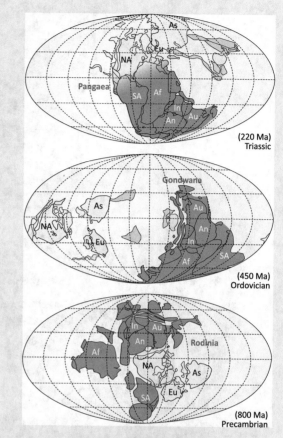

■ **Fig. 6.7** Reconstruction of the supercontinents Rodinia in the Late Precambrian, Gondwana in the Early Paleozoic, and Pangaea in the Early Mesozoic (Hoffman 1991). Af = Africa, An = Antarctica, As = Asia, Au = Australia, Eu = Europe (corresponds to Baltica), In = India, NA = North America, SA = South America

Summarizing these complex relations, it appears that hot zones of the LLSVP and the basal mantle layer correspond with hot zones of the core and that these coincidences produce mantle plumes. Consequently, more heat is transferred into the LLSVP resp. PGZ above the hot bulges of the core than is transferred by the core outside the bulges. Moreover, there is an indication that some amount of iron from the core is transferred to the LLSVP zones in such places. This transfer is actually suggested by relatively high osmium contents and a high $^{187}Os/^{188}Os$ ratio seen in some basalts that are derived from hot spots (Walker et al. 1995). The complex processes and features of the core and the adjacent LLSVP zones suggest that both heat and limited amounts of material are transmitted from the former to the latter.

6.3 Hot Spot Tracks in the Ocean

Hot spots penetrate oceanic lithosphere relatively easy whereas the transfer of magmas through the thicker continental lithosphere is retarded. Therefore, hot spot magmas form tracks on the ocean floor in response to the motion of the ocean plate relative to the fixed hot spot. The tracking the hot spots and dating the volcanoes along the track allow the velocity and direction of plate motion to be accurately calculated. The most famous example for such a track is the Hawaii–Emperor chain in the Pacific (■ Fig. 6.8). The mantle plume is presently located below Hawaii. The 6000 km-long track, which runs towards WNW and then bends abruptly NNW, consists of extinct volcanoes of increasing age to the northwest. The pattern demon-

6

Fig. 6.8 The Hawaii–Emperor chain with volcanic islands (yellow names) and submarine seamounts and guyots (white names). The age of the volcanoes increases from Hawaii towards NW. At 42 Ma a kink of 60° occurs that indicates a change in the plate movement direction. Ages given in millions of years

strates how the Pacific Plate drifted over the hot spot and during the Eocene at 42 Ma, changed its direction of movement by 60°. The oldest remaining volcanoes in this chain are older than 80 Ma. However, the hot spot is somewhat older because the northernmost volcanoes are already subducted along with the northwestern part of the Pacific Plate beneath Kamchatka.

Similar tracks with a kink are also found in other regions of the Pacific (■ Fig. 6.1) and reveal a consistent picture of the drift of the Pacific Plate since the Late Cretaceous with a change of the movement direction in the Eocene. Other tracks in the Pacific do not show a kink and are thus younger than 42 Ma. The kink is most probably caused by global plate tectonic patterns, specifically a plate reorganization that took place during the Alpine–Himalayan orogeny in the Eocene. The resulting mountain belt extends through Southern Europe and South Asia across more than 10,000 km and was initiated by the collision of Africa

and India with Eurasia. Such a major plate tectonic event caused a re-orientation of the plate drift pattern and is reflected on the other side of the world by the abrupt movement change of the Pacific Plate.

Linear volcanic chains that represent tracks of mantle plumes are also found in the Atlantic and the Indian oceans. In the Atlantic, the hot spot of Tristan da Cunha produced two volcanic chains, the Rio Grande Ridge on the South American Plate and Walvis Ridge on the African Plate. Both ridges have northerly components of movement in addition to their west and east motions, respectively. This pattern evolved because the hot spot spent most of its life on the Mid-Atlantic Ridge, ca. 125–30 Ma, and the volcanoes were thus partly built on each plate (■ Fig. 2.6). Approximately 30 Ma in the Oligocene, the drift of Africa slowed significantly although the spreading velocity remained unchanged. Therefore, the Mid-Atlantic Ridge was pushed westward resulting in a stronger westward drift of South America. The hot spot of Tristan da Cunha remained stable and was thus completely enclosed by the westward extending African Plate (■ Fig. 6.9). Presently, Tristan da Cunha is located about 400 km east of the Mid-Atlantic Ridge and produces its track only on the African Plate.

In the Indian Ocean the Maldives Ridge and the Ninetyeast Ridge, which got its name from its position on the 90th east line of longitude, were generated by the hot spots of Réunion and the Kergueles, respectively. Each displays kinks in plate direction, also ca. 40 Ma (■ Fig. 6.10), and both produced voluminous basaltic rocks, especially early in their respective histories. After significant early basaltic production, each continued with sufficient magmatism to leave prominent and rather continuous tracks of plate drift in the ocean.

The hot spot of the Kergueles, the older one of the two, built a plateau with a volume of 20 million km³ of rocks across an area of 2 million km². Over its geologic history that began 120 Ma, the hot spot has had a great influence on the greater Indian Ocean and adjacent continents. Volcanic rocks were especially abundant during its first 30 million years of history and are presently located in the Kergueles Plateau, the Antarctic, the Broken Ridge Plateau, Western Australia, and Rajmahal, India. The basalts of the Rajmahal are not widely distributed and were separated from the occurrences further south by the formation of the early Indian Ocean.

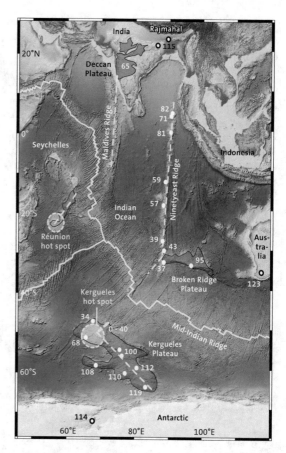

■ **Fig. 6.10** Tracks of the hot spots of Réunion and the Kergueles. Ages of dated basalts (in Ma) show the series of events that separated India, Australia, and Antarctica (Zhao et al. 2001). At ca. 120 Ma, the three continents were adjacent to the Kergueles–Broken Ridge hot spot. From 115 to 40 Ma, India moved rapidly northward and the Ninetyeast Ridge marks the track of the hot spot on the Indian Plate. At ca. 40 Ma, Australia and Antarctica separated and the hot-spot plateau was severed into two major pieces. The Maldives Ridge records a similar though younger history of movement of the Indian Plate. Note that both hot spots were separated from their tracks on the Indian Plate by the establishment of the present Mid-Indian Ridge around 40 Ma. Their most recent tracks are only on the Antarctic and African plates

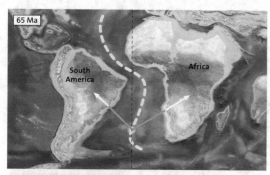

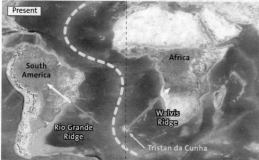

■ **Fig. 6.9** Early Tertiary and present maps that demonstrate the drifting of Africa and South America and location of the hot spot of Tristan da Cunha. Following a period of symmetric drift when the hot spot lay on the mid-ocean ridge, the plate boundary shifted 30 Ma, and the hot spot tracked across only the African Plate; therefore, the most recent track is restricted to this plate (compare ■ Fig. 2.6; Burke and Wilson 1976)

The track of the hot spot is found in the Ninetyeast Ridge and records when India and the Antarctic drifted apart. At 40 Ma, the present spreading axis between the Indian Ocean was formed and separated the Broken Ridge Plateau from the Kergueles Plateau. This reorientation of the plate drift pattern in the Eocene caused the kink in the track. The recorded hot spot history of Réunion began with a huge bang as it generated the Deccan Traps of the Deccan Plateau at the Cretaceous–Tertiary boundary 65 Ma. Again, the hot spot track was generated in two episodes, before and after 40 Ma. The shorter tracks of the two hot spots on the African and Antarctic Plate, respectively, resulted from slow motion of these plates relative to the hot spots since the establishment of the actual Mid-Indian Ridge in the Eocene. In contrast, the long hot spot tracks north of the ridge reflect the rapid drift of India northward during the Late Cretaceous and Early Tertiary.

A Guyot Evolves

One of the most important American geologists in the middle of the twentieth century was Harry H. Hess, one of the principle formulaters of the theory of plate tectonics. Hess was a naval officer in the 2nd World War where he was assigned to submarines. Through voyages in the Pacific, he discovered volcanic seamounts in the Pacific. Some seamount summits are 2000 m below sea level, are remarkably flat, and have fringe reefs on their tops (Hess 1946). He called these peculiar seamounts guyots, after Arnold Guyot, a Swiss geographer who emigrated to America and who, like Hess, taught at Princeton University but a century earlier. Hess concluded from his observations that the summits of these seamounts must have been at sea level earlier in their history. Marine erosion beveled the peak of the volcano and created a flat abrasion platform. During the gradual subsidence of the seamounts, shallow marine sediments and coral reefs were formed on the planed summits. Eventually, the summits subsided too rapidly for coral growth to keep up and the reefs drowned and sank to the depths.

Plate tectonic processes readily explain the evolution of flat-topped seamounts (◻ Fig. 6.11). A volcano that develops above a hot spot on ocean crust may grow, as in the case of Hawaii, above the sea level and form an island. As the volcano drifts away from the hot spot area, volcanic activity becomes extinct and erosion exceeds volcanic production. Continued drifting causes the island to slowly submerge because its plate basement also subsides with increasing age. Marine processes bevel the volcanic island to near sea level and create an abrasion platform. In tropical regions, the flat top may be rimmed with coral reefs, usually in a circular pattern. Because the vertical growth of coral can keep up with all but the most rapid rates of subsidence, the fringing coral atoll continues its upwards growth. However, eventually the subsidence, perhaps coupled with rapidly rising sea level, wins out and the reef drowns and subsides passively below the sea. With continuously increasing age of the crust, the eroded, flat-topped volcano sinks deeper into the deep sea.

Seamounts that formed above hot spots in the deep sea but never rose above the sea level maintain their summit as a peak. Volcano chains formed above hot spots may consist of various volcanic islands, guyots, and seamounts—each records the balance between volcanic production, subsidence, and sea level.

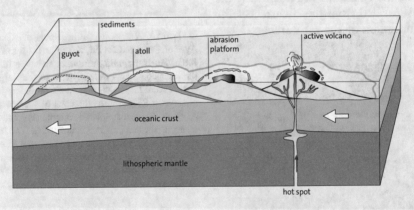

◻ **Fig. 6.11**   Evolution of a guyot. Stage 1—the hot spot builds a volcano above sea level that in tropical regions may be fringed by reefs. Stage 2—the ocean plate moves off of the hot spot and the volcano erodes by marine processes to a flat-topped extinct volcano. Stage 3—the fringing reef builds around the margin of the truncated volcano. Stage 4—the volcano subsides and the reef builds upward; eventually subsidence lowers the volcano to depths at which reefs can no longer grow. The now flattened volcano becomes a guyot whose summit can be in water depths as great as 2000 m

Deviations from the regularity of a plume track are caused by various reasons. A large plume can produce magma at different locations. If a fault zone drifts over the hot spot, the magma might be displaced along this fault zone and diverted across the track. It is also conceivable that the magma might find its way back along the plume track that has been weakened by fractures and older supply channels. Magma chambers can form and drift within the lithosphere and extrude at a later time. Such behavior could disrupt the linear sequence of volcanoes. The three long and kinked hot spot tracks in the Pacific (◻ Fig. 6.1) do not coincide exactly. The two southern tracks display irregularities in their volcanic sequence that may be related to one or the other of the mechanisms described above.

6.4 Hot Spot Tracks on the Continent

The tracks of hot spots are more difficult to follow across continental crust where they are less distinctive and commonly not marked by volcanic chains.

Fig. 6.12 Track of the hot spot of Trindade that today is in the South Atlantic adjacent to the Brazilian coast (Crough et al. 1980). The track is marked in Brazil by occurrences of kimberlite and alkaline intrusions that decrease in age from west to east. The dashed red line tracks the position of the hot spot as reconstructed from the global plate drift pattern. Ages are given in million years

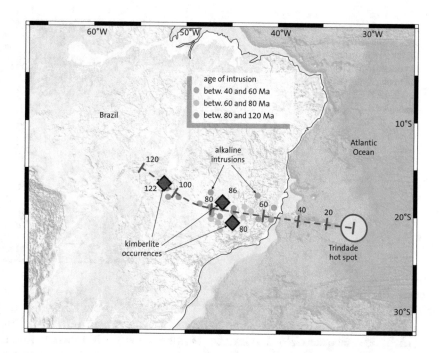

The thick continental lithosphere is difficult for the magmas to penetrate and disruptions of magma paths and changes of magma composition make detection more difficult. However, examples of hot spot tracks across continental terrain do exist. The hot spot of Trindade in the South Atlantic left a track during the Cretaceous when the Brazilian Shield drifted over the hot spot (■ Fig. 6.12). Subsequent erosion has removed the volcanoes but has left a track of originally shallow alkaline intrusions, some diamond-bearing kimberlite occurrences amongst them. Kimberlites are formed in diatremes (kimberlite pipes) that originate in the sub-lithospheric mantle; many are thought to be related to hot spots. The magmatites of the Trindade track decrease in age towards the present position of the hot spot. The positions of the hot spot below the South American continent as reconstructed from the plate drift history and the age of the intrusions coincide well even though the magmatism above hot spots in continental areas scatters across a larger radius.

The track of the Mesozoic Great Meteor Plume was imprinted across the North American continent where it is marked by kimberlites and other alkaline magmatic complexes. Only magmatic rocks that formed deep in the crust are preserved. As North America moved across the hot spot for 100 m. y. during the Jurassic and Early Cretaceous, a 4000 km-long track was formed that extends from Hudson Bay to New England (■ Fig. 6.13). As it migrated off shore, it produced the New England Seamount Chain in the NW Atlantic. Some of these volcanoes were islands earlier in their history but today their summits are in water depths of more than 1500 m. The Great Meteor Plume is presently located at the Mid-Atlantic Ridge.

Two mantle plumes located beneath the European continent, the hot spots of the Central Massif in France and the Eifel in Germany (■ Fig. 6.1), did not form a track. Both are considered to be the product of a shallow source in the upper mantle and not of deeper mantle origin (Granet et al. 1995); each was relatively short-lived. The activity in the Central Massif culminated in the formation of two volcanoes, the Cantal (11–2.5 Ma) and the Mont Dore (4–0.3 Ma). The last eruption occurred 4000 years ago. Although the volcanism is considered to be extinct, this may turn out to be a misjudgement in the long term.

The volcanoes of the Eifel were active in the Early Tertiary, especially between 40 and 25 Ma. Activity recommenced in the Quaternary ca. 0.6 Ma and the youngest eruptions are dated ca. 10,000 years old. The formation of NW–SE striking rows of volcanoes corresponds to the alignment of fissure systems (Schmincke 2004) that are related to the NE–SW extension of the Lower Rhine Embayment (Ch. 3). The alkaline melts are dominantly of basaltic composition and only slightly differentiated. Contacts of the ascending magmas or the already solidified hot magmatites with ground water led to explosive events (phreatomagmatic resp. phreatic eruptions; *phreas, Greek* fountain, water container) accompanied by thick deposits of tuff. These volcanic deposits comprise mostly cinder cones and highly explosive maars and are well exposed in the numerous quarries of the Eifel region. These are phreatic pipes (diatremes) with ring walls of pulverized rock material that was ejected around the explosive crater without the eruption of magma. The craters are the sites of scenically attractive lakes. Presently the volcanism is dormant but it is not considered extinct as testified by frequent escape of magmatic carbon dioxide.

Fig. 6.13 Track of the hot spot of the Great Meteor Plume on the North American Plate as reconstructed from dated magmatic rocks (Van Fossen and Kent 1992; Heaman and Kjarsgaard 2000). Ages in million years

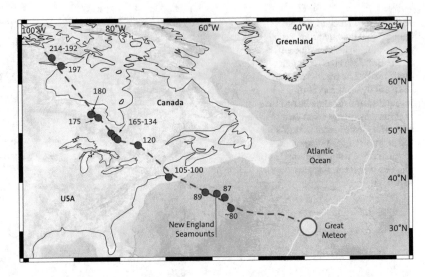

6.5 Flood and Trap Basalts

Flood basalts, also called large igneous provinces (LIP), are characterized by layered basaltic deposits several kilometers thick that form huge plateaus on land as well as below sea level. On land the basaltic plateaus typically erode at their edges into distinctive stair steps. The steps are caused by erosion of the intersecting horizontal bedding planes between lava flows and the vertical contraction joints that develop on cooled lava sheets. Erosion takes advantage of the horizontal and vertical planes to form the distinctive landscape. The term "traps" or "trap basalts" was introduced by Carl von Linné, Axel Fredrik Cronsted, and Johan Gottschalk Wallerius for the Deccan basalts, India, in the eighteenth century—*trappa* is *Swedish* for stair (Faujas de Saint-Fond 1788).

Typically huge eruptions involve millions of cubic kilometers of basaltic rock that is extruded during an interval of only one or several million years. The hot basalts that emanate from fissures or erupt as lava curtains have a low viscosity that enables successive flows to cover areas as large as several hundred-thousand square kilometers. Multiple layers of basalts several meters thick stack up to form widespread plateaus. During brief periods of magmatic quiescence, soils can develop on top of the basalts to form prominent stratigraphic horizons within the sequence.

Large eruptions of flood basalts have occurred throughout Earth history (■ Fig. 6.14). Notable examples include:

- The Siberian trap basalts formed 250 million years ago at the Permian/Triassic boundary with a volume of 2.5 million cubic kilometers.

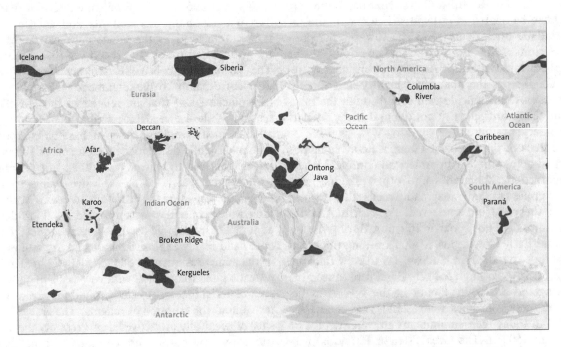

Fig. 6.14 Global distribution of large igneous provinces in the oceans (submarine plateaus) and on land (flood basalts)

- The hot spot of Tristan da Cunha (see above) that contributed to the opening of the South Atlantic, created the Paraná flood basalts in South America and the Etendeka basalts along the Southwest African coast. Each is associated with alkaline magmatic complexes that formed before they were separated by continental break-up in the Early Cretaceous.

- The Deccan trap basalts in India are related to the break-up of part of Pangaea 65 Ma where the hot spot of Réunion produced the enormous eruptions that now form the Deccan Plateau (◻ Fig. 6.10). These eruptions splintered India from a continental remnant that today is represented by the Seychelles Islands in the Indian Ocean. Paleomagnetic investigations indicate that the Deccan traps were formed more than 30 degrees farther south than their present position which confirms the location of origin as close to the hot spot of Réunion. The Deccan traps erupted in less than one-half million years with a volume greater than 2 million cubic kilometers.

When continental crust is strongly thinned by the rifting process, hot asthenosphere rapidly ascends. If magma was already enriched in the head of a mantle plume, additional melting of the mantle occurs because of pressure release. Huge quantities of basalt pour out across the continental margin and the adjacent oceanic depression in a geologically very short time interval. The eruptions during this process occur in phases. Calculations indicate that many cubic kilometers of basalt may be produced per year. Therefore, the magma production of one hot spot may exceed the total production along the global mid-ocean ridge system that is presently ca. 20 km^3 per year.

Large submarine basaltic plateaus are formed within intraplate oceanic settings. Basaltic extrusions at intra-ocean hot spots can thicken the basalt/gabbro oceanic crust to thicknesses up to 30 km. The resulting large plateaus rise by up to 3000 m above the surrounding deep sea floor. The largest extant oceanic plateau is the Ontong–Java Plateau east of the Bismarck Archipelago in the Western Pacific (◻ Fig. 6.14). It covers over two million square kilometers and includes up to 30 million cubic kilometers of gabbro and basalt that ascended from a plume (Coffin and Eldholm 1993). Additional examples of large submarine basaltic plateaus include the Kerguelen Plateau in the Indian Ocean (see above) and the Caribbean Large Igneous Province. All three of these plateaus were formed during the superplume event in the Cretaceous (see below) during several episodes that individually lasted for only a few million years.

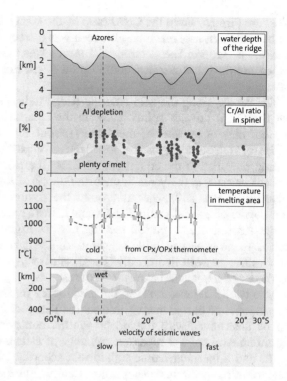

◻ **Fig. 6.15** Topographic profile along the Mid-Atlantic Ridge between 60° N and 30° S latitude and conditions in the mantle below (Bonatti 1994). Beneath the Azores, the formation of basaltic melts is high; the temperatures, however, are low in the area of melting. Seismic investigations show that the mantle below the Azores is rich in watery fluids. Cpx = clinopyroxene, Opx = orthopyroxene

6.6 The Azores—Hot, Cold or Wet Spot?

The Azores are a group of volcanic islands near the Azores Triple Junction where the transform fault representing the plate boundary between Eurasia and Africa branches off from the Mid-Atlantic Ridge towards east (◻ Figs. 1.5 and 6.1). For years the Azores were considered to be a hot spot. Detailed investigations indicate that the mantle immediately below the Azores is not hotter but rather cooler than other segments of the Mid-Atlantic Ridge (Bonatti 1994).

Peridotites were examined from the lithospheric mantle directly below the crust along the Mid-Atlantic Ridge from the equator to northern latitudes (◻ Fig. 6.15). The basalt formed from upper mantle peridotite by partial melting (Ch. 5). Clinopyroxene (diopside) is easier to melt than the other minerals of the peridotite and thus preferentially goes into the melt. Also, the elements iron and aluminum preferentially go into the melt whereas magnesium and chrome mostly remain in the peridotite. The peridotite is thus depleted by partial melting in Fe and Al and enriched in Mg and Cr. This process is recorded in the mineral spinel that occurs in low quantities in peridotite. Generally, mixed crystals between chrome and aluminum

6

spinel are formed. From the Cr/Al ratio of the spinel, it can be estimated how much basaltic melt was extracted from the peridotite. The electron microprobe was used to measure the exact chemical composition of the minerals.

The profile along the Mid-Atlantic Ridge indicates that the Cr/Al ratio of spinel is low in the equatorial area but high in the area of the Azores (about 39° N; ◘ Fig. 6.15). This confirms that low degrees of partial melting prevailed in the equatorial Atlantic (therefore, the oceanic crust at the Vema Transform Fault is of low thickness; ◘ Fig. 5.12) whereas around the Azores, a high degree of partial melting is assumed. In the area of the Azores, a 25% partial melting of the original rock is to be expected; other segments of the profile show values between 10 and 20%. These results support the theory of a hot spot below the Azores.

Meanwhile, temperature of the peridotites during the processes of partial melting and subsequent cooling can be determined with the aid of geothermometers. Within mineral pairs in rocks, an exchange of elements occurs and a thermodynamic equilibrium according to the current temperature develops. For example, the distribution of Mg and Fe in orthopyroxene and clinopyroxene can be used as geothermometer. Corresponding measurements in peridotites along the Mid-Atlantic Ridge indicate that the area of the Azores is not hotter than the other ridge segments but has slightly lower temperatures (◘ Fig. 6.15). Is this a cold spot? But how does one explain the high percentage of partial melting that is responsible for the high elevation of the ridge in the area of the Azores?

The explanation is that the melting point of rocks is not only dependent on the rock composition and temperature but also on the pressure and the content of fluids. As already discussed, the degree of partial melting increases when the mantle current ascends because the pressure release reduces the melting point (◘ Fig. 6.16). Fluid phases included in the rock act in the same direction. Fluid phases are liquid or gaseous components that occur in practically all rocks; however, they occur in different amounts, and are dissolved in magmas. Magmatic rocks containing very little fluid phases are described as "dry", and those with a lot of fluids (in particular water) are classified as "wet". According to chemical analyses, basalts of the Azores contain three to four times the amount of water as normal basalts of mid-ocean ridges indicating a "wet" melting region. This is corroborated by seismic investigations (◘ Fig. 6.15). The seismic waves are relatively slow in the uppermost mantle in the area of the Azores, a characteristic attributed to high contents of fluids.

Therefore, the substantial production of basaltic melts in the area of the Azores Triple Junction is not related to a hot spot. But this area is not a cold spot either! Low temperatures can not be responsible for the high magma production. The critical feature is the high content on fluid phase in the mantle rocks that effectively reduces the melting point of the peridotite. At given temperatures between 1000 and 1100 °C, large portions of basaltic melt develop that are otherwise only possible with higher temperatures (◘ Fig. 6.16). Additionally, with melting, a loss of heat develops in the remaining peridotite and may explain the somewhat lower temperatures according to the geothermometer. Wet areas are not rare in the mantle. Water is permanently transported into the mantle in subduction zones as water-containing sediments or within basalts whose original dry minerals are converted into water-containing minerals. Water may even come from the mantle itself by methane degassing from the deeper mantle that is oxidized to water and carbon.

◘ **Fig. 6.16** Amounts of partial melting of the ascending mantle peridotite if it is dry and hot, dry and cool, cool and "wet" (Bonatti 1994). The percentages indicate the portion of basaltic melt extracted from the peridotite; melt proportion increases during the ascent of the peridotite because of pressure release and related lowering of the melting point. Watery fluids ("wet" mantle) reduce the melting point of peridotite and thus increase the degree of partial melting

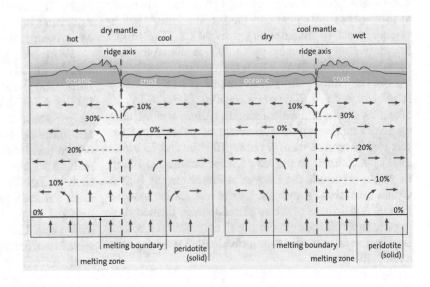

6.7 Hawaii—A Typical Oceanic Hot Spot

The main island of Hawaii, the locus of most recent volcanism in the Hawaiian chain, has been connected to the hot spot plume for less than one million years (Fig. 6.8). The resulting volcanic complex rises at the peak of Mauna Loa to more than 4000 m above the sea level and has its base at a water depth of more than 5000 m below sea level and thus, with nearly 10,000 m of relief forms the tallest mountain on Earth measured from its base. Its diameter at the ocean floor is ca. 1000 km, roughly corresponding the diameter of the plume head below. In spite of the fast drift of the Pacific Plate, the magma production was rapid enough to produce the present mountain with a volume of 40,000 km^3 (Schmincke 1981). This volume would cover Switzerland with a 1 km thick layer of basalt; however, such an event probably would be prevented by a referendum! Geophysical investigations indicated that a pear-shaped magma chamber is located between 3 and 10 km below the summit of the Kilauea. This magma chamber is fed by a 30 km-long vertical, tube-like feeder channel that terminates in the mantle.

The Island Hawaii will drift out of the area of influence of the mantle plume within the near geological future because the Pacific Plate drifts over the plume at a rate of ca. 100 km in one million years. Approximately 30 km southeast of Hawaii is a new volcano, Loihi, currently under construction (Fig. 6.8). Therefore, the plume is not currently located directly beneath Hawaii. The summit of Loihi is presently at ca. 1000 m below sea level.

Islands such as Hawaii typically evolve in a predictable fashion. Incipient submarine volcanoes are built on the sea floor above hot spots and comprise pillow lavas and massive dikes and sills. As the lavas build vertically and gradually approach the surface of the water, gas in the magma can blast through water to the surface and result in volcanic explosions that intensify as seawater penetrates into fissures and joints and is mixed with magma in shallow magma chambers. The explosions generate volcanic ashes that solidify to tuff or are transported into deeper areas where they form aprons of volcanic fragments. If the production of lava continues, the volcano builds a large oceanic island that is composed of a sequence of tuffs and lavas with diagonally cutting dikes that represent the solidified feeder channels. The island of Hawaii is currently in this mature stage of development; however, as the island moves off the locus of the hot spot, erosion processes will take over and eventually reduce the island to a low-lying pile of volcanic rocks and debris.

6.8 Iceland

Slow plate motions over active hot spot generate large amounts of basalt. This is the case in Iceland where the hot spot is directly at the Mid-Atlantic Ridge and the spreading rate is low (ca. 1 cm/year in each direction; Fig. 6.17). The coincidence of a large production of basalt from the hot spot and a slow plate drift results in an elevated midocean ridge that in Iceland is exposed on land for greater than 300 km. Iceland is cur-

 Fig. 6.17 Map of the Greenland–Iceland–Faroe–Swell. This topography evolved because the Iceland hot spot was located below the young mid-ocean ridge of the North Atlantic 40 Ma. Red shading marks basaltic rocks produced by the hot spot. Numbers indicate the spreading rates in cm/year

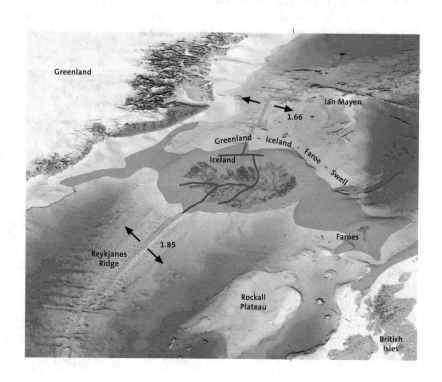

rently one of the most active hot spots on Earth with a plume head of probably 1000 km diameter. The crust was thickened to more than 25 km by the intense magmatism and thus raises the question about the possibility of a continental splinter below Iceland. A negative gravity anomaly appeared to support this assumption. Seismic investigations demonstrate that the Icelandic crust is basaltic although acidic volcanic rocks (rhyolites) also occur. The rhyolites are differentiates of the basaltic magmas and not developed from the melting of continental crust as it is frequently observed at continental hot spots. Approximately 8% of rhyolitic melts may form from a basaltic magma by differentiation during cooling (Schmincke 2004). The negative gravity anomaly of Iceland is thus not related to lighter continental crust but to the anomalously hot and molten, and thus relatively light, uppermost mantle, the head of the plume and the MORB source. The thick basaltic crust and the large plume head explain the high elevation of the 100,000 km^2 island.

Iceland is a product of the activity of a mantle plume, which ca. 70 Ma (Late Cretaceous) was located below Western Greenland and the Canadian Arctic and probably contributed to the aborted separation of North America and Greenland (◘ Fig. 4.9). Greenland then drifted over the hot spot, which 55 Ma was located below Eastern Greenland. At that point, the opening of the Atlantic between Greenland and Scandinavia initiated. Eruption of basalts occurred in pulses that lasted one to three million years and recurred at intervals of 5–10 m. y. (O'Connor et al. 2000). Basalts derived from the hot-spot activity are found in Eastern Greenland, along the Greenland–Iceland–Faroe–Swell in the North Atlantic, at the northwestern edge of the British Isles, and at the continental margin of

Norway (◘ Fig. 6.17). These basaltic centers make up the North Atlantic Volcanic Province. The young mid-ocean ridge came under the influence of the mantle plume in the Eocene ca. 40 Ma. Since that time, the Greenland–Iceland–Faroe–Swell was formed as a diagonal ridge by the hot spot on both sides of the spreading axis.

6.9 Yellowstone

The Yellowstone hot spot appeared ca. 17 Ma in the Miocene and has left a track as the relatively slow-moving, WSW-moving North American Plate drifted over it at 4 cm/yr. The track has left major volcanic centers that shift to the ENE. The earliest record of volcanism occurs in the Columbia River Plateau and comprises chiefly basalts but important rhyolites as well (◘ Figs. 6.18, 6.19). This acidic volcanism arises because the upwelling magmas pass through thick continental crust.

In fact, the Yellowstone hot spot is currently the most prolific producer of acidic magmatic rocks on Earth. Compared to basalts, rhyolitic magmas are highly explosive and produce violent eruptions. The explosive nature is a result of the high viscosity of the magma and water dissolved in the magma-during the explosion, water is abruptly released in a gaseous state. These types of volcanic eruptions drain very large magma chambers in a very short time span; the emptied chamber immediately collapses to produce *calderas* (*Spanish* cauldron), bowl-shaped craters that commonly fill with water to form a lake. Because of the explosive nature of the volcanism, the vast majority of the rhyolites are erupted as tuffs.

◘ **Fig. 6.18** The bimodal (basaltic and rhyolitic) volcanic rocks along the track of the Yellowstone hot spot (Christiansen et al. 2002). Because of the plate drift of North America towards WSW, the volcanic rocks decrease in age towards ENE. Collapse craters (calderas) with diameters of several tens of kilometers developed in areas of large rhyolite eruptions

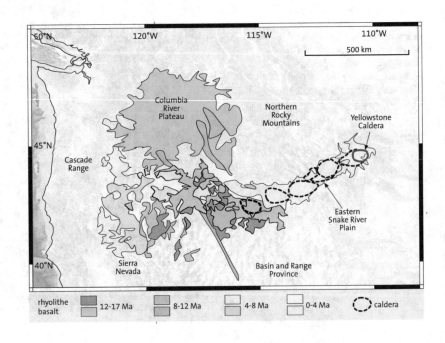

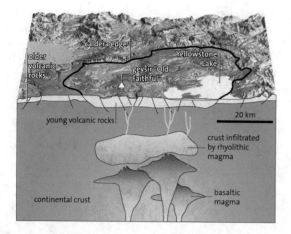

☐ **Fig. 6.19** Schematic block diagram of the Yellowstone Caldera and the magmas below

6.10 The Superplume Event in the Cretaceous

The Cretaceous from ca. 125 Ma to 85 Ma was a time of extremes. According to a theory of Roger L. Larson, extreme conditions were caused by the exceptionally high activity of mantle plumes that were expressed in many large and unusually productive hot spots on the surface. This event is called the "Cretaceous superplume event" (Larson 1995). Large igneous provinces (LIP) generated by superplume events apparently occur at irregular intervals throughout geologic time, although two large LIP, the Siberian Traps at the Permian/Triassic boundary and the Deccan Traps at the Cretaceous/Tertiary boundary, are not associated with superplume events.

The high activity of mantle plumes in the Cretaceous is the result of a heat accumulation in the boundary zone between the core and mantle, the place of origin of the large mantle plumes. Overheating was caused by convection currents in the outer core. A persistent and stable convection current pattern could be the reason that the boundary zone to the mantle was extensively heated at different locations (at the LLSVP zones) and finally led to heat accumulation at various global locations to subsequently ascend from hot mantle material in the form of plumes (☐ Fig. 6.20). As the plume heads reached the base of the lithosphere, they were converted into particularly large mushroom-like heads that provided the intense, abrupt magmatic activity.

The superplume activity not only increased the activity of hot spots within the plates, but also those at mid-ocean ridges where the magma production increased significantly. As a consequence of the strong magmatic activity within the plates, as well as at the mid-ocean ridges, new oceanic crust, highly elevated oceanic plateaus, oceanic islands, and seamounts were rapidly formed. In the mid-Cretaceous, the Pacific Ocean was a place of enormous expansion of basaltic crust. Both these trends continue to the present. Globally, almost all submarine high plateaus and a large number of seamounts were formed in the Cretaceous superplume period. Approximately 125 Ma, the total production of oceanic crust increased, within a time interval of only a few million years, from around 20 km^3 per year (this corresponds more or less to the present value) to about 35 km^3 per year (☐ Fig. 6.20).

The intense magmatic activity had global effects on other geologic and climatic processes. Large amounts of gases that were dissolved in the magma were released to the atmosphere, especially the well-known greenhouse gas carbon dioxide (CO_2). This caused an additional increase of the global mean annual temperature that already was at a higher level than today. The

The heat supply from the Yellowstone mantle plume has repeatedly generated large volumes of rhyolite melts that have resulted in violent eruptions and the formation of huge calderas. During the history of the hot spot, nearly 150 highly explosive eruptions have been recorded. The youngest, Yellowstone Caldera, is oval and has dimensions of 75 45 km. The last three large eruptions occurred in regular intervals at ca. 2.1 Ma, 1.3 Ma, and 0.64 Ma with a production of up to 2500 km^3 of tuff during one event. It is easy to calculate that another such catastrophic eruption is to be expected in the geologically near future and the devastating consequences will not be restricted to the direct surroundings of the center of eruption.

During the eruption at 0.64 Ma, the volcano edifice collapsed several hundreds of meters. The caldera was filled with volcanic lavas and lake deposits. Collapse was followed by the formation of domelike bulges, an indicatation that magma chambers at depth were again filled. At present, the surface continues to pulsate suggesting magmatic agitation in the depth. Because of the permanent heat transfer from the hot basaltic magma chambers to the surrounding crustal rocks, increasing amounts of explosive acidic magmas develop. If the internal pressure becomes excessive, a devastating eruption will occur because the pressure release during the explosion abruptly releases all gases within the melt. According to seismic investigations, the actual magma chamber is swollen over a distance of 40–50 km and a thickness of up to 10 km (☐ Fig. 6.19). In addition to volcanic destruction across a wide area, a new eruption would have global climatic consequences. A similar huge eruption at Toba, Sumatra 0.74 Ma formed a caldera 100×60 km. A global drop in temperature of ca. 5 °C drastically reduced the population of mankind at that time. However, Toba is not a volcano located above a hot spot but rather above a subduction zone.

Fig. 6.20 The superplume event in the Cretaceous and some of its global effects (Larson 1995). Shown are rates of formation of ocean crust (at hot spots and mid-ocean ridges), high-latitude temperatures, and sea level elevation, each compared to the present. Also shown is the long normal magnetic polarity period that lasted 40 m. y. This period may have been caused by a stable convection current pattern in the liquid outer core that was related to the superplume event

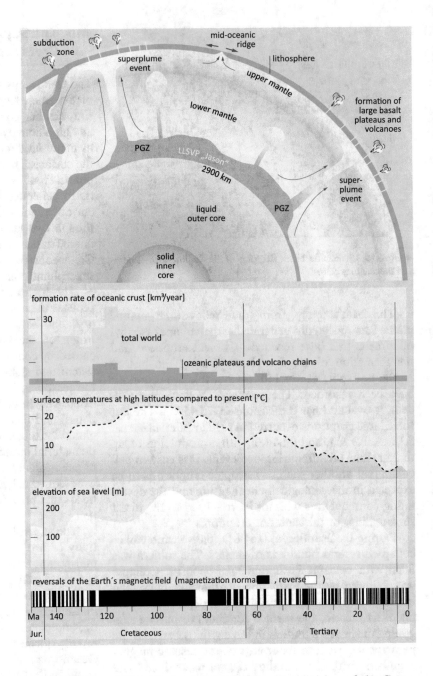

polar regions were particularly affected by the warming (Fig. 6.20). The time interval between 115 and 90 Ma was presumably the warmest period since the Precambrian.

Rapid sea-floor spreading generated broad, large-volume mid-ocean ridges. The ridges, the oceanic plateaus and the hot-spot volcanoes caused a decrease in volume of the ocean basins thereby forcing sea-water onto the continents. Also, few or no glaciers were present so little water was tied up in ice (today's sea level is depressed some 80 m by extensive Antarctic and Greenland glaciation). The result was the highest sea level since the Ordovician, as much as 250 m above present sea level (Fig. 6.20). Shelf areas expanded as coastlines retreated towards the land. Approximately

93 Ma during the Early Turonian division of the Cretaceous, the zenith was reached.

The broad shelf areas and the warm, equitable climate led to an enormous increase in the biomass of organisms that produce calcareous shells. The calcareous production is particularly high in shallow warm water. Therefore, shallow water limestones from the Cretaceous period are especially widespread. The increased amount of carbon to build calcite ($CaCO_3$) in the form of limestone from the organic carbon in living things was provided by the carbon dioxide that was released to the atmosphere and the oceans by the basaltic volcanism. Among the most impressive and beautiful landscapes built from this sedimentary formation are the Cretaceous chalk rocks of Dover (Great

Britain) and Rügen (Germany), and the chalk bluffs from the Dakotas to Texas (USA); each is composed of the remnants of calcareous microfossils. But the rich habitat also produced organic matter that could not be decomposed completely because of the lack of oxygen in the sediments; this led to the formation of bitumen-rich clays that matured to black shales rich in petroleum. Mid-Cretaceous black shales were formed repeatedly and represent important petroleum host rocks. Oil deposits thus derived from Cretaceous black shales comprise nearly half of the known global petroleum reserves. Another giant storehouse of Cretaceous carbon occurs in the huge coalfields of the Western Interior of North America. Coal formed in coastal plain settings when the rapid accumulation of plant material greatly exceeded the rate at which it could be oxidized; the coals now comprise one of the greatest energy reserves in the world.

The increased formation of oceanic crust at mid-ocean ridges must have been compensated by a faster removal of ocean floor in the subduction zones. Subduction produces a characteristic magmatism above the subduction zone. The faster the subduction rate, the greater are the amounts of subduction-related magmatites. Therefore, thick bodies of Cretaceous plutonic and effusive rocks are found in the core of the Andes, the North American Cordillera and other mountains around the Pacific. The 110–80 Ma magmatic event in the Sierra Nevada and adjacent batholiths represents one of the most concentrated plutonic events in the geologic record.

Curiously, coincidentally with these other extreme events, most of the diamond deposits of the world evolved. Diamonds are, except from the pressure developing at meteorite impacts, only stable under the high pressure of the mantle and occur at depths greater than 100 km. Diamonds were formed in the mantle over eons of time and were transported to the surface by the Cretaceous plume magmatism. They reached the surface when magma from the upper mantle rapidly ascended through pipes to form kimberlites, the host rock of diamonds. Diamonds are commonly found in river placer deposits because they can withstand long river transportation due to their hardness.

Yet another extreme phenomenon that seems to have a causal relation to the superplume event is the mid-Cretaceous magnetic normal period. The reversals of the Earth's magnetic field occur successively in very irregular time intervals from some thousand to several tens of millions of years. The magnetic field of the Earth is generated within the outer core that is mostly composed of iron. Convection currents in this liquid layer and electric currents caused by the convection play the decisive role. The reversals of the magnetic field are not understood completely, but turbulent currents from the outer core may cause more frequent reversals than convection currents that are stable over long time periods. The intense and fast convection currents demanded for superplume periods also stabilize the magnetic field over long time periods and reversals do not occur (Larson 1995). The removal of hot mantle material by the plumes maintains the steep temperature gradient across the core/mantle boundary which in turn stabilizes the convection in the outer core.

The mid-Cretaceous is the time interval with the longest span without a reversal in the Phanerozoic (◘ Fig. 6.20). Over some 40 million years, the magnetic field maintained a polarity that was oriented in the same sense as the current normal magnetization. During the Late Cretaceous and Early Tertiary, the frequency of reversals increased gradually until such events became very frequent in the last 30 million years, typically occurring several times within one million years. The frequent reversals indicate slow and turbulent currents in the outer core. This increase in reversal frequency is coincident with the onset of icehouse conditions in the Early Oligocene. This is in sharp contrast to the greenhouse conditions of the Cretaceous.

Subduction Zones, Island Arcs and Active Continental Margins

Contents

© The Author(s), under exclusive license to Springer Nature Switzerland AG 2022
W. Frisch et al., *Plate Tectonics*,
Springer Textbooks in Earth Sciences, Geography and Environment,
https://doi.org/10.1007/978-3-030-88999-9_7

7

Subduction zones are created when two lithospheric plates move against each other and one of the two plates descends under the other through the process of subduction. However, only oceanic lithosphere is able to sink deeply into the Earth's mantle to become reincorporated there. Continental crustal material is generally too light to be subducted to great depth. The interaction of the subduction zone and the asthenosphere of the mantle generates the melts that rise to feed the volcanism typical of island arcs and active continental margins.

Subduction zones are critical to the dynamics of the Earth because they represent the essential driving force behind the movement of plates. Moreover, magmatism initiated by subduction is responsible for the creation of continental crust through a series of complex processes. The continental crustal material generated in this fashion has a low specific weight and remains at the outer rind of the Earth and is not reintegrated into the mantle. Without this lightweight continental crust, which forms high topographic features on the Earth, our planet would have a completely different face. The total length of global subduction zones sums to more than 55,000 km, a length only slightly shorter than the total length of the mid-ocean ridges (60,000 km).

Four types of convergent plate boundaries are recognized (Figs. 7.1 and 7.2).

The first type occurs when ocean lithosphere is subducted below other ocean lithosphere ("intra-oceanic subduction zone") to create a volcanic Island arc system built on oceanic crust ("ensimatic island arc"; *sima*—artifical word first used by Wegener made from silicon and magnesium to characterize ocean floor and Earth's mantle). Examples for intra-oceanic, ensimatic island arc systems include the Mariana Islands in the Pacific and the Lesser Antilles in the Atlantic.

The second type occurs where oceanic lithosphere is subducted beneath continental lithosphere and an island arc underlain by continental crust forms ("ensialic island arc"; *sial*—silicon and aluminum for continental crust). The island arc of this system is generally separated from the continent by a marine basin underlain by oceanic crust. An ensialic island arc is either the splitting off of a continental splinter through the formation of a marginal basin as a result of the subduction of oceanic lithosphere or an island arc system with subduction activity lasting more than 100 million years, in which an originally ensimatic island arc is gradually converted into continental crust. Examples for island arc systems underlain by continental crust are the Japanese Islands and the eastern Sunda Arc.

The third type of convergent plate boundary represent the active continental margins where oceanic lithosphere is subducted beneath continental lithosphere without a marine basin behind the volcanic arc; rather, the arc is built directly on the adjacent continent. The continental margin is connected directly to the hinterland, although a shallow marine basin may exist behind the volcanic arc. Examples for active continental margins are the Andes, SE Alaska, and the western and central Sunda Arc that includes Sumatra and Java.

The forth type of convergent margin occurs along zones of continent–continent collision. If two continental masses collide during continuous subduction, they eventually merge. Telescoping of the two plates and the buoyancy of the subducting continent eventually leads to a standstill of subduction within the collision zone. The oceanic part of the subducting plate tears off and continues to drop down, a process referred to as "slab breakoff". Continent–continent collisions ultimately result in the formation of mountain ranges like the Himalayas or the Alps.

 Fig. 7.1 Convergent plate margins of the Earth characterized by ensimatic island arcs (underlain by oceanic crust), ensialic island arcs (underlain by continental crust), and active continental margins

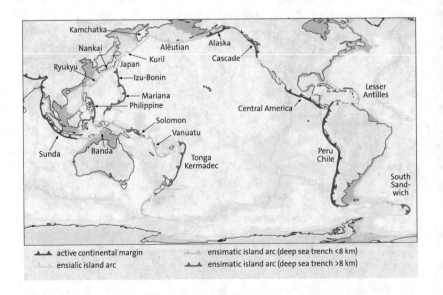

Fig. 7.2 Examples of different types of plate margins with subduction zones. The island arc of the Marianas developed on oceanic crust, that of Japan on continental crust. The volcanic zone of the Andes is built on the South American continent (active continental margin). The collision of two continents produces a mountain range like the Himalayas—subduction wanes, leading to slab breakoff

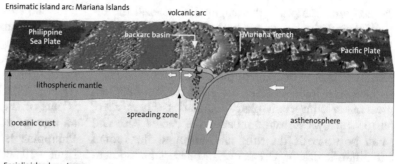

Ensimatic island arc: Mariana Islands

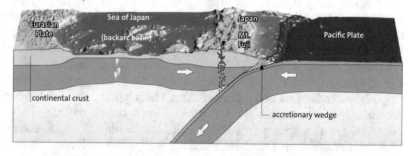

Ensialic island arc: Japan

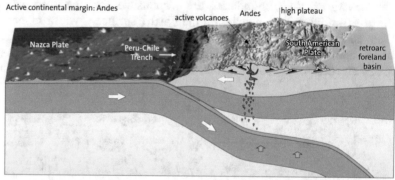

Active continental margin: Andes

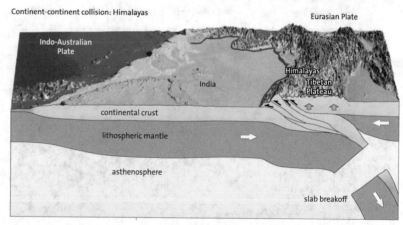

Continent-continent collision: Himalayas

7.1 Structure of Plate Margin Systems with Subduction Zones

Systems of convergent plate boundaries are characterized by a distinct topographic and geologic subdivision. Although the plate boundary itself is only represented by a line at the surface, commonly within a deep sea trench, a zone several hundreds of kilometers wide is formed by processes which are related to subduction. The volcanic zone above the subduction zone, in many cases expressed as an island arc, is the dominating element of this plate boundary system. We use arc as shorthand for the terms volcanic arc or magmatic zone. The arc is the point of reference for the convergent boundary that is usually divided into three parallel zones: from the trench to the arc is the forearc zone, the arc zone comprises the magmatic belt, and the region behind the arc is the backarc zone (■ Fig. 7.3). This generally agreed upon subdivision of three parts turned out to be practical in order to describe the complex structures of convergent plate boundary systems.

Deep-sea trenches form the major topographic expression at convergent plate boundaries. These deep, narrow furrows surround most of the Pacific Rim and small portions of the rims that surround the Indian and Atlantic oceans. Oceanic lithosphere is bent downward under the margin of the "upper plate" and dives into the asthenosphere. The result is an elongated deep trench that is located between the abyssal plains and the border of the upper plate. The deepest trenches, with water depths of about 11,000 m, are known from the Challenger and the Vitiaz deep (named after an English research vessel and a Russian researcher) in the southern Mariana trench. Water depths of more than 10,000 m are also known from the Kurile, Izu–Bonin, Philippine and Tonga–Kermadec trenches (■ Fig. 7.1).

Convergent plate boundaries are responsible for the greatest differences of relief on the Earth's surface. Differences in altitude of 10 km between the deep sea trench and volcanoes of the magmatic arc are not unusual. A relief of 14,300 m across a distance of less than 300 km can be observed between Richards Deep (−7636 m) and Llullaillaco (+6723 m) in the Chilean Andes, the highest active volcano on Earth.

In a transect from the trench onto the upper plate, the following morphological features generally occur (■ Fig. 7.3). The landward side of the deep sea trench is part of the upper plate and consists of a slope with an average steepness of several degrees. In front of the Philippine Islands the angle exceeds 8° where a rise from −10,500 m to −200 m occurs over a distance of 70 km. The outer ridge follows behind the slope. In most cases, the ridge remains substantially below sea level; however, in several cases, islands emerge above sea level (Sunda Arc: Mentawai; Lesser Antilles: Barbados). The outer ridge is not always distinctive. Next in the transect, directly in front of the volcanic arc, lies the forearc basin, another prominent morphological element.

Collectively, the deep sea trench, outer ridge, and forearc basin comprise the forearc region. The distance between the plate boundary and magmatic zone has a width of between 100 and 250 km. This region is also called the arc-trench-gap, a magmatic gap that with very few exceptions is void of magmatic activity. Low temperatures in the crust are caused by the coolness of

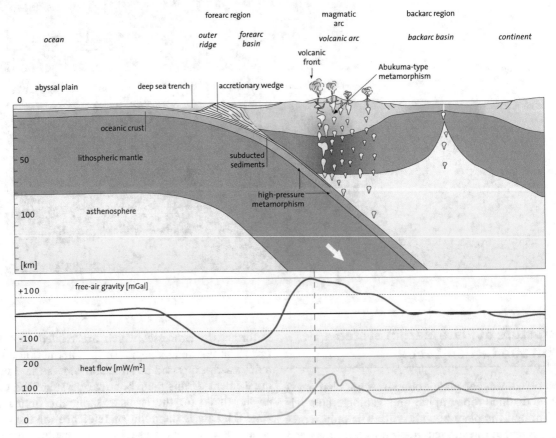

■ **Fig. 7.3** Structure of a plate margin system with subduction zone and ensialic island arc. Gravity and heat flow data show typical pair of negative and positive anomaly. 1 Gal (galilei) = 1 cm/s^2 (unit of acceleration). 1 mGal = 10^{-3} Gal. mW/m^2 = mW per square meter

the subducting plate underneath and prevent the formation of magma by melting or the rise of magma from deeper sources.

The volcanic arc, with an average width of 100 km, is the central part of the island arc or the active continental margin and is characterized by significant magmatic activity. Approximately 90% of all active volcanoes above the sea level, most of which are in or around the Pacific Ocean, are subduction-related volcanoes. The zone of volcanism has a sharp boundary on the forearc margin and a gradual one on the backarc margin (■ Fig. 7.3).

Island arcs are separated from adjacent continents by a marine basin underlain by ocean crust, the backarc basin (■ Figs. 7.2 and 7.3). Active continental margins, synonomous with continental arcs, have a backarc region that may have thinned continental crust and/or a zone of compressional structures.

7.2 Spontaneous and Forced Subduction: Mariana- and Chile-Type Subduction

Subduction zones can be subdivided into two types, based on characteristics of the subducted plate (Uyeda and Kanamori 1979). The first type is referred to as a Mariana-type subduction zone and is characterized by old and dense lithosphere that is subducted and sinks

into the sublithospheric mantle by its own weight. Therefore, it is generally the steepest-dipping of the two. The second type is referred to as a Chile-type subduction zone and is characterized by younger, hotter and less-dense lithosphere that dips at a shallower angle (■ Fig. 7.5).

Because oceanic lithosphere becomes denser with increasing age, it can achieve a density greater that that of the underlying asthenosphere and thus be more easily subducted; this is spontaneous subduction (Nicolas 1995) or free subduction (Frisch and Loeschke 1986), meaning that it arises from internal cause. Theoretically this inversion of density occurs when the oceanic lithosphere reaches an age of ca. 30 million years. In order to create new subduction, the aging process must proceed further because an excess of density is necessary to create the vertical forces that are capable of tearing off the lithosphere and initiating subduction. As mentioned above, oceanic lithosphere may reach ages of up to 200 millions of years (oldest oceanic crust today: ca. 185 Ma, in front of the Marianas and ca. 175 Ma, at both margins of the Central Atlantic Ocean). Therefore, old ocean crust can undergo spontaneous subduction. Old ocean crust has a lithospheric mantle that is approximately 2% denser than the asthenosphere directly beneath it. However, if the age of the oceanic lithosphere is young, subduction can only be initiated by compressional forces; this is forced subduction.

What is the Reason for the Arcuate Shape of Island Arcs?

In principle the arcuate shape of island arcs is easy to explain. A thumb pushed against a rubber ball causes the normal convex bulge of the ball to become a concave dent and the line of bending marks a circular line on the surface of the ball. Before a plate enters a subduction zone it possesses a curvature according to the curvature of the Earth. When it dives into the subduction zone, this curvature inverts and convex becomes concave. Adjacent arcs commonly display a catenary-like pattern as observed in the Western Pacific where one island arc drapes next to the other (■ Fig. 7.1). The radius of an arc, r, can be calculated using the formula

$$r = \frac{1}{2} R \, \text{arc} \, \alpha$$

where R is the radius of the Earth (6370 km) and a the inclination angle of the subduction zone (■ Fig. 7.4).

Subduction zones have angles of inclination between 30° and 90° and average approximately 45°. The radius of an island arc (r) is 2500 km using the formula above with a subduction angle of 45° and 3335 km with an an-

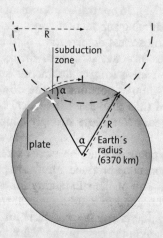

■ **Fig. 7.4** Geometric relation between the angle of a subduction zone (α) and the curvature radius (r) of a deep sea trench-island arc system (Bott 1982)

gle of 60°. However, most island arcs have radii that do not match these calculations. For example, below a depth of 100 km, the subduction zone of the Mariana Arc dips nearly vertical (■ Fig. 7.2). However, The radius of bending is less than 1500 km; its southern margin is even more stronger bent (◘ Fig. 7.1). There are several possible explanations for this discrepancy. One may be attributed to a delay factor as the angle of inclination varies over certain periods of time and the geometry of the upper plate does not respond in immediate fashion. A second reason may be related to inhomogeneous structure and stiff components in the upper plate. In the case of the Mariana Arc, it appears that some other tectonic feature has also bent the arc. The strong curvature in its southern part is related to drag along a transform fault.

In the case of active continental margins, the influence of the subduction zone upon the shape of the upper plate and that of the overall plate boundary becomes minimal. The subducted plate is not able to substantially cut or deform the thick and stiff edge of the continental lithosphere. The long, linear convex and concave shape of the plate boundary to the west of the Andes is determined by the shape of the South American continent and the various older plates that make it up. Subduction zones commonly occur along old passive continental margins at the zone of weakness where the oceanic lithosphere has been welded to continental lithosphere and fractures formed due to their different buoyancies, especially where the old oceanic lithosphere is cold and dense. This scenario appears to have occurred along the western margin of South America.

7

The western rim of the Pacific Plate is characterized by lithosphere with ages greater than 100 Ma. Therefore, spontaneous subduction dominates. However, younger lithosphere typical of the eastern margin of the Pacific, mostly less than 50 Ma (◘ Fig. 2.12), as well as that of the Philippine Sea Plate and the western Sunda Arc is too buoyant to subduct spontaneously, but rather is forced underneath the upper plate by compressional forces. Such conditions promote shallow subduction and a strong coupling to the upper plate. In other words, the horizontal compressional force is transferred from the subducting plate to the upper plate and compressional structures such as folding and stacking of crustal units evolve far into the upper plate (◘ Fig. 7.5). In contrast, spontaneous subduction with steeply dipping subduction zones generates extensive decoupling so less deformation of the upper plate occurs and extensional structures form. Forced subduction may evolve from spontaneous subduction as increasingly younger portions of the downgoing plate are subducted thus changing the vertical component at a plate boundary to a more strongly coupled horizontal force.

Steeply dipping subduction zones such as the *Mariana-type subduction* produce significant consequences. A strong slab pull of the subducting lithosphere forces the subduction zone to roll-back oceanward—i.e., towards the subducting plate or easterly into the edge of the Pacific Plate in the case of the Marianas. The locus of the hinge zone, where the ocean floor is bent down into the subduction zone and which marks the plate boundary, migrates backwards, away from the locus of the arc. This process is termed subduction or slab roll-back (◘ Fig. 7.5, insert). The roll-back causes extensional forces to develop at the plate boundary and the edge of the upper plate is extended. The roll-back causes the island arc to migrate towards the outside of the arc system which in turn results in a strong exten-

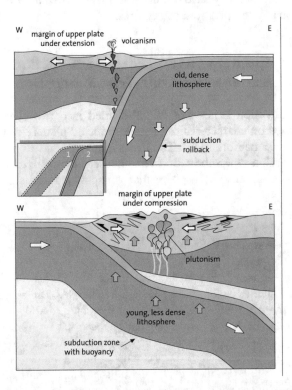

◘ **Fig. 7.5**   Mariana-type (above) and Chile-type (below) subduction zones. At the Marianas, subduction occurs easily and the subduction zone rolls back to the east (insert). At Chile, young and specifically light lithosphere pushes upwards; subduction is forced

sion of the backarc region (◘ Fig. 7.5). If the arc was originally built on a continent, this process leads to the separation of the island arc from the upper plate continent. The resulting separation behind the island arc generates a backarc basin and with sufficient extension, new oceanic crust may form (example: Sea of Japan). The western Pacific is characterized by numerous island arc–backarc systems (◘ Fig. 7.1). The garland- or drape-like alignment of the island arcs results from their separation from the continent and the formation of backarc basins. Typically, the garlands have lengths from 2000 to 2500 km. Because the Mariana Arc comprises an intra-oceanic subduction zone, the formation of the backarc basin occurred solely within oceanic crust (◘ Fig. 7.2).

Subduction roll-back that forms during Mariana-type subduction has an effect on the topography of the plate boundary system. The mean topographic elevation on the edge of the upper plate is low because of the effect of suction and extension. On the other hand, the depth of the deep sea trenches is significantly deeper on average because the subducting plate bends downward very steeply and the island arcs do not deliver enough sedimentary material to even partially fill the trench. The world's deepest trenches, all with depths greater than 8000 m, occur along westward-directed subduction zones (including those of the Atlantic Ocean) that involve old oceanic lithosphere (◘ Fig. 7.1). The roll-back of the westward directed subduction zones may be enhanced by an eastward directed asthenospheric current as calculated from global plate drift compared to flow of the sublithospheric mantle (LePichon 1968; Doglioni et al. 1999).

A curious situation at subduction zones with backarc basins (Mariana-type subduction) is the juxtaposition of strong convergence and divergence. Although relative plate movement velocities at convergent plate boundaries in the western Pacific approach 9 cm/yr, a wide area at the edge of the upper plate is under extensional stress. Compressive structures are restricted essentially to the tip of the upper plate.

The subduction zones and convergent plate boundaries of *Chile-type subduction* are fundamentally different in many ways. The subducting plate is intermittently coupled to the upper plate because of the buoyancy of the former; this causes compression and thickening of the upper plate (◘ Fig. 7.5). Thickening of continental crust leads to orogenesis and the generation of a high mountain range. Deep sea trenches are shallower and volcanic zones are characterized by significantly higher elevations than volcanoes associated with Mariana-type subduction. The highly elevated hinterland is more likely to deliver sedimentary material into the trenches, and the sediment supply is generally not hampered by intervening ridges. Even more consequential than the topographic height is the structural height, the total amount of uplift within the volcanic zone, which may exceed 20 km. Crustal structures originally formed at depth are subsequently uncovered by erosion. In fact, metamorphic and intrusive magmatic rocks at the surface in the Andes can be used to estimate the amount of uplift and subsequent amount of erosion.

Along Chile-type subduction zones the compressional forces are transferred far into the upper plate. Therefore, earthquakes in this area are particularly strong and frequent. Regionally, compressional structures with overthrusts evolve that are oceanward-directed in the forearc and continentward-directed in the backarc region (◘ Fig. 7.5). Cosequently, the backarc area is also under compressional stress. Crustal shortening occurs across the entire magmatic zone, which in turn has an effect on the magmatism itself. Crustal melts typically become trapped in the crust and crystallize as intrusions or melt adjacent crustal rocks and feed highly explosive acidic volcanoes (see below). The collective processes of tectonic stacking and magmatic accretion have formed a crustal thickness of 70 km in the Andes, one of the areas of thickest continental crust on Earth.

The transfer of forces at convergent plate boundaries varies. Stress analyses in the forearc region of the upper plate along Mexico and Central America revealed changing states of stress through time (Meschede et al. 1997). Phases of compression that reflect coupling of forces between both plates may be followed by extensional phases during which decoupling occurs. During the extensional phases the crustal stack at the edge of the upper plate becomes unstable, and thus collapses. Reverse and thrust faults that evolved during the compressional phases can be reactivated as normal faults. Subsequent phases of coupling and decoupling are also observed at the plate boundary between the Juan de Fuca Plate and North American Plate. During periods of plate coupling, the edge of the upper plate is vaulted until there is a spontaneous decoupling because of the high accumulation of energy. As a result, the upper plate abruptly steps forward towards the subducting plate and its margin is accompanied by subsidence (◘ Fig. 7.6).

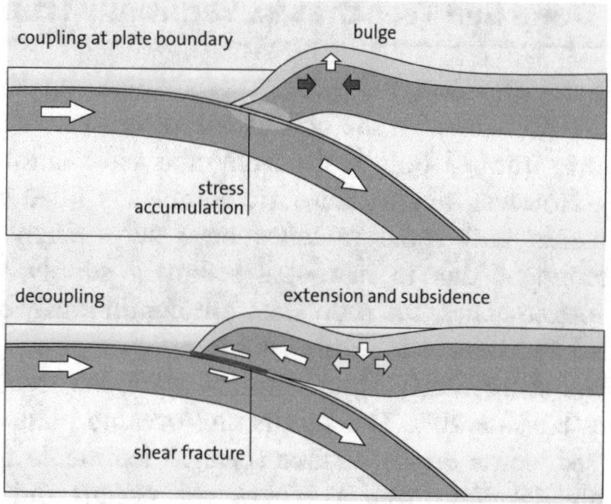

◘ **Fig. 7.6** Stress accumulated during coupling at the boundary between Juan de Fuca and North American Plate followed by rapid decoupling (Hyndman 1996). A similar decoupling combined with an abrupt movement of the edge of the upper plate towards the trench was responsible for the earthquake in the subduction zone of Sumatra that caused the devastating tsunami on 26 December, 2004

7.3   **Deep Sea Trenches as Sediment Traps**

Deep sea trenches, the greatest depressions on Earth and deepest zones in the oceans, are places where sediment is trapped both tectonically and sedimentologically. However, few trenches are completely filled with sediments and some trenches have surprisingly little sediment due to restricted sediment supply; and in most trenches, the sediments are continuously subducted and thus tectonically removed.

The sedimentary input into the deep sea trenches varies considerably. The subducting oceanic plate carries sediments on its surface (pelagic sediments from the abyssal plain; *pelagos, Greek* sea, ocean) and this material can be carried tectonically into trenches. Pelagic deposits can also settle directly into trenches from the thick water column above. Other sediments come from the adjacent island arc, forearc region, and continental margin as suspension deposits, turbidity current deposits, and various slump and landslide deposits. These terrigenous sediments (*terrigenous, Latin–Greek* of the land) are carried into the trench across the continental slope and typically transported over large distances along the trench axis by trench-parallel currents. These sediments are the so-called contourites. Therefore, most trenches are filled with mixtures of pelagic and terrigenous deposits as well as sediments and sedimentary rocks transported into the trench on the subducting lower plate.

The amount of sediment in a trench, regardless of origin, is related to the balance between the sediment supply and the slow tectonic removal of trench fi 11. Thick trench deposits are generally favored by a low subduction rate and accompanying slow tectonic removal of trench-fill material. If trench fill is rapid under these conditions, the trench fills and trench morphology is flat and the surface grades into the adjacent abyssal plains. Such a condition exists today in the trench off Oregon and Washington, USA. In contrast, trenches are deep and more irregular if the sedimentary supply is low or if swells adjacent to it such as the forearc outer ridge intercept and block sediment entry. The lack of sediment in the deep trenches of the Marianas, the Tonga and Kermadec Islands, and the Kuril Islands can be explained by these conditions.

A comprehensive example of the factors that are relevant to sedimentation in deep sea trenches is found in the Northern Pacific Trench system (◻ Fig. 7.7). Comparing the Washington–Oregon Trench, the Aleutian Trench, and the Kurile–Kamchatka Trench, a systematic change of the sedimentary balance can be observed. The sedimentary cover tectonically transported into the trench by the ocean floor (pelagic sediments) increases significantly in thickness from east to west. One of the oldest parts of Pacific oceanic crust occurs in the northwestern Pacific. Accordingly, the sediments formed on the abyssal plains are thicker here than fur-

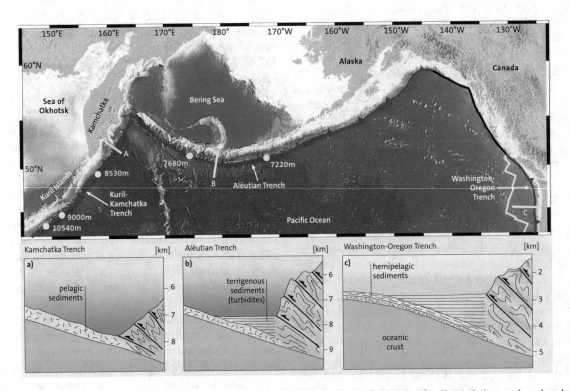

◻ **Fig. 7.7**   The deep sea trench system of the northern Pacific (Scholl 1974). A systematic change of sediment balance takes place between the Kurile–Kamchatka Trench (**a**), Aleütian Trench (**b**), and Washington–Oregon Trench (**c**)

ther to the east and these deposits are carried into the trench by the subduction process.

Thick turbidite sequences have been deposited in the Washington–Oregon Trench during the Pleistocene ice age. Sea level was lowered by more than 100 m during glacial maximums, and a lack of barriers enabled free transport of sediments from the land to the shelf margin where turbidites directed the sediment into the deep sea. The trench was completely filled with turbidites and consequently no longer morphologically evident. Therefore, turbidity currents could reach the open ocean and the pelagic sedimentary layer was thickened considerably by distal turbidites (hemipelagic sediments: ◘ Fig. 7.7). Turbidites are described as distal (as opposed to proximal) when they are far away from their origin and contain only fine-grained, silty to clayey material. Presently, the huge Columbia River continues to direct sediment into the trench.

The Aleütian Trench is not directly influenced by terrigenous sediments because the outer basin and outer ridge intercept the sediment supply. However, canyon-like gaps in the outer ridge exist adjacent to Alaska and in the westernmost part near Kamchatka. There sediments are transported as turbidity currents into the trench where they are diverted parallel to the trench axis. At some locations a slope angle of 0.2° is sufficient to transport the finest fraction of sediments along the trench axis. Trench-parallel transportation has been documented over a distance of 1400 km. In this manner, sediments brought into the trench at local points are spread longitudinally over long distances.

The Kurile–Kamchatka Trench generally lacks trench turbidites (sediment input at one location is diverted into the Aleutian Trench). The depth of the trench with respect to the adjacent abyssal plain reflects the different sedimentation regimes of these three areas: The Kurile–Kamchatka Trench is approximately 3–4 km deeper; the Aleutian Trench is 1–2 km deeper, and the Washington–Oregon Trench does not show notable trench morphology or difference in depth.

7.4 Accretionary Wedge and Outer Ridge

The morphology of the outer ridge forms as sediments are scraped off of the subducting plate during subduction. As the subducting plate transports its sedimentary fill from the trench towards the arc, some portion of the sediment is subducted and transported down to great depths where it becomes an important factor in the feeding of subduction-related magmas. The remaining portion, or in some cases the entire sedimentary layer and parts of the oceanic crustal basement, can be scraped off at the frontal tip or the bottom of the upper plate and added to a so-called accretionary wedge on the upper plate (◘ Fig. 7.8).

An accretionary wedge grows from below. The scraped-off sedimentary layers are stacked and continuously uplifted by renewed underplating from below, a process that results in morphological elevation of the outer ridge. The more material that is scraped off, the higher the elevation of the outer ridge. The process of accretion has been modeled in sandbox experiments so the evolution of the accretionary wedge is well understood (◘ Fig. 7.8). The underplating process resembles large-scale nappe thrusts in mountain ranges. Previously juxtaposed layers of sediment are stacked during the shortening process and at each overthrust, older sediments are placed over younger ones. The process and sequence of events can also be viewed from the opposite perspective—younger units are forced below older ones by underthrusting.

Eight overthrust planes that display repetitions of the sedimentary layers have been drilled at the Vanuatu accretionary wedge in the SW Pacific (◘ Fig. 7.8). Because the sedimentary layer at the subducting plate has a thickness of slightly more than 100 m, the much greater thickness of the accretionary wedge is a result of intense tectonic stacking. In contrast, south of Japan, where the Philippine Sea Plate subducts beneath the Eurasian Plate, a thick sedimentary layer greater than 1000 m is entering the Nankai subduction zone. Here, the decollement zone remains in the sedimentary layer and does not cut through the underlying oceanic crust as is the case in Vanuatu. Approximately the lower third of the sedimentary layer is being subducted and does not contribute to the growth of the accretionary wedge.

Increased pressure and temperature occurs in the deeper parts of accretionary wedges. The rocks are metamorphosed and deformed by dynamic processes acting within the wedge. Deformation structures like scaly fabric in shear zones form at shallow depths shortly after entering the subduction zone (◘ Fig. 7.10). Such a fabric is characterized by a large number of polished fault planes (slick-ensides) on small, cm-sized rhombohedra-shaped blocks of claystone. With continuing deformation penetrative narrow cleavage planes develop ("slaty cleavage") and with increasing pressure and temperature, metamorphism begins.

Large submarine slides may occur along the steep slope of deep sea trenches. Such large-scale slides have been studied by bathymetric investigations along the Pacific coast of Costa Rica and along the Kermadec deep sea trench north–northeast of New Zealand. It is hypothesized that the slides are triggered by larger earthquakes that are frequent in a subduction zone. Competent layers such as sandstones are torn and then embedded as fragments into softer clay-rich layers to produce a mixed layer of sandstone and claystone with a "block-in-matrix" structure. Such rocks are called

7

◻ **Fig. 7.8** Upper: structure of the accretionary wedge in front of Vanuatu, southwest Pacific (Meschede and Pelletier 1994). Each thrust plane is characterized by younger rocks that underlie older rocks. Oceanic crust has been scraped-off from the subducting plate and added into the accretionary wedge. Limestones and basalts were strongly sheared at the thrust planes and tectonic breccias were formed. Middle: map of subduction zone and stratigraphic columns obtained from Deep-Sea Drilling Project. Lower: tectonic stacking of the sedimentary sequences has been modeled in a sandbox experiment (construction of the experiment and photo: Nina Kukowski, GFZ Potsdam, Germany)

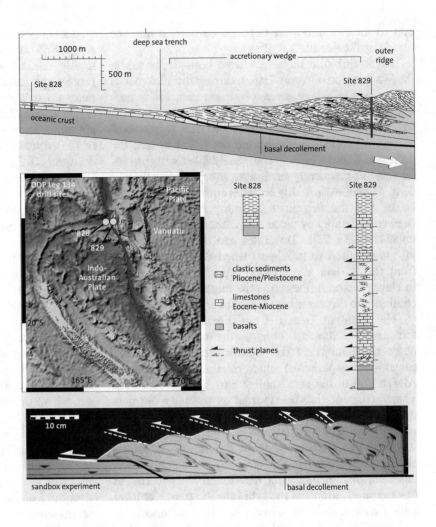

mélanges (mixture of rocks), and more specifically sedimentary mélanges because they formed primarily by sediment-gravity gliding processes (◻ Fig. 7.11).

The slide masses, including some isolated large pieces of rocks (*olistolites, Greek* sliding rocks), may eventually reach the deep sea trench and then enter the subduction zone (◻ Fig. 7.11). Therefore, off-scraped rocks of the oceanic crust (ophiolites) and various rocks transported into the deep sea trench are juxtaposed within the tectonic zone of friction between the two plates where all components are mixed by the strong tectonic movement. Pieces of all magnitudes (up

to kilometers in size) of disrupted competent rocks like basalt and indurated sedimentary rocks within a soft matrix of clay or slate comprise a tectonic mélange. Parts of this mélange may have been mixed initially by sedimentary and later by tectonic processes (sedimentary-tectonic mélange; ◻ Fig. 7.11) although the processes can be difficult to distinguish. If ophiolitic material forms the dominant component, it is called an ophiolitic mélange. Such ophiolitic mélanges are commonly alined along the suture zone between plates after a collision.

The Accretionary Wedge of the Sunda Arc

The Sunda Arc provides an instructive example demonstrating how the shape of the accretionary wedge depends on the amount of sediments transported into the subduction zone (◻ Fig. 7.9). Fueled by an enormous monsoon-controlled supply of sedimentary material, the Ganges and Brahmaputra rivers constructed the huge several-kilometer-thick submarine Bengal fan (◻ Fig. 4.16) that extends to the southernmost point of Sumatra. The

sedimentary fan lies on the Indo-Australian Plate and is transported towards the NNE where it is being subducted in the Sunda Arc. Therefore, the deep sea trench along the northwestern part of this island-arc system is mostly masked by the high sedimentation rate while the outer ridge, built by the tectonic stacking of fan-supplied sediments, is emergent several hundred meters above sea level. It forms islands such as the Andaman and Nicobar Islands and the Mentawai Ridge in front of Sumatra. Far-ther SW away from the influence of the Bengal fan and adjacent to Java, the trench is distinctive with depths of almost 7500 m. Accordingly, the outer ridge is not well expressed and lies mostly below a water depth of 2000 m. Along Sumatra, the slope of the accretionary wedge fall-ing from the outer ridge to the deep sea trench typically displays distinctive subdivisions. Here active thrusts pro-duce elongate flat areas and depressions that are called slope basins (■ Fig. 7.9, insert). They are common along Mentawai Ridge where some are exposed above sea level. During their complex history, they acted as sediment traps that recorded the uplift history of the outer ridge. Analyses of the microfauna indicate uplift from deep wa-ter to shallow water conditions. The youngest sedimen-tary rocks include the formation of reefs in very shal-low water, followed by uplift above sea level (Moore et al. 1980).

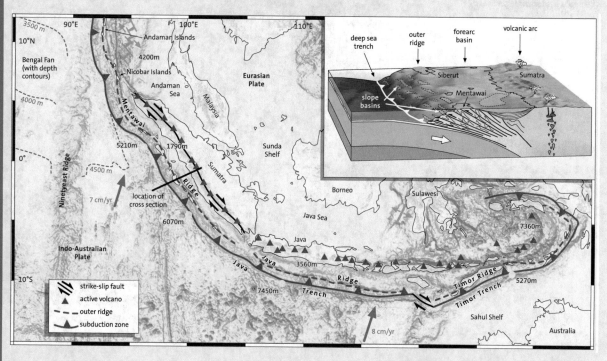

■ Fig. 7.9 Map showing contrasting plate-tectonic condi-tions along the Sunda Arc. In front of Sumatra, sediment of the thick Bengal fan are scraped-off and incorporated into the accretionary wedge. This causes the outer ridge to emerge from the sea at this location (Mentawai Ridge—see insert). In front of Java, the deep sea trench and the outer ridge are significantly deeper. In front of Australia, the continental crust of the Sahul shelf is being subducted beneath the Sunda Arc; this causes a particularly strong uplift of the outer ridge (Timor Ridge) and marks the initial stage of orogenesis

■ Fig. 7.10 Development of scaly fabric and slaty cleavage along thrust planes in an accretionary wedge

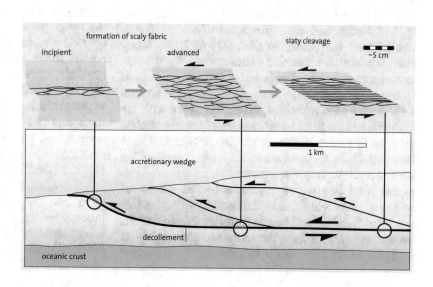

7

■ **Fig. 7.11** Formation of mélange in an accretionary wedge and subduction zone (Cowan 1985). Inserts show variations across different portions of the mélange. The outcrop photo shows a typical mélange structure. Mélange is characterized by a block-in-matrix structure (photograph by M. Meschede). Mud-volcanoes develop by explosive dewatering of sediments under high pressure

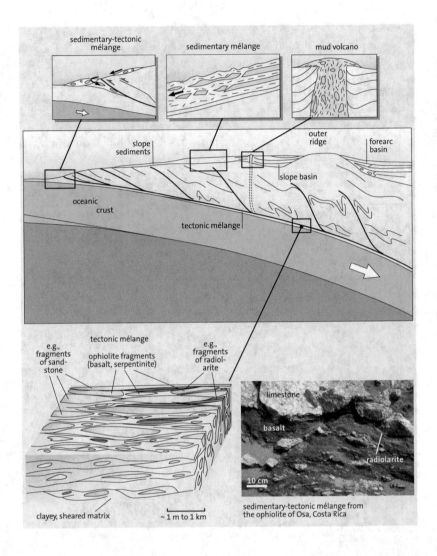

7.5 Subduction Erosion Instead of Accretion

Sedimentary accretionary wedges occur in many subduction zones including large portions of the Gulf of Oman (Makran subduction zone, ■ Fig. 5.16), Sumatra, SW Japan, and in smaller areas of western North America and the Lesser Antilles (■ Fig. 7.13). Many other subduction zones are characterized by the opposite process, subduction erosion. During subduction erosion, rock material is scraped off from the bottom of the upper plate and transported downward with the subducting plate. Such basal erosion of the upper plate is known from the Marianas, the Tonga Islands, Costa Rica and Chile. In these cases, an accretionary wedge has not evolved. It is assumed that many modern convergent plate boundaries have accretion at the tip of the upper plate and erosion farther down the subduction zone. Subduction erosion transfers a considerable amount of sediment from the surface to greater depths. In places where subduction erosion occurs, the outer

ridge is not a significant feature because sedimentary material does not accumulate in an accretionary wedge. The forearc basin commonly grades directly into the trench with little or no topographic rise between the two.

Subduction erosion is particularly effective when the lower plate has a roughly textured surface and is only covered by a thin sedimentary layer (■ Fig. 7.14). As the lower plate enters the subduction zone and is bent, the upper part of the plate is extended; a series of trench-parallel horst and graben structures are subsequently formed. The roughness caused by these structures acts like a grater on the basal part of the upper plate (basal erosion, ■ Fig. 7.14), much like a cheese grater removes cheese from the base of the cheese block. Material that is scraped off the upper plate enters the subduction zone where it is transported downward into deeper parts of the subduction zone. This process is currently happening in the Mariana Arc. Seamounts on the subducting plate create a somewhat different kind of subduction erosion. Single seamounts scrape material from the frontal tip of the accretionary wedge along the upper plate and carry it into the subduction zone (frontal erosion). A superb example of this setting occurs off the Pacific coast of Costa Rica where seamounts currently buried by the tip of the up-per plate leave obvious bulges followed by indentations in the slope of the upper plate. Subsequently, sediment gravity slides may fill up or obscure these indentations (◻ Fig. 7.14 lower, ◻ Fig. 7.15).

Mud Volcanoes

Mud volcanoes are widespread features that occur on the top of accretionary wedges. They are volcano-like features which are not fed by molten rocks but rather by mud rich in water and clay. The mud originates at the shallow plate boundary and forces its way through the margin wedge (◻ Fig. 7.11). Water originates from porous sediment along the decollement zone between the upper and lower plate where it becomes overpressured because of lateral pressure in the subduction zone. Additionally, the pore fluid pressure increases due to additional water contained in opal that is released from siliceous organisms, and in clay minerals (primarily smectite). Methane and other volatile hydrocarbons are generated from decomposing organic material and add to the increase in pore fluid pressure (Kopf 2003). Overpressure occurs when the internal pressure of a rock exceeds that of the surrounding confining pressure, a value determined by the weight of the overlying rocks. Overpressure generates cracks that are used by low-density, water-rich masses of mud to rise in a diapir. If the mud extrudes at the surface it may form fountains and mud-volcanoes either on subaerial or submarine surfaces.

The rise of a mud diapir happens quickly. For example, salt diapirs and mantle diapirs have rising velocities of millimeters to centimeters per year whereas mud diapirs that lack gas, have velocities of about 2 mm/s or more than 60 km/yr. At these velocities, they can cut through an entire accretionary wedge of several kilometers thickness within a couple of weeks. If methane gas is present, velocities can reach 100 m/s or 360 km/h (Kopf 2003). Gas-charged eruptions of mud volcanoes produce an explosion. Mud volcanoes without gas produce cones that rapidly form hard slopes with angles steeper than 5°. Eruptions with gas as a component form flat, dough-like structures that can reach diameters of tens of kilometers. In the Mediterranean region, the African Plate subducts south of Crete beneath the Aegean Sea. Adjacent to Crete, mud volcano fields occur on the sea floor with some cones more than 500 m high and 40 km wide. They occur as much as 150 km away from the plate boundary. The dewatering of minerals, a chemical conversion of water-containing (hydrated) minerals into water-free minerals, plays an important role. The water, which originally occurred as pore fluids and in hydrated minerals, is forced out of the highly strained subduction zone upward through the accretionary prism.

Mud volcanoes can also occur in collision zones following the subduction that led to continental collision. The

◻ **Fig. 7.12** Mud-volcano at the foot of the Apen-nine Mountains at Maranello, south of Modena, Italy (photo kindly provided by Achim Kopf, University of Bremen, Germany). The steep tip of the cone is approximately 1 m high

principle of evolution is the same as discussed above. Mud volcanoes in collision zones are known from the Caucasus, Trinidad and Pakistan, and locally in the Apennine Mountains (◻ Fig. 7.12) and Romania. Eruptions of these mud volcanoes occur in intervals of years to decades.

A special situation is present in the forearc area of the Marianas in the western Pacific where serpentinite mud-diapirs occur. These mud-volcanoes are up to 2000 m high and 30 km wide and contain abundant fragments of serpentinite (mantle peridotite transformed into serpentinite by water absorption), gabbro, and basalt. The imbedded rock fragments are probably scraped off from the upper oceanic plate by the rising mud masses that are generated in the decollement zone between the two plates. However, blueschist, i. e., high-pressure metamorphosed basalt rock fragments are also present among the components; such rocks can only be derived from the subducting plate after it is buried to a depth of more than 15 km. Serpentinite mud-volcanoes are only known from the intra-oceanic subduction zone of the Marianas; this suggests that they only form where oceanic lithosphere is present in the outer part of the upper plate.

7

Fig. 7.13   Map of circum-
Pacific showing location of
subduction zones with and
without accretionary wedges (von
Huene and Scholl 1991)

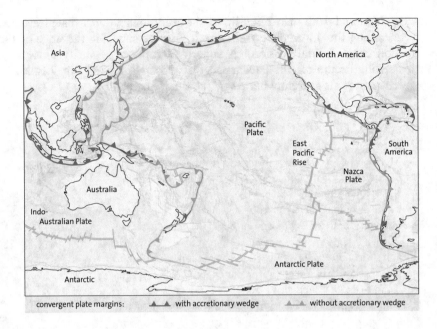

Fig. 7.13   Map of circum-Pacific showing location of subduction zones with and without accretionary wedges (von Huene and Scholl 1991)

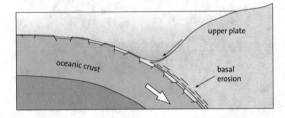

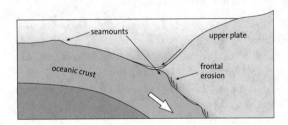

Fig. 7.14   Subduction erosion upper: at the base of the upper plate caused by subducted horsts and grabens; lower: at the front of the upper plate caused by subducted seamounts

Basal erosion on the upper plate causes the plate boundary to migrate towards the arc, which itself is moving backward in response to the migrating subduction zone. This process occurred along Costa Rica and in parts of the Andes between the Miocene and present (■ Fig. 7.16). A tectonic mélange, consisting of parts of the scraped off upper plate, has evolved along the decollement zone. The mélange is transported into the depths of the subduction zone along with the plate being subducted.

Subduction erosion can occur both where the subduction zone angle is relatively low and there is a strong coupling between the two plates (Chile type) and where the subduction angle is high and subduction roll-back occurs (Mariana type). Examples for the first case include Central America (see below) and parts of South America, and for the second case the Marianas, Tonga, and NE Japan. In Chile-type subduction, subduction erosion occurs because of the strong contact pressure caused by strong plate coupling. In Mariana-type subduction, erosion is caused by the rough, saw-tooth-like morphology on the subducting plate.

Pore fluids have important consequences in subduction zones and on subduction erosion. Recall from above that pore fluids originate both from dewatering of hydrous minerals and from expulsion of water from sediments. Pore fluids in the high-stress zone along the plate boundary decollement actually increase mobility and slippage in this zone by acting as a lubricant. During explosive expulsions of fluids, framework grains break and loose their connectiveness, and the rocks become ground by shearing. Such rocks are called cataclasites. *Cataclasis* (*Greek* strong fracturing) is a common process associated with subduction erosion. Fluids that escape upwards through the upper plate may generate mud volcanoes (see box; ■ Fig. 7.11).

Subduction erosion can be proven only indirectly as the process occurs at depth and the eroded rocks become subducted to even greater depth. In contrast, accretionary wedges can be directly studied. An active plate margin that may be the site of subduction erosion is the Central American subduction zone adja-

Fig. 7.15 Subduction of seamounts at the plate margin off Costa Rica (von Huene et al. 2000). The indentation of the seamount into the subduction zone leaves deep furrows in the upper plate (figure after bathymetric data of the research vessel Sonne, made by Wilhelm Weinrebe, IfM–Geomar Kiel, Germany)

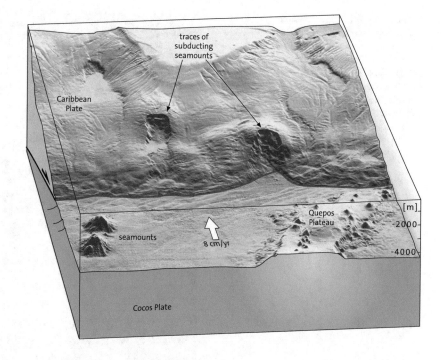

cent to Costa Rica. Here, the Cocos Plate is being subducted beneath the Caribbean Plate with a velocity of about 9 cm/yr. The volcanic front has shifted 40 km backwards (towards the Caribbean) during the last 15 million years. This shift probably corresponds to the amount of shortening formed by subduction erosion (■ Fig. 7.16).

As subduction erosion thins the upper plate, the outer part subsides. Sediments deposited in the forearc area can be used to document this subsidence. For instance, benthic foraminifera, single-celled organisms which live within the sediment, indicate the approximate sedimentation depth based on their shell shape and design. If foraminifera and other sedimentological features in layers of decreasing age at a convergent margin indicate increasing water depth, subduction erosion is a likely explanation. At a drill site of the Ocean Drilling Program offshore Costa Rica, shallow water sediments occur at the deepest part of a hole whereas near the top of the well core, deep water sediments occur (Meschede et al. 2002). Accretionary sequences produce the opposite pattern because the gradual buildup of the accretionary wedge causes a shallowing-upward sedimentation pattern over the wedge. This pattern is present at the Mentawai Ridge in the Sunda Arc (see box above).

7.6 The Forearc Basin

Forearc basins are zones of crustal subsidence that lie between the outer ridge and the mountain range of the volcanic arc. They are generally dominated by

great thicknesses of marine sediment and sedimentary rock that form in depths that range from abyssal to shallow. Present examples of foreaerc basins include the Sandino Basin of offshore Nicaragua and the basin that lies between the Mentawai Ridge and Sumatra. The former contains deposits approximately 10 km thick and the latter consists of sediment 4–5 km thick that displays a general shallowing–upward sequence (■ Fig. 7.9). Although the deposits in the Sumatra example are dominated by sediment derived from the adjacent arc, they also contain appreciable amounts of marine biogenic carbonate, especially within the younger, shallower deposits.

Forearc basins are underlain by either thinned continental or normal oceanic crust. The distribution of the two is related to the origin of the subduction zone. Typically new subduction zones evolve at the juncture between continental and oceanic crust, along a passive continental margin, because this zone is generally structurally weak. Continental margins typically show irregular shapes with alternating salients and embayments; subduction zones tend to form smooth, arcuate plate boundaries. Therefore, the margin of the upper plate along the new plate boundary is likely to contain embayments of oceanic crust (■ Fig. 7.17). Hence, the type of crust in the forearc region may be continental, oceanic, or both in a given section. Such oceanic entrapments are called remnant or residual ocean basins.

Forearc basins on oceanic crust may feature great water depth. The oceanic crust merges into thinned continental crust towards the volcanic arc. If an accretionary wedge and an outer ridge develop at the convergent margin, the basin will be topographically con-

7

■ **Fig. 7.16** Subduction erosion at the plate margin of Costa Rica where the Cocos Plate subducts beneath the Caribbean Plate (Meschede et al. 1999). Backward shift (towards the backarc) of the volcanic arc is caused by subduction erosion of the upper plate since the Miocene

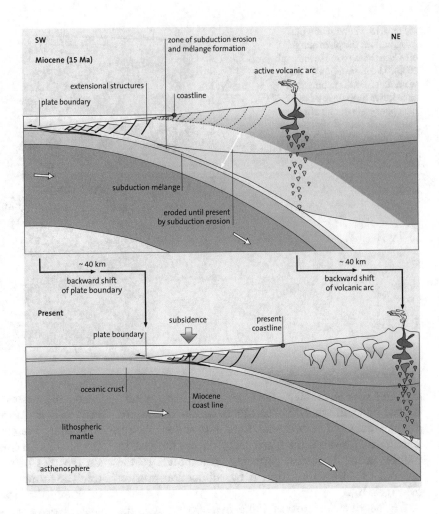

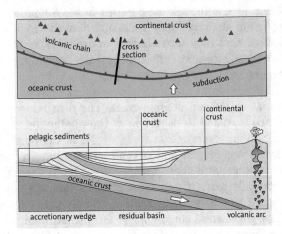

■ **Fig. 7.17** Development of a residual forearc basin caused by the formation of a subduction zone at an irregular continental margin

stricted and fill rapidly with sediment (see box). During collision, ophiolites from the basement of forearc basins may be accreted onto the adjacent continen-

tal crust. As opposed to subducted ophiolites from the lower plate, they do not experience high-pressure metamorphism unless they are scraped off and dragged down.

7.7 Earthquakes and Benioff Zones

Subduction zones are easy to detect because they form long, linear belts with intense earthquake activity. These seismically active zones are responsible for 95% of all earthquakes on Earth and are referred to as Benioff zones after the geophysicist, Hugo Benioff, who recognized and systematically investigated such zones in the Pacific Ocean and below the Sunda Arc (Benioff 1954), or Wadati–Benioff zones (in 1935 Kiyoo Wadati documented earthquake epicenters along a down plunging surface below Japan). The earthquakes associated with Benioff zones reach a maximum depth of almost 700 km (■ Fig. 7.19).

The Shigatse Flysch in Tibet

A well exposed, extraordinary example of a fossil forearc basin is the Shigatse Basin in Tibet, which prior to the collision of India and Central Asia, was filled with flysch (Einsele et al. 1994). *Flysch* (*Swiss–German* to flow, derived from the tendency of weathered sandy-clayey layers to flow downhill) sequences comprise thick units of turbidite deposits that typically evolve pre- to synorogenically in subduction zones. Turbidites are frequently deposited in deep sea trenches and subsequently scraped-off and deformed in accretionary wedges. Turbidites are also common in deep-water forearc basins flanked by steep submarine slopes along the volcanic arc, especially if water-saturated sediments are periodically shed of the arc over long spans of time. Proximal turbidity currents and distal suspension currents carry sediment far into the basin; the periodic events are triggered by earthquakes, a common phenomenon in forearc basins such as the Shigatse Basin.

From the mid-Cretaceous to Early Tertiary, oceanic crust of the rapidly northward-drifting Indian Plate was subducted beneath the Tibetan Block. The Gangdese Belt, which today forms mountains 7000 m high in the Transhimalaya, evolved over the subduction zone as a volcanic arc. In the forearc area, the Shigatse Basin developed as a residual ocean basin (■ Fig. 7.17); it transitioned into thinned continental crust towards the Gangdese Belt (■ Fig. 7.18).

Along the northern edge of the basin, sedimentation commenced s with shallow-water carbonate deposits in coastal areas, whereas farther south, the sedimentary rocks reflect deep-water environments. The increased sedimentary load generated rapid subsidence and thick flysch sediments, derived from the north, that steadily filled the basin (■ Fig. 7.18). The turbidite rocks are mostly arc-derived volcanic-rich clastics that were deposited in water depths between 1000 and 3000 m during several cycles of sedimentation. Each cycle reflects early rapid subsidence accompanied by deep-water deposition followed by shallowing-upwards deposits that mark basin filling. Smaller-scale cycles are defined by individual fining- or coarsening-up sequences within the larger cycles. They reflect the complex subduction-related dynamics of the basin. Increased basin subsidence rates and/or hinterland uplift caused by episodes of enhanced magmatic activity led to coarsening-upwards trends. Tectonic quiescence or decreased rates of basin subsidence lowered the relief and led to finer-grained deposits and fining-upwards cycles.

In areas where large rivers entered the flysch basin, coarse-grained gravels extend far into it. Conglomerate layers are concentrated throughout given measured sections suggesting that river courses remained constant through time. Rivers transported abundant volcanic material that was derived from the active arc as well as plutonic detritus derived from more deeply dissected parts of the arc. Additional non-arc-derived material was supplied by the distant hinterland, the present terranes of the Tibetan Plateau, indicating that some rivers headed north of the arc and subsequently dissected through it in their journey to the forearc flysch basin.

The southern, more distal portions of the basin are characterized by finer-grained and thinner deposits. Farther seaward, sediments were scraped off the subducting plate along the plate boundary and contributed to an accretionary wedge. Although the edge of the upper plate was bent to form an outer ridge, it remained below the sea level (■ Fig. 7.18).

Collision of the Indian continent with Central Asia occurred in the Early Tertiary ca. 55 Ma. During the collision, the character of the Shigatse forearc basin was completely transformed as it was underplated by the Indian continent that was partially subducted below Central Asia (■ Fig. 7.18). The new basin became entirely emergent above the sea level. It was surrounded by complex tectonic elements including the now inactive volcanic chain to the north and the stacked nap-

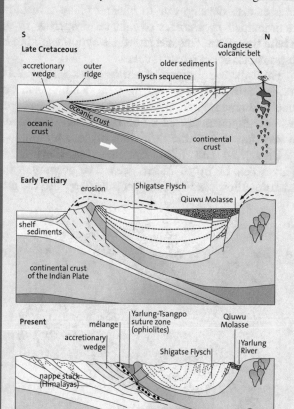

■ **Fig. 7.18** Development of the Shigatse Flysch in Tibet. Subduction of the Indian Plate beneath the Eurasian Plate created a forearc basin that evolved as a residual basin partly on oceanic crust. Subsequent continental collision has deformed the basin as shown

pes of the Himalayan thrust belt to the south. As sedimentation continued, the basin evolved into a non-marine molasse basin or successor basin. Rivers supplied coarse-grained gravels and sands and fine red muds were deposited in lakes. These deposits comprise the Qiuwu Molasse. The shortening during and after the collision reduced the Shigatse Basin to one third of its original width of 70 km or more. The collision between the plates is marked by the Yarlung–Tsangpo suture line, which farther west is called the Indus suture line; Yarlung Tsangpo derives its name from the upper course of the Brahmaputra river in Tibet.

Foredeep basins in the foreland adjacent to an emerging mountain range following continental collision are called foreland, molasse, or successor basins (see ▶ Chap. 11). Molasse deposits consist of thick, shallow water, marine to non-marine, sedimentary deposits of sandstone, conglomerate, and mudstone that typically form syn- to post-orogenically in foreland/foredeep basins. The term molasse apparently is derived from the use of harder sandstones as millstones (*Latin mola*). The Qiuwu Molasse of the Shigatse Basin is equivalent to the Kailash Molasse further to the west that is named after the 6700 m-high holy mountain of the buddhists and hindus.

Continued tectonic shortening folded both the Shigatse Flysch and the Qiuwu Molasse. Presently, the eroded layers of folded mudstone, sandstone, and conglomerate form an attractive landscape along a 20 km-wide strike valley with characteristic low, parallel ridges. The rocks are exceptionally well exposed because of the dry conditions in the Tibetan highlands; the mountainous area ranges from 4200 to 5000 m in height. In more humid landscapes, mud-dominated, thin-bedded flysch sequences tend to be mostly concealed by soils and vegetation.

Subduction zones are characterized by earthquakes with three different depth ranges: (1) shallow earthquakes, earthquakes with epicenters from near the surface to depths of 70–100 km, the normal thickness of the lithosphere; (2) intermediate earthquakes with depths between 70 and 400 km; and (3) deep earthquakes with depths between 350 and 700 km (◻ Figs. 1.8, 7.20). This pattern of earthquake depth occurs only in subduction zones; in fact earthquakes with epicenters deeper than 100 km are not known outside of subduction zones because hot asthenospheric mantle flows plastically when subjected to stress. In most regions outside of subduction zones, earthquakes with epicenters deeper than 20 km are rare because at this depth, continental crust behaves ductilely with stress. Subduction zones suffer such deep earthquakes because cool, rigid oceanic lithosphere is transported to great depths relatively rapidly. These conditions influence these rocks and allow large stresses to accumulate that are spontaneously released into earthquake energy. Stresses also develop due to mineral phase transitions accompanied by changes of mineral properties in the subducting plate. This mechanism is hypothesized to be responsible for deep earthquakes.

Fault-plane-solutions of earthquakes (◻ Fig. 2.14) are indicative of their origin. The bending zone of the subducting plate is characterized by low-magnitude earthquakes caused by horizontal tensional stress in accordance with the formation of graben structures (◻ Figs. 7.14 and 7.20). Epicenters of these earthquakes can be up to 25 km deep. The accompanying fractures allow sea water to penetrate deep into the lith-

◻ **Fig. 7.19** Distribution of earthquake epicenters in various modern subduction zones. Such features are known as Benioff zones. These zones penetrate to the boundary between upper and lower mantle to approximately 660 km depth

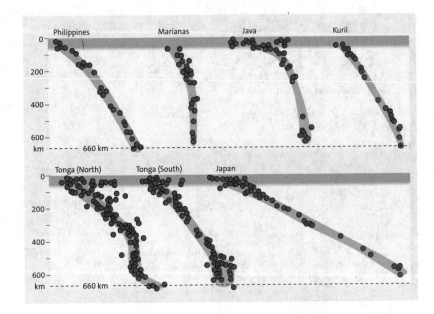

◘ Fig. 7.20 Cross section through a subduction zone showing the different types of earthquake mechanisms (Green, 1994). The uppermost area is dominated by shallow earthquakes caused by horizontal compression along thrust planes. Serpentine dewatering causes intermediate earthquakes to depths of 400 km. Deep earthquakes are probably created by the mineral phase transition from olivine to spinel and occur in depths between 350 and 700 km

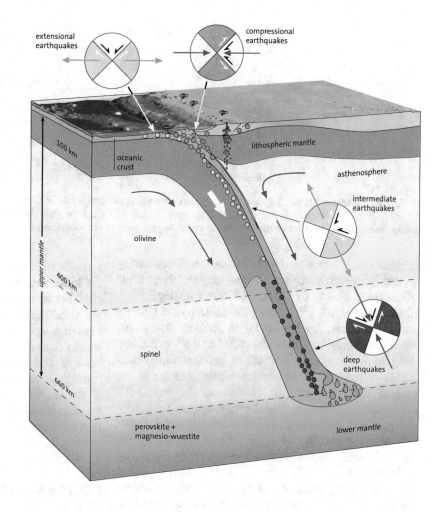

ospheric mantle and cause it to change partly to serpentinite. The serpentinite may rise to the surface in serpentinite diapirs as discussed previously.

In contrast, shallow earthquakes along the subduction zone are generated by horizontal compression and the friction between the two converging plates. They are released directly along the plate boundary or along accompanying fractures parallel to the down-plunging plate boundary (◘ Fig. 7.20). They can reach a depth corresponding to the thickness of the upper-plate lithosphere, typically 70–100 km, although there is a clear decrease in frequency below 50 km. The largest earthquakes, like of the ones that struck Japan in 1923, Kamchatka in 1952, Chile in 1960, Alaska in 1964, and Sumatra in 2004, originate within this zone. These earthquakes all (except Japan) had moment magnitudes between 9.0 and 9.5 and they represent the strongest ever registered. In each of these cases, the downgoing plates subduct with relatively high velocities and at a shallow angle beneath the overlying continental crust. These large, shallow earthquakes are responsible for approximately 90% of the total earthquake energy worldwide. The periodic occurrence of large earthquakes in intervals of several decades to centuries

demonstrates that the movement in the zone of friction between the two plates occurs in an irregular way.

Shallow earthquakes in subduction zones can cause two types of devastation, (1) destruction directly from the physical shaking and vibration and (2) the generation of tsunamis (*Japanese harbor wave*). The subduction-generated massive earthquake of the 26th December 2004, adjacent to the northern tip of Sumatra demonstrated this to the entire world. The subduction of the Indo-Australian Plate beneath the Eurasian Plate moves with a velocity of 7 cm/yr at this location. Relative displacement of blocks at the focus of the earthquake at a depth of 20 km was approximately 20–30 m. This instantaneous physical movement is equal to the amount of movement that would occur during an average subduction creep over an interval of 300–400 years.

Tsunamis are difficult to detect in the deep, open ocean where they form long (more than 500 km), low (ca. 1 m), rapidly moving (more than 800 km/h ◘ Fig. 7.20 Cross section through a subduction zone showing the different types of earthquake mechanisms (Green 1994). The uppermost area is dominated by shallow earthquakes caused by horizontal compression along thrust planes. Serpentine dewatering causes inter-

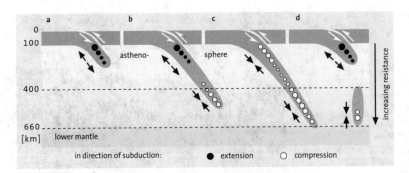

Fig. 7.21 Intermediate and deep earthquakes indicate extension or compression in direction of the downward plate movement (Isacks and Molnar 1969). This results from the resistance of the upper mantle into which the plate plunges. If the plate reaches the boundary of the lower mantle, compressive stress can be traced upwards into the upper parts of the subducting plate

7

mediate earthquakes to depths of 400 km. Deep earthquakes are probably created by the mineral phase transition from olivine to spinel and occur in depths between 350 and 700 km. above 6000 m deep ocean floor) waves. However, their behavior in shallow water and along coastlines is drastically different and extremely dangerous. Here, their velocity of propagation is dramatically reduced in the shallow water because of increased friction caused by contact of the wave with the sea floor. Concurrently, the wave length drastically shortens and the wave amplitude increases until it may rear up to more than 30 m high along the coast. This wall of water rushes across low landscapes creating widespread havoc.

Earthquakes that occur at greater depths are consequences of internal stress within the subducting plate. The zone of intermediate earthquakes is generally separated from the zone of deep earthquakes by a gap of low seismic activity. Intermediate earthquakes are mostly tensional whereas deep earthquakes are compressional, relative to the direction of the plate movement (■ Fig. 7.21a, b). This condition results from the medium into which the subducting plate sinks. Outside of the subduction zone, the Earth's mantle has its lowest density and viscosity in the asthenosphere, the zone between the base of the lithosphere down to 250–300 km. Cooled and dense lithosphere of the subducting plate will sink down into the asthenosphere relatively rapidly, without substantial resistance, because of its higher density. This leads to downward-directed tensional stress within the plate and tension-caused earthquakes (■ Fig. 7.20). The higher density of the subducting plate relative to its surrounding material has been described above as an essential factor in the driving force of plate motion.

At approximately 400 km in depth, the surrounding mantle material becomes denser because of a pressure-caused mineral phase transition; this leads to a higher resistance towards the subducting plate. Therefore, compressive stresses develop parallel to the direction of downward movement (■ Figs. 7.20, 7.21b). At

a depth of approximately 660 km, the boundary with the lower mantle, another mineral phase transition occurs in the mantle and strengthens it. Subsequently the subducting plate as a whole becomes less dense than the lower mantle. Additional compressive stress within the subducting plate is generated by the resistance of the stronger and denser lower mantle and is transferred upwards to shallower depths (■ Fig. 7.21c). If slab-breakoff occurred, such stress transfer is impossible and the subduction zone throughout the asthenosphere remains in a state of tensional stress (■ Fig. 7.21d).

7.8 The Secret of Deep Earthquakes

All rock material at depth incurs a lithostatic pressure (the equivalent to the hydrostatic pressure in water) that is determined by the weight of the overlying pile of rock. Tectonic movements add an additional, directed pressure that leads to an overall deviatoric stress where the pressure on a single particle of rock is greater in one direction than in all other directions. Earthquakes are the result of directed tectonic pressure. If a certain limiting value, which depends on the strength properties of the rocks is exceeded, the rocks react with shear fracturing where frictional gliding occurs between the two separated blocks. In cases of large earthquakes, the movement between the blocks in the area of the focus may be greater than 10 m.

Shear fractures in a rock have been investigated experimentally. When subjected to compression, numerous microcracks, thin, short hairline fractures, develop that are oriented parallel to the direction of maximum compressional stress; there is little extension of the rock parallel to the direction of minimum compressional stress. The result is a decrease in the strength of the rock. If tectonic stress continues in the same direction, failure occurs and shear planes oblique to the main direction of pressure develop that join the existing microfractures (■ Fig. 7.22a).

Fig. 7.22   a Microcracks form in response to tectonic pressures that weaken the rock; shear fractures follow. **b** Below a depth of approximately 350 km, so-called anticracks form in the subducting plate due to mineral phase transition from olivine to spinel. Connected spinel lenses create continuous bands of weak material that favor abrupt movement (Green 1994)

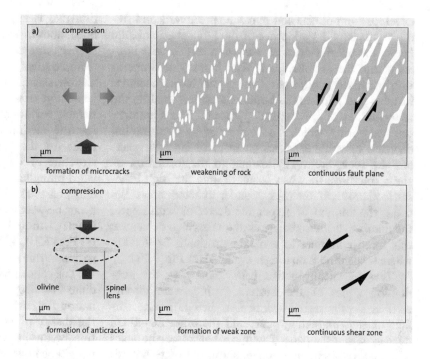

a)   compression

μm

formation of microcracks

μm

weakening of rock

μm

continuous fault plane

b)   compression

olivine   spinel lens

μm

formation of anticracks

μm

formation of weak zone

μm

continuous shear zone

However, at greater depths, where lithostatic pressure is increased, such displacements along fractures would not be possible because rocks under these conditions react plastically by ductile deformation when subjected to tectonic stress. Moreover, the high lithostatic pressure does not allow the formation of microfractures because they are coupled with extension. Nonetheless, 20% of earthquakes in subduction zones occur at intermediate depths and approximately 8% occur as deep earthquakes. Within the subducting plate, they are concentrated along a belt in the internal part of the plate (◘ Fig. 7.20). Frequency of earthquakes decreases from a maximum in the upper lithosphere and reaches a minimum at 300–400 km depth. Deep earthquakes are most common across a depth range of 550–600 km (◘ Fig. 7.23).

The zone in which deep earthquakes occur corresponds to the stability range of spinel within the subducting plate. Outside of the subduction zone, olivine, the main component of the mantle rock peridotite, compacts at a depth of 400 km and changes into the more densely packed mineral structure of spinel with the same mineral formula as olivine (Mg_2SiO_4). The structure of spinel is stable to a depth of about 660 km; below this boundary it again changes into a denser-structured mineral, perovskite ($MgSiO_3$) plus magnesio-wuestite (MgO). Within the subducting plate, the stability range for spinel exists between 350 and 700 km, the depth at which deep earthquakes occur. Consequently, it seems likely that the mineral phase transition of olivine to spinel is responsible for the deep earthquakes.

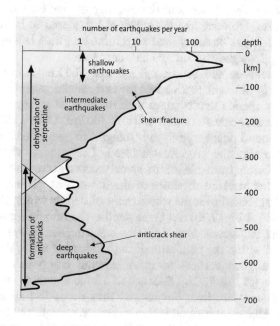

number of earthquakes per year

1   10   100   depth

0

[km]

shallow earthquakes

— 100

intermediate earthquakes

shear fracture

— 200

dehydration of serpentine

— 300

— 400

formation of anticracks

anticrack shear

— 500

deep earthquakes

— 600

— 700

Fig. 7.23   Frequency of earthquakes in a subduction zone (Green 1994). Deep earthquakes are differentiated through their different mechanism of formation

Intermediate and deep earthquakes also are released by spontaneous displacements within rocks; however, the mechanisms differ from those of shallow earthquakes (Green 1994). Intermediate earthquakes are related to dewatering processes with increasing depth of the mineral serpentine. Serpentine is a magnesium silicate mineral with a high water-content ($Mg_6[Si_4O_{10}(OH)_8]$; the water-content is 13 wt%). Serpentine develops from olivine and pyroxene precursors and frequently occurs in peridotites of the lithospheric mantle below the oceanic crust as well as in serpentinite bodies within the oceanic crust. With increasing temperature and increasing pressure, the crystal lattice becomes instable and releases the bound water; this enables the re-formation of olivine and pyroxene. The spontaneous dewatering, which is accompanied by a substantial decrease of volume, causes microfracturing and weakens the rock, eventually leading to the formation of shear fractures; they can occur, despite the high lithostatic pressure, if the rock is under directed stress. Because the amount of serpentine decreases with depth, the number of intermediate earthquakes also decreases; this trend continues to a maximum depth of 400 km where serpentine is no longer available (◘ Fig. 7.23).

Two seismically active belts, mostly in the subduction zones of the western Pacific, were discovered at depths near 100 km. One belt is at the surface of the subducting plate, and the other 20–40 km deeper, within the plate. The upper belt of earthquakes is caused by shear movements along the plate boundary and thus belongs in the category of shallow earthquakes. However, the lower belt is considered to be related to the dewatering of serpentine and thus belongs to the category of intermediate earthquakes.

Between the depths of 350 and 700 km, a completely different mechanism for earthquake generation must operate that overlaps with the zone of intermediate earthquakes between 350 and 400 km. Once again, a mineral phase transition is thought to be a critical in explaining the processes involved. The mineral phase transition from olivine to spinel occurs according to the temperature structure of the downgoing plate and is distributed over the whole range of depths within the plate (◘ Fig. 7.20). At these depths, the remaining olivine is "metastable". Initially, small lenses of spinel form within the peridotite of the subducting plate; they are aligned vertical to the maximum compressional stress (◘ Fig. 7.22b). Because spinel has a denser mineral structure than olivine and thus needs less space, this change corresponds to a reduction of volume. The spinel lenses can be interpreted as "anticracks" because they are rotated 90° compared with normal microcracks. If enough anticracks are formed, the rock becomes weakened. Finally the anticracks are joined and an instantaneous shear movement of the rock develops. However, the shear movement does not occur at a fracture plane where blocks of rocks break off, but rather through a process called superplasticity.

Superplasticity occurs at high temperatures and is favored where rock-forming minerals are very fine grained. The mineral grains slip past each other along their boundaries. Although this may seem to be an apparently trivial process, it leads to fast plastic deformation and is apparently unique to the conditions outlined above. The amalgamated spinel lenses are extremely fine-grained and with appropriately oriented tectonic stress, result in rapid grain reorientations along narrow shear zones in the rocks. And this process may generate earthquakes. The characteristic of such "anticrack earthquakes" is not discernible from normal fracture earthquakes because in both cases, blocks of rocks are displaced against each other. The thermal structure of the subducting plate, with a cold downward plunging finger in its interior, indicates that mineral phase transitions from olivine to spinel gradually occur between 350 and 700 km (◘ Fig. 7.20). The maximum earthquake occurrence, at depths of 550–600 km (◘ Fig. 7.23), is related to frequent phase transitions within in this zone.

The above-mentioned mechanisms for inter-mediate and deep earthquakes give a satisfactory explanation for earthquakes in subduction zones. However, the theory of deep earthquakes also raises contradiction because the metastability of olivine and its gradual change over a large depth range is controversial. A contrasting explanation for deep earthquakes suggests that thermal instabilitiy within the plate may lead to high stress that spontaneously produces films of melted rocks at fault planes (Wiens 2001).

7.9 High-Pressure or Subduction Metamorphism

Rocks in subduction zones are metamorphosed under conditions of high pressure at relatively low temperature, hence the terms high-pressure (properly high-pressure/low-temperature) metamorphism or subduction metamorphism; ◘ Fig. 7.24). Such metamorphism results because cool rock material is rapidly subducted to great depths. Normally, at a depth of 50 km that corresponds to a pressure of 1.5 GPa (1 GPa = 10 kilobar), the temperature is approximately 800–900 °C. In contrast, oceanic crust that reaches the same depth in a subduction zone reaches temperatures between 200 to more than 500 °C. The lower value is valid for old, cool, rapidly subducting lithosphere and the higher value is for young, and warm, more slowly subducting lithosphere; the NE and SW margins of Japan, respectively provide examples of each (Stern 2002).

The contrast in temperature within and outside of subduction zones becomes even more noticeable at depth. This is indicated by the patterns of isotherms (lines of equal temperature) within subducting plates. The isotherms are strongly declined downward and under the upper plate because rocks are poor heat conductors and they need millions of years to ad-

◻ Fig. 7.24 Pressure–
temperature (PT) diagram
showing different types of
metamorphism and metamorphic
facies. High-pressure
metamorphism is typical for
subduction zones and Barrovian-
type regional metamorphism
typifies continent–continent
collisions. Abukuma-type
regional metamorphism
and contact metamorphism
occur in the magmatic belts
above subduction zones.
Diagenesis occurs in the realm
of rock consolidation and is
considered a non-metamorphic
process. The insert shows a
coësite crystal, formed at the
conditions of ultrahigh-pressure
metamorphism, enclosed in
a garnet. Because of pressure
reduction during the uplift of
the rock, the coësite was partly
changed into quartz. This caused
a volume increase in the quartz
and produced the cracks in the
garnet crystal

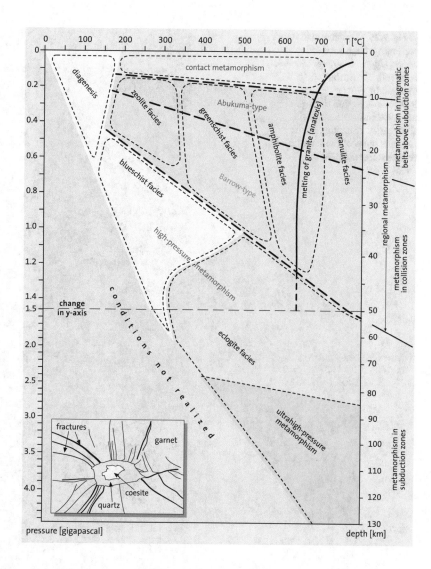

just to the higher temperatures normally seen at depth
(◻ Fig. 7.25). However, the pressure within the subducted plate increases immediately, without delay,
based on the given depth because it is dependent on the
burden of the overlying rock pile. Therefore, the rocks
and minerals within the subducted plate experience the
extreme conditions of subduction metamorphism—
rapid deep burial causes high pressure with relatively
low temperatures (◻ Fig. 7.24). Rocks that have been
buried more than 100 km within subduction zones and
subsequently exhumed during orogenic processes can
be observed directly on outcrop.

Metamorphic conditions in a subduction zone
transform basalts, dolerites and gabbros of the oceanic
crust into glaucophane-schists (blue-schists) and eclogites (◻ Fig. 7.24). Glaucophane is a blue-violet Na-am-
phibole that forms at a pressure of about 0.6 GPa (corresponding to about 20 km burial depth). It remains
stable over a large pressure range as long as the temperature remains below 500 °C. Eclogite, on the other
hand, forms at a pressure of about 1 GPa (corre-

sponding to about 35 km burial depth) and its stability field is open towards higher pressure and temperature. Eclogite is mainly composed of omphacite (Na–
Al pyroxene) and pyrope-rich garnet (Mg–Al garnet).
Plagioclase (Na–Ca feldspar), one of the major components of oceanic crust, is no longer stable under these
conditions. Denser mineral phases generally form under high pressure and thus high-pressure metamorphic
rocks have higher specific weights (eclogite: about 3.5 g/
cm^3) compared to their non-metamorphic equivalents
(about 3.0 g/cm^3).

Oceanic crust is dewatered during subduction. Basaltic rocks contain abundant water because interaction with seawater near the mid-ocean ridge forms water-containing minerals (zeolite, chlorite, epidote, amphibole; ▶ Chap. 5). Under conditions of subduction
metamorphism, the water-containing minerals become
unstable and are replaced. Water set free by metamorphism of these minerals is partly bound in other water-containing but high-pressure resistant minerals (e.
g., lawsonite, zoisite, glaucophane); however, the major

Fig. 7.25 Diagrammatic cross sections through subduction zones. Upper: computer-modeled temperature distribution, in °C, of a subduction system and, lower: cross section through an active continental margin showing an estimation of the isograds (lines of the same temperature) (Schubert et al. 1975)

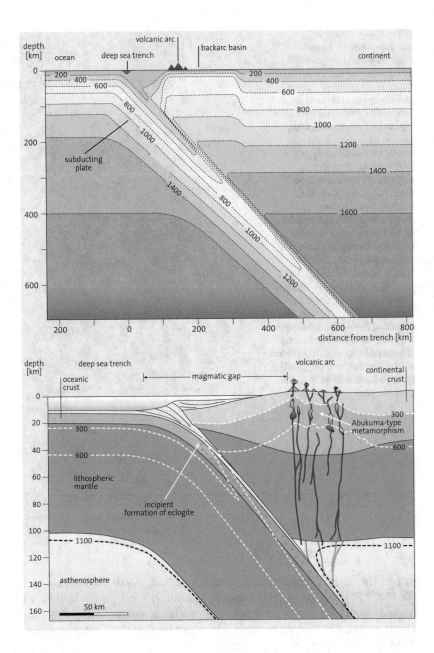

portion is emitted into the mantle above the subduction zone. Subducting rocks are thus gradually dewatered and that is the reason why eclogite, to a large extent, is a "dry" rock.

In addition to basalt, dolerite and gabbro, sedimentary rocks and splinters of continental crust carried with the subducted material experience high-pressure metamorphism. However, the mineral composition in these rocks is not easy to identify as having formed under high-pressure conditions. Many groups of minerals (garnets, amphiboles, micas, chloritoids, etc.) have special members formed under high-pressure conditions which are chemically different from their relatives formed under different pressure–temperature conditions. Only a detailed microchemical analysis of the minerals using an electron-microprobe can detect

whether they were formed under high-pressure or normal conditions of metamorphism. Under conditions of high-pressure metamorphism, iron–magnesium silicates mostly have compositions similar to those of the Mg end-members because the small ion radius of magnesium supports denser packing of atoms.

Sediments contribute significantly to the dewatering in a subduction zone although following high-pressure metamorphism they continue to include water-containing minerals such as phengite (a type of mica), karpholite, chloritoid and lawsonite. Water released in the subduction zone causes intense changes in the mantle above and, from a certain depth, melting processes which are responsible for magmatism above a subduction zone.

It is difficult to explain how rocks that have been metamorphosed at great depths in subduction zones are exhumed at the surface. Such high-pressure metamorphic rocks occur in mountain ranges in lenses or in larger bodies within ophiolite sequences and their adjacent rocks. Typically they occur within long narrow zones that contain ophiolites and mark the suture zone of collisions between two continental masses.

Commonly, high-pressure mineral associations are only preserved as relics. This is because high-pressure minerals adapt to decreasing pressure and temperature conditions that accompany slow isostatic ascent. Only rapid uplift accompanied by relatively low temperatures or rapid cooling maintains high-pressure minerals in rocks. Investigations of textures in eclogites suggest that regions of extreme crustal extension above high-pressure metamorphic rocks generate the conditions in which high-pressure minerals can be preserved; such conditions permit rocks from deep-seated subduction zones to rapidly reach the surface (Platt 1986). Processes associated with erosion by water and ice occur much too slowly to remove the overlying rocks rapidly enough.

The geologic setting of the Western Alps clearly illustrates this last point. The mineral assemblage with pyrope and coesite there (see box) suggests that 100 km of rock had to have been removed above the ancient subduction zone rocks. Yet regional relations show that only 20–25 km of rock were removed by erosion (England 1981). What happened to the other 80 km of rock? Large detachment faults have been mapped in the Dora-Maira massif. These faults document considerable extension in the region. In fact, mapping shows that a 50 km-thick package of rock and another 10 km-thick package of rock are missing from the top of the coesite-bearing metamorphic rocks. These rocks must have been removed by tectonic, not erosional processes. The missing rocks were removed by rapid horizon-

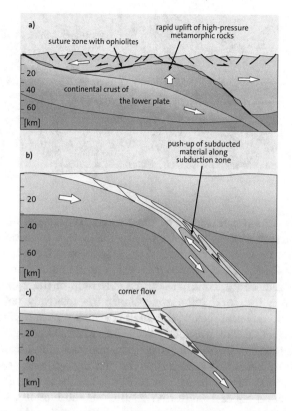

Fig. 7.26 Conceptual models of rapid uplift of high-pressure metamorphic rocks by **a** lateral migration of crustal wedges, **b** push-up along a subduction zone, **c** corner flow in a subduction zone

tal displacement during orogenic collapse of the orogen (Fig. 7.26a). Such features were first described from the North American Basin and Range Province in structures called metamorphic core complexes or metamorphic domes (▶ Chap. 3). Here, deeply buried rocks are rapidly brought to the surface by tectonic processes.

Ultrahigh-Pressure Metamorphic Rocks

Eclogites and glaucophane-schists are important indicators of extinct subduction zones that are exhumed at the surface by orogenic processes (▶ Chap. 11). In some mountain ranges, the rocks were buried to a depth of more than 100 km and pressures of 3–4 GPa. Coesite and diamond, which have been detected at various locations mostly as very small crystals less than 0.1 mm in diameter, are indicators of such deep burial. Metamorphism within the range of stability of coesite and diamond is also called ultrahigh-pressure metamorphism (Fig. 7.24).

Coësite is a high-pressure modification of quartz that occurs at pressures greater than 2.5 GPa (at 500 °C), which corresponds to a depth of 80 km. Before its discovery in metamorphic rocks of the Dora-Maira massif in the Italian Western Alps (Chopin 1984), coesite was

only known naturally in deposits related to meteorite impacts such as Nordlinger Ries, Germany. The coesite of the Western Alps was preserved as an armored relic in pure pyrope crystals up to 10 cm across. The pyrope shield prevented the conversion of the denser coesite into the less dense quartz because it prevented the related increase of volume. High-pressure garnets with radial tension fractures around quartz inclusions document the conversion of coesite (density 2.93 g/cm^3) into quartz (2.65 g/cm^3); the volume increase during the conversion forced the garnet to break. Commonly small relics of coesite are preserved adjacent to the quartz (Fig. 7.24, insert).

Diamonds form from carbon in organic material that was preserved in sediments that were later subjected to high pressures. At approximately 300 °C, organic material is transformed into graphite (density ca. 2.2 g/cm³) and subsequently, at a pressure of about 3.5 GPa (corresponding to a depth of nearly 110 km), into diamond (3.5 g/cm³). Both, graphite and diamond are composed of carbon, but the atomic structure of the latter is much more compact.

Diamonds have yet to be discovered in the Dora-Maira massif. Apparently, the maximum pressure of 3.5 GPa, the stability field of diamond, was not attained. Both diamond- and coesite-bearing metamorphic rocks are present in ancient subduction zones in the Caledonian Mountains of Norway, the Dabie Shan of Eastern China, the Kokchetav massif in Kazakhstan, the Erzgebirge in Germany, and other mountain ranges. Because metamorphic diamonds have diameters in the range of micrometers, they are difficult to detect.

Crust that is thickened by underplating during the subduction process is gravitationally unstable, a condition that initiates crustal extension by an orogenic collapse (▶ Chap. 11). If subduction, which produces compression at depth, continues during extension in the upper plate, the subduction zone transports relatively cold rock material below the high-pressure metamorphic rocks that have previously been sheared off. This condition uplifts the metamorphic zone and prevents it from being heated. It is also thought that high-pressure sequences can be pressed in reverse upwards along subduction zones even during continuous subduction and compression. This mechanism has been proposed for high-pressure units in the Swiss Alps (◘ Fig. 7.26b; Schmid et al. 1996).

Corner flow is another mechanism that can rapidly uplift high-pressure metamorphic rocks. Rocks of the blueschist facies within the accretionary wedge are forced upwards because the wedge consists of scraped-off sediments that becomes narrower at depth. The subducted and metamorphosed rocks from the lower part of the wedge are returned towards the surface along the frontal part of the upper plate (◘ Fig. 7.26c).

High-pressure minerals are destroyed, especially if uplift of the rocks occurs slowly or the water content increases during the uplift process (e. g. thrusting interleaved with water-bearing sedimentary rocks). Water may also be released by the collapse of water-containing minerals. Water is an important reagent for mineral reactions. If the amount of available water is low, adaptation to the changed pressure and temperature conditions is only partial and eclogite may still be present as a relict mineral assemblage.

An example of re-metamorphosed eclogites occurs in the area around Zermatt, Switzerland (◘ Fig. 7.27; Bearth 1959). Pillow lavas, former ocean-floor remnants from the zone of Zermatt–Saas Fee, were subjected to high-pressure eclogite metamorphism, yet the pillows remained amazingly well preserved. This indicates that despite high lithostatic pressure, directed stress was rather low during subduction of the rocks. Lithostatic pressure alone does not cause deformation of rocks; deformation only occurs when directed stress is applied. When the metamorphosed sequence bearing the pillow lavas was uplifted, water appeared; initially while the rocks were under less extreme pressure and slightly decreasing temperatures, glaucophane was formed. The water did not penetrate the inner protected core of the pillows where the eclogitic mineral assemblage was preserved. However, the edges of the pillows were changed into blueschist. At shallower depth, the blue glaucophane of the spandrels between the pillows was changed into green amphibole; blueschist was converted into greenschist. Therefore, these petrographically complex rocks contain mineral assemblages from three different metamorphic pressure–temperature fields that document decreasing pressure and uplift of the rock assemblage (◘ Fig. 7.24).

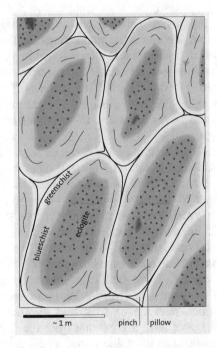

◘ Fig. 7.27   Schematic sketch of pillow lavas near Zermatt, Switzerland. The mineral composition was partly adjusted to lower pressure conditions during uplift from the subduction zone. Therefore, the central parts in the core of the pillows, which are protected against circulating fluids, remain as eclogite, the edges as glaucophane-schist, and the pinches where fluids could corrode easily, as greenschist

Rapid Burial, Rapid Uplift

Subduction rates calculated today mostly range between 3 and 9 cm/yr indicating that rocks in subduction zones rapidly reach great depths. Depending on the inclination angle of the subduction zone (most are between 30 and 60°), the depth of burial of a given mass of rock increases 15–80 km per million years and reaches a depth of 100 km in 1 to 7 million years. Temperatures rise in subduction zones at 5–10 °C/km, compared to normal geothermal gradients in continental crust of 30 °C/km.

The eclogites of the Dora-Maira massif in the Western Alps were subducted to approximately 100 km depth during the Early Tertiary. It is uncertain how long this process lasted, but with slow subduction rates of 2 cm/yr as is assumed for the Alpine orogenesis, this process required less than 10 m. y. According to age determinations on the critical minerals, exhumation from 100 to 20 km depth was in the range of 5–8 m. y. (Gebauer et al. 1997); this corresponds to an average exhumation rate of 10–16 km/m. y. During the same time interval, temperatures decreased from approximately 800 to 300 °C, a rate of 60–100 °C/m. y. The high rates of exhumation occurred because of extensional processes in the overlying rock pile, but the rapid cooling rate is unclear. The low heat conductivity of rocks suggests that rapid uplift of rocks should occur without an accompanying large loss of heat.

Pressure and temperature conditions are documented in the mineral parageneses. Para-geneses are assemblages of minerals in equilibrium within a rock that formed during defined pressure, temperature and fluid conditions. Conditions of parageneses are well known from experiments and calculations on the thermodynamic equilibrium. Because mineral suites are confined to specific conditions during burial or uplift of a rock, the pressure–temperature path (P–T path) can be reconstructed when the older parageneses have not completely been destroyed. If some minerals in a paragenetic sequence are suited to radiometric age dating, it is possible to extend the P–T path with the age datings to construct a pressure–temperature-time path (P–T–t path). However, P–T–t calculations are marked with certain errors.

Because of its low density, continental crust cannot be subducted and incorporated in significant amounts into the upper mantle. However, coësite-and diamond-bearing rocks indicate that they can be dragged rather deep into a subduction zone. Eventually, they will move upwards even though they increase in density during their high-pressure modification; their buoyancy is less than that of surrounding mantle material. It is unclear if crustal material can move deeper into subduction zones more than 200 km, perhaps to eventually disappear into the Earth's mantle. If this process occurs, the amount transferred is certainly minor.

7.10 Subduction-Related Magmatism—A Paradox?

At first glace, it would seem paradoxical that subduction zones, some of the coldest tectonic zones at given depths produce prolific amounts of magmatism. How can locations where cold lithosphere is rapidly subducted, in turn, cause magma to rise and feed intense plutonism and volcanism? About a quarter of the magmatic rocks on Earth, nearly 9 km³/yr, are formed above subduction zones. This compared to 21 km³/yr, more than 60% of all magmatism, at mid-ocean ridges (Schmincke 2004). The rising melts in the area of the magmatic zone cause a high geothermal gradient with heat flow values of 100 mW per m² or more (■ Fig. 7.3).

Volcanism commences abruptly at the volcanic or magmatic front. Magmas develop above subduction zones when the downgoing slab plunges into the asthenosphere and reaches a depth of 80,100 km (■ Fig. 7.28). Melting above the subduction zone is a complicated process but can be described in a simple way: Sediments and ocean crust, along with serpentinites within it, release water upwards. The water-bearing fluids rise into the hot asthenosphere wedge between the lower and upper plate and cause melting. Melting is initiated because the solidus temperature, slightly greater than 1000 °C for "wet" mantle, is exceeded at this location below the volcanic belt. When the solidus temperature is exceeded, the melting process begins—rocks are solid below this temperature. Without the fluids, "dry" conditions prevail and the solidus would be at approximately 1400 °C. Water depresses the melting point of rocks.

The critical role of water in the evolution of subduction-related magmatic rocks is mirrored in their water content, generally comprising several percent by weight (up to 6% H_2O). For comparison, basalts at mid-ocean ridges have typical water values below 0.4% and basalts at hot spots below 1.0%.

The most violent volcanism occurs directly behind the magmatic front where the large amount of water derived from the subduction zone enters into the asthenospheric wedge. With increasing depth of the subduction zone, the magma sources become deeper (■ Fig. 7.29). As distance increases from the magmatic front towards the backarc, less water is released from the subduction zone and magmatic activity

7

□ **Fig. 7.28** Processes of melting below volcanic arcs (Stern 2000). Dewatering from the subduction zone causes formation of serpentine in the lithopheric mantle of the upper plate and melts in the hot asthenospheric wedge at greater depths

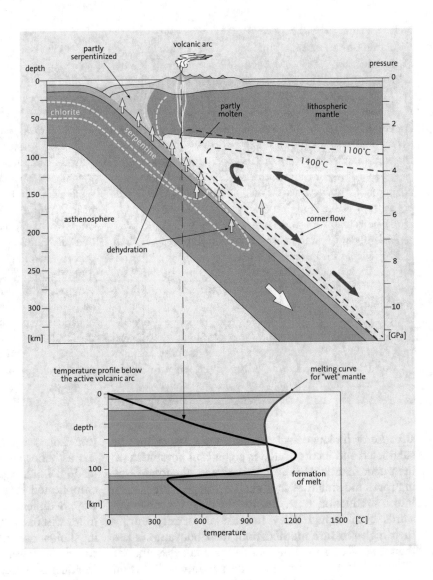

□ **Fig. 7.29** Cross section of the Japanese Islands showing the change in melting depth (Schmincke 2004). The lower diagram shows the volume of magma produced at various zones across the subduction zone

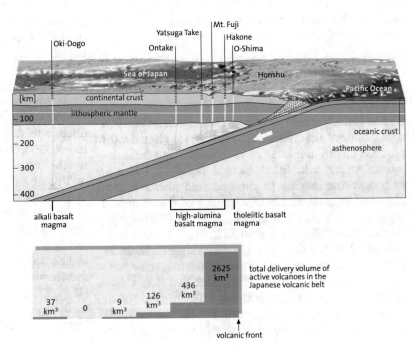

decreases. Most magmatism fades when the subduction zone reaches a depth of 200–250 km. In SW Japan, volcanoes most distant from the volcanic front are located in the backarc basin area, where the subduction zone reaches a depth of more than 300 km. The downward motion of the subduction zone induces a downward current in the asthenospheric wedge above. As a result, hot asthenosphere is drawn in from behind. In this manner, the asthenospheric wedge is continuously supplied with new, hot mantle material. The result is a corner flow between the two plates (Fig. 7.28).

Water released from the subduction zone is already generated before the top of the subduction zone reaches the asthenosphere, and thus, coincides with the zone where the subducting plate glides along the upper plate. A conversion occurs within the lithospheric mantle of the upper plate above the subduction zone in which large amounts of water-containing minerals like amphibole and serpentine are formed (Fig. 7.28). The serpentinization weakens the mantle in the upper plate causing the plate boundary to deform and be drawn downwards along the subduction zone. Amphibole and serpentine become unstable at increasing depth and temperature and are transformed back to water-free pyroxene and olivine. The additional water released by the mineral transformation rises and enriches the water content of the subduction zone and thus enhances to the melting process. Merging magma blobs eventually become large enough to form diapirs that rise against the downward-directed mantle currents.

Subduction zones have extreme contrasts in crustal temperature in very close proximity. Consequently, metamorphic processes contrast across fairly tight zones. For example, rising magmas generate dramatically differnt types of metamorphism within the upper plate as compared to processes in the subducted plate immediately below.

7.11 Rocks of the Magmatic Zone

Melted rock material rising in the magmatic belt above the subduction zone forms huge intrusive bodies, called *batholiths* (*Greek* deep bodies of rocks); melts that reach the surface form volcanoes. Volcanic rock compositions include andesite, volumetrically most important and named after the Andes, basalt, dacite and rhyolite (names of rocks are presented on the inner jacket at the back of the book). Collectively, these rocks are referred to as calc-alkaline rocks because of their high concentration of calcium as well as the importance of alkali elements. Calc-alkaline rocks have distinctive chemical characteristics and typically occur above subduction zones. Tholeiitic magmatic rocks can occur in minor amounts near the magmatic front and certain alkaline rocks can be important in the backarc basin area. Recall that most tholeiites are produced at mid-ocean ridges and above large hot spots whereas alkaline rocks occur dominantly in intra-plate domains, especially hot spots and continental graben structures.

Much of the magma produced above a subduction zone does not reach the Earth's surface but rather gets trapped in the crust below the volcanoes as plutons (intrusive bodies). Diorite, the intrusive equivalent of andesite is the most important plutonic rock. Other important rock types include, tonalite, granodiorite and granite (the equivalents of dacite and rhyolite); gabbro, the equivalent of basalt, is rare.

Andesites are almost exclusively the product of subduction zones and thus are strong indicators of previous subduction activity in old mountain ranges. Presently, 339 of 342 (99%) active andesitic volcanoes are located above subduction zones. Generally, the chemical composition of magmas produced above subduction zones reflects their complex mechanism of formation. Their composition includes elements from the subducted plate that are transported by the released aqueous fluids and give a characteristic imprint to the magmatic rock. The source of most of the melted substance is the process of partial melting of as-thenospheric material in the mantle wedge above the subduction zone. Additional control of magma composition comes from the crust of the upper plate through which the magmas penetrate. This crustal influence varies depending on what crustal type is present within the upper plate. Continental crust and thick island arc crust especially influence magmatic composition. Contribution from the different sources is reflected by a number of characteristic elements.

Melts generated in the asthenospheric wedge above the subduction zone show the composition of basalts or basaltic andesites. They are enriched in mobile elements that are easily transported by aqueous fluids and include potassium, rubidium, strontium, barium, thorium and uranium, all of them *lithophile* (*Greek* "liking rocks", tending to concentrate in continental crust). These elements, which are concurrently mantle-incompatible because of their large ionic radii, are mainly derived from subducted metamorphosed sediments, according to their concentration pattern (Stern 2000).

Other elements with high ion potential, a high relation of charge to ionic radius, are difficult to mobilize and remain in the metamorphic rocks of the subduction zone. They also preferentially remain in the residual mantle peridotite during partial melting of the asthenospheric wedge; only limited amounts transfer into the basaltic melt. Therefore, subduction-generated melts are poor in these "immobile" elements such as the heavy rare earth elements, niobium, tantalum, hafnium, zirconium, yttrium, phosphorus or titanium.

7

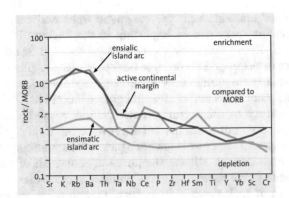

Fig. 7.30 Spider-diagrams for calc-alkaline basalts produced above subduction zones (Pearce 1983). The x-axis shows elements with decreasing mantle incompatibility: the initial five elements are strongly incompatible or lithophile, the last two are compatible, and those in between are the partially compatible elements. Concentration of every element in the rock is divided by an average value for MORB (mid-ocean ridge basalt-MORB-normalized values). The scale of the y-axis is logarithmic

Compared to mid-ocean ridge basalt (MORB) the subduction-generated melts are partly depleted in these elements, for instance in niobium and tantalum. The depletion in Nb and Ta can be attributed to early crystallization and stay back of magnetite which incorporates these elements in its crystal lattice.

From the different behavior of lithophile and immobile elements subduction-related magmatites can be detected. Compared to MORB, the lithophile elements are strongly enriched in subduction-related magmatites, the immobile elements show similar or lower concentrations as in MORB. To visualize the distribution pattern in the rocks the concentrations of the elements are normalized against MORB. They are calculated in relation to an average composition of MORB and the enrichment or depletion is depicted in the so-called spider diagrams (□ Fig. 7.30). This kind of diagram offers a clear view because any number of elements can be used and the normalization allows a direct comparison. Trace elements are generally useful in characterizing magmatic rocks because their strictly defined concentrations commonly indicate specific conditions of formation. The most important character of the subduction-related magmatic rocks are their high concentrations in potassium and rubidium relative to MORB and their negative tantalum and niobium anomalies (□ Fig. 7.30).

The evolution of subduction magmas is not yet fully understood in detail but agreement exists that the following processes are crucial:

— Some elements originate from the subducted metamorphosed sediments and basalts. The subduction component is considered to be derived mostly from the sediments, in particular water, the cosmogenic beryllium-10 (see box below), and lithophile elements.

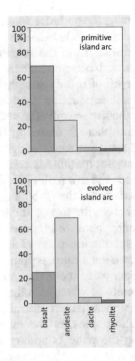

Fig. 7.31 Frequency diagram showing different volcanic compositions in primitive and evolved intra-oceanic, ensimatic island arcs

— Primarily, the melts are formed within the asthenospheric wedge between the subduction zone and the upper plate. They rise through the crust because of their lower density.

— If the crust of the upper plate in the area of the magmatic arc is oceanic, the Marianas, for example, the magmas are not altered substantially and remain mainly basaltic to andesitic in composition (□ Fig. 7.31). Island arc tholeiites and boninites (andesites rich in Mg, named after the Bonin Islands north of the Marianas) are formed.

— Magmas that penetrate continental crust are altered by two factors: assimilation of crustal material and gravitational differentiation, both of which produce compositions enriched in SiO_2 and K_2O. Calc-alkaline suites are generated that include aluminum-rich basalt (see below), andesite, dacite and rhyolite; andesite is clearly more common than basalt.

Acidic rocks (dacites to rhyolites) are most abundant where the continental crust is thick and subduction is long lasting; the composition of the magmas evolves through time. Mature island arcs and active continental margins with thick continental crust display an increase in acidic rocks (rich in SiO_2) compared to more primitive island arcs that are built on oceanic crust or thin continental crust. For example, volcanic rocks of the Marianas and those of the Izu and Bonin Islands to the north are located on oceanic crust; they have an average of 53–55% SiO_2 only 2–3% greater than levels in normal oceanic crust. In contrast, volcanic rocks of the Andes are clearly more acidic, ca. 63% SiO_2 a figure that approaches the average composition of the upper continental crust of the Andes, ca. 66% SiO_2. Globally, continental crust has an average value of 57% SiO_2. Therefore, the lower crust must have a basic composition.

Newly formed melts periodically rise upwards. During their ascent, the composition of the magma is changed through differentiation. Minerals with high melting points crystallize and are removed from the magma through fractionated crystallization and gravitational differentiation. Crustal rocks can become melted and assimilated in the magma; or a chemical exchange occurs with crustal rocks and "contaminates" the magma. Through these processes, the composition of the melt increasingly differs from that of the primary melt. The result is that subduction-related magmas within continental crust are more enriched with lithophile elements (◻ Fig. 7.30). Minerals with more basic composition are removed from the melt and remain in the lower continental crust, thus accounting for its basic composition compared to the more acidic composition of the upper crust. On average, the upper crust is 1.5 times thicker than the lower crust. The above processes explain why magmas generated by subduction processes are much more complex and variable at locations where continental crust is involved (e. g., Andes, Japan) than at the locations of intra-oceanic island arcs like the Marianas.

If the subducting oceanic crust is very young and hot, the subducted basaltic crust can partially melt in the presence of water. A distinct rock type is formed from these magmas, known as adakite after the Adak Island in the Aleutians. These rocks are rare and currently form in only several places. Besides the Aleutians, where young oceanic crust is currently being subducted, adadakites are also present in Costa Rica. Here, the Cocos Ridge, a young oceanic ridge formed at the Galápagos hot spot, is being subducted beneath the Central American isthmus at a flattened angle of approximately 30°.

7.12 Zonation of Magmas in Space and Time

Although magmatic belts above subduction zones are dominated by calc-alkaline rocks, tholeiitic and alkaline magmas also occur. These three suites of magmas display a zonation in space and time. The zonation in space is related to magmatic variation that occurs in parallel belts of increasing depth across the subduction zone. At the frontal part of the volcanic belt, tholeiitic magmas occur and are generally expressed as basaltic rocks. The broadest and central belt comprises calc-alkaline magmas that include the complete calc-alkaline suite of rocks from basalt to rhyolite, but are dominated by andesite. Calc-alkaline rocks are commonly modified by crustal assimilation and differentiation. Potassium-rich calc-alkaline rocks increase in abundance towards the backarc region and eventually pass into shoshonites that are a special type of alkaline rocks. The volume of rock produced clearly decreases into the backarc basin (◻ Fig. 7.29). The changes in character of the magmas (tholeiitic, calc-alkaline, alkaline) are controlled by the high production of partial melt beneath the magmatic front and the decreasing magma generation and decreasing extraction of elements from the subducted rocks with increasing depth of the subduction zone.

A similar trend can be observed in the potassium content that consistently increases across the arc. Island arc tholeiites, the outer belt, are depleted in K (mostly <0.7 wt% K_2O), calc-alkaline basalts have K_2O contents of about 1% (acidic differentiation products are correspondingly richer in K since this element is enriched during differentiation), and shoshonites, named after the Shoshone River in Wyoming, have K_2O contents of 2% and more. The increase in potassium with increasing distance from the magmatic front is most likely a result of the gradual extraction of the element from the metamorphic sediments in the subduction zone where K-containing minerals, especially mica, dehydrate and yield K to aqueous fluids.

Because the amount of partial melt in the asthenospheric wedge decreases towards the backarc region, potassium experiences relative enrichment in the melts although less K is extracted from the subducted material. Moreover, the SiO_2 content of melts generally decreases towards the backarc region and thereby emphasizes the alkaline character of the rocks, because less SiO_2 is available to saturate the potassium in potassium feldspar ($KAlSi_3O_8$). Therefore SiO_2-undersaturated minerals like leucite ($KAlSi_2O_6$) are formed, typical of shoshonitic rocks. Shoshonites are indicative of the rear side of subduction-related magmatic zones. In Japan they are found in the backarc basin, the Sea of Japan (◻ Fig. 7.29). The chemistry of alkaline rocks of hot spots and continental rifts differs from that of shoshonites as expressed by the dominance of sodium over potassium.

Calc-alkaline basalts are "high-alumina basalts" that are characterized by their enrichment in aluminum, 16–20 wt% Al_2O_3 compared to 12–16% in tholeiite. Probably the aluminum is derived from the subducted sediments. It is transported into the asthenospheric wedge, where melting occurs, along with the fluids and other elements enriched in calc-alkaline magmas. Clay-rich sediments, common on ocean floors and in deep-sea trenches, are rich in Al_2O_3.

A change from tholeiitic to calc-alkaline character of the magmatic rocks can also develop over time as young, intra-oceanic island arcs evolve to mature island arcs. Young island arcs such as the Marianas, with underlying oceanic crust, only produce basalts and basaltic andesites. The basalts have a tholeiitic composition and the andesites are boninites. The composition of the boninites indicates that they are not derived from the asthenospheric wedge above the subduction zone but rather derived from partially melted harzburgites of the lithospheric mantle of the upper plate; normally, harzburgites do not reach sufficiently high temperatures to melt (Wilson 1989). Through time, the intra-oceanic island arc evolves into a typical calc-alkaline magmatic complex, dominated by andesitic rocks, above the subduction zone (◘ Fig. 7.31). The primary melts are basalts, with an SiO_2 content ca. 48–53 wt%, and basaltic andesites with 53–57% SiO_2; both tend to differentiate towards more acidic rocks. During the evolutionary process of becoming an evolved island arc, the 10- to 20-km thick mostly basaltic crust of the primitive island arc is thickened to greater than 20 km as it chemically evolves towards the andesitic chemical average composition of continental crust.

In cases of island arcs built on continental basement or arcs on active continental margins, the rising basaltic to basaltic-andesitic magmas have to penetrate the older continental crust. However, the lower density continental crust forms a physical barrier for the magma rising from the mantle. Especially at Chile-type continental margins, the rising magma is retarded over long time periods by the thick crust. The magmas in the crustal chambers differentiate at depth and evolve towards more acidic compositions. Through this process, substantial amounts of andesitic (57–63% SiO_2), dacitic (63–68% SiO_2) and rhyolitic (68–75% SiO_2) magmas evolve. The melts accumulate into granodioritic batholiths that get stuck in the crust or rise to the surface to form mostly explosive volcanoes.

Subduction magmatism is the most important process in the formation of continental crust. Partial melting of mantle peridotite produces basalts but no acidic rocks. Only the multiple recycling of eroded island arc material in subduction zones and repeated melting gradually lead to the evolution of acidic magmatic rocks. Repeated volcanic erosion and transport of rock particles and minerals into sedimentary basins enriches the content of quartz and clay. If these minerals are supplied to a subduction zone, magmatic rocks rich in silica and aluminum can develop above the subduction zone. If this cycle is repeated numerous times, the amount of acidic rocks in the continental crust gradually increases. Because the crust in the magmatic arcs does not move horizontally relative to the magma centers as it does at mid-ocean ridges or at hot spots, substantial crustal accretion is the result. Not surprisingly, the bulk composition of continental crust is similar to andesite, the most important rock type of the volcanic belts above subduction zones.

7.13 Explosive Stratovolcanoes as Indicators for Subduction Magmatism

Basaltic magmas are low in viscosity and contain relatively few volatiles. Therefore, basaltic volcanoes are characterized by smooth-flowing, low-viscosity lava and enjoy the reputation of being non-explosive. Acidic magmas are more viscous and may contain large amounts of volatiles, especially water. High-level magma chambers can build up enormous gas pressures that may lead to spectacular explosive eruptions. Intermediate to acidic volcanoes above subduction zones have high-risk explosive potential.

Typically, explosive volcanoes above subduction zones are andesitic to dacitic in composition and form stratovolcanoes, so named because the stratification comprises alternating layers of lava flows and ash layers. The volcanoes have steep slopes, typically inclined to 40°, and are the prototype of the "common volcano". Examples occur around the Pacific: Mount Pinatubo (Philippines), Mount Fuji (Japan), Mount Saint Helens (USA), Popocatépetl (Mexico; ◘ Fig. 7.33), Chimborazo (Ecuador). The explosive activity produces volcanic ashes, ejected loose material solidifying to volcanic tuffs, whereas calmer eruptions produce lavas. Lavas are defined as extruding magma that has been degassed during the ascent because pressure has been released.

The eruption of Mount Saint Helens in northwestern USA in 1980 was magnificently documented. The volcano is a product of the subducting Juan de Fuca Plate beneath the western coast of North America. Surface measurements, seismic and other geophysical investigations provided detailed information about the location of the high-level magma chamber and thus, it was possible to predict an eruption in principle but not exactly as to when it would occur. Much preparation was made to record the eruption and to avoid loss of life.

Isotopic Signatures and the Influence of Continental Crust

The influence of continental crust can be detected by isotopic analyses. Prolonged mixing in the mantle has caused strontium isotopes to have a fairly consistent isotopic ratio. The $^{87}Sr/^{86}Sr$ ratio for basalts of the upper mantle at midocean ridges is presently ca. 0.703 and therefore reflects the isotopic composition of the mantle source (■ Fig. 7.32). In the geological past this ratio was lower because ^{87}Sr forms by the radioactive decay of ^{87}Rb. Rubidium, an element that is incompatible in the mantle, primarily enters the acidic rocks of the continental crust. Cosequently, the production of ^{87}Sr from ^{87}Rb in the continental crust is high. Throughout Earth history, this has caused acidic crustal rocks to increase their $^{87}Sr/^{86}Sr$ ratio with respect to the mantle.

Magmas directly derived from the mantle have a concurrent Sr isotopic ratio, ca. 0.703 for MORB and its asthenospheric source. However, magmas formed by melting of older continental crust have high $^{87}Sr/^{86}Sr$ ratios with values of 0.710 and higher. But crustal contamination of mantle-derived material creates intermediate isotopic ratios. Island arc magmas typically have $^{87}Sr/^{86}Sr$ isotopic ratios between 0.704 and 0.710 (■ Fig. 7.32) because the amount of crustal Sr is generally low. However, it is not possible to discern whether the crustal component was derived from the subducted sediment or the continental basement of the volcanic zone.

In similar fashion, neodymium isotopes can be used to separate mantle and crustal origin. However, ^{147}Sm that decays to ^{143}Nd is enriched faster in the mantle than in continental crust because samarium is a mantle-compatible element that remains in mantle rocks when partial melts are removed. The mantle origin is thus revealed by higher $^{143}Nd/^{144}Nd$ ratios. ^{143}Nd $^{87}Sr/^{86}Sr$ Island arc magmatites generally have $^{143}Nd/^{144}Nd$ ratios of 0.5124–0.5130 (■ Fig. 7.32).

Based on isotopic ratios and other chemical and mineralogical characteristics, granitic rocks have been subdivided into I-type and S-type granites (granite is used sensu lato or "granitoids": granites, granodiorites, tonalites). I-type granites ("I" stands for "igneous"), are typically formed at island arcs and active continental margins and generate large batholiths in the magmatic zone. They have $^{87}Sr/^{86}Sr$ ratios typical of subduction-related magmas, mostly 0.706–0.710, and reflect a restricted crustal component. They evolved from magmas primarily generated from the mantle. Batholiths with I-type granitoids are exposed over thousands of square kilometers in the Sierra Nevada (USA) and the Andes. The batholiths formed the roots of

■ **Fig. 7.32** $^{87}Sr/^{86}Sr$ and $^{143}Nd/^{144}Nd$ relations for volcanics from island arcs as compared to basalts of midocean ridges (MORB). 0 Earth: average values for bulk Earth composition

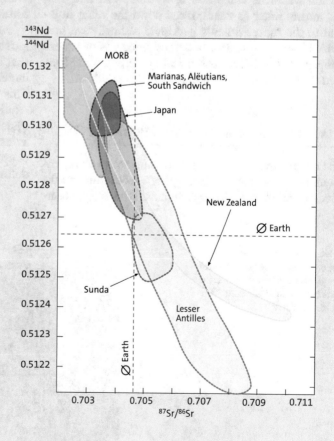

subduction-related volcanic belts, the deep crustal material later exposed by erosion. They testify to former, extensive subduction activity as the large Farallon Plate was subducted under North and South America.

S-type granites have a higher $^{87}Sr/^{86}Sr$ ratio (>0.710) and indicate crustal melting processes, especially from metamorphosed sedimentary rocks. "S" stands for "sedimentary" origin. Principally, they evolve during orogenic processes related to the collision of continental crust and not as subduction-related magmas.

Some radioactive isotopes that have relatively short half-lives can be used to identify the contribution of subducted sediments in subduction-related magmatic rocks. The isotope beryllium-10 (^{10}Be) is formed through cosmogenic radiation in the atmosphere, flushed by rain, and then bound into marine sediments. If these sediments are subducted, they release the beryllium to the subduc-tion-related melts that soon erupt to form volcanic rocks. Because this process develops over a time period of only a few million years, some ^{10}Be is not radioactively decayed (the half-life of ^{10}Be is 2.6 m. y.) when the magmas reach the surface.

Volcanic rocks that contain ^{10}Be indicate that young sediments were subducted and that they partly fed the melts. The youngest sediments are always in the uppermost layer of the subducted sedimentary pile and they would be the first to be scraped-off during accretion. Therefore, the occurrence of ^{10}Be in the volcanic rocks demonstrates that the youngest sediments were subducted rather than being scraped-off into an accretionary wedge. If large-scale accretion occurs, the young sediments are added to the accretionary wedge and no detectable ^{10}Be will be present in the volcanic rocks above the subduction zone. In older sediments, most ^{10}Be is decayed and is no longer detectable.

The following events characterized the eruption of Mt. Saint Helens: (1) Rising magma caused the 3000 m-high volcano to bulge at its top; (2) the peak region started to glide, related pressure release inside the mountain generated a violent explosion that asymmetrically blasted away a huge part of the top and side of the volcano; (3) an avalanche composed of volcanic ash and gasses, mostly water as steam, raced down the valley adjacent to the ruptured flank, generated a shockwave, and leveled trees over an area of several hundreds of square kilometers; (4) the explosion and ejected material lowered the elevation of the mountain by about 400 m.

7.14 Metamorphism in the Magmatic Belt

The magmatic zones of island arcs and active continental margins are characterized by high geothermal gradients generated by the rising magmas. Whereas the tem-perature increases downwards in the subduction zone and in the forearc area by less than 10 °C per kilometer, the geothermal gradient in the upper crust at the magmatic front jumps to 35–50 °C/km (■ Fig. 7.25). Consequently, rocks of the crust below the volcanic belt experience a completely different metamorphism than rocks in the accretionary wedge or in the subduction zone.

Metamorphism in the magmatic zone is characterized as low-pressure/high-temperature; this means that at a given depth, the temperature is substantially increased compared to subduction metamorphism. This temperature-dominated metamorphism is named Abukuma-type metamorphism after the Abukuma Plateau in Japan (■ Figs. 7.24 and 7.34) and is characterized by characteristic mineral assemblages. Because of the high temperatures, the rocks are easily deformed.

■ **Fig. 7.33** Aerial view of the Popocatépetl in Mexico, a typical stratovolcano above a subduction zone (photograph taken in 1987 by M. Meschede)

Paired Metamorphic Belts

As first identified in Japan, pairs of elongate, parallel belts with significantly different metamorphic rocks are indicative of former subduction activity. One metamorphic belt consists of rocks formed by high-pressure metamorphism and the other consists of rocks of the Abukuma facies high-temperature metamorphic rocks. These two belts of metamorphic rocks, which were formed under extremely different pressure/ temperature relations, occur in direct association and are called "paired metamorphic belts".

Orogenic processes involve uplift that raises portions of the metamorphic rocks towards the surface and into the zone of erosion. From the spatial arrangement of the two contrasting belts at the surface it is possible to detect not only former subduction activity but also the polarity of the subduction zone (⬛ Fig. 7.34). The high-pressure belt (subduction metamorphism) marks the area of the subduction zone and the high-temperature belt (Abukuma-type metamorphism) marks the location of the magmatic arc. In Japan the paired metamorphic belts indicate subduction beneath the Asian continent began in the Permian. Since the Permian, the subduction zone has shifted twice towards the present Pacific Ocean. The presently formed metamorphic rocks are still hidden at depth.

⬛ **Fig. 7.34** Paired metamorphic belts in Japan (Miyashiro 1973). The different ages of the belts indicate that subduction lasted for more than 250 million years, each with subduction directed towards N or NW. The subduction zone shifted stepwise outward (southeastward) and is today offshore the Japanese coast in the Pacific

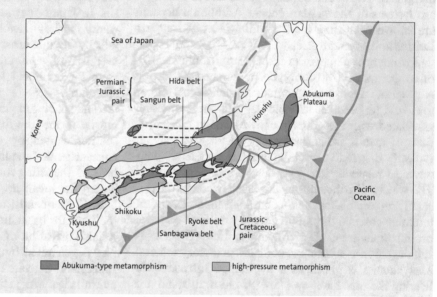

■ Abukuma-type metamorphism   ■ high-pressure metamorphism

Due to the high geothermal gradient, the temperature in the magmatic zone increases to generate the highest grades of metamorphism, 650–900 °C, present at crustal depths. Rocks may experience partial melting (anatexis). Sedimentary rocks begin to melt at a temperature of approximately 650 °C, if water is present (⬛ Fig. 7.24), and produce migmatites, a mixed rock, formed from granitic melts that contain highly metamorphosed but not melted residuals. Dry magmas and sedimentary rocks, which have already expelled their water, are not melted under these conditions; instead granulites develop. In general, the formation of granulites occurs at the base of the continental crust.

Abukuma-type metamorphism occurs below 7–10 km depth and reflects regional metamorphism. In contrast, intrusions that rise into shallow depths of only a few kilometers create a zone of contact along their margins, the metamorphic aureole. During this contact metamorphism (⬛ Fig. 7.24), heat of the intrusive body reacts with the cold country rock and mobilizes aqueous fluids that form numerous mineral changes. This process is mainly visible in sedimentary rocks where clay-rich sediments are metamorphosed to knotted schists or hornfels and carbonatic rocks are metamorphosed to skarns. Knotted schists are slaty rocks in which newly grown metamorphic minerals form either specks or nodules. Hornfels is a product of higher temperatures in which the rock is recrystallized into a fine-grained mix of minerals that creates the horny appearance. In carbonatic rock the supply of SiO_2 from the pluton produces calcsilicate minerals. Skarn is a Swedish word for miners of calcsilicate rock. Contact and Abukuma metamorphism are transitional in magmatic arcs.

7.15   Ore Deposits in the Magmatic Belt

A number of important ore deposits, especially in the circum-Pacific region, occur in the magmatic belt above the subduction zone. Mostly these are porphyry-type copper and molybdenum deposits that are related to high-level (subvolcanic) intrusions of granodiorites (I-type granites; see box above; ▣ Fig. 7.35). The name refers to the porphyric structure (large single crystals in a fine-grained matrix) of the granodiorites. The ore deposits are also called "disseminated ores"—the copper or molybdenum ore particles are distributed throughout the rock and not concentrated in veins. Large porphyry-type deposits occur at the active continental margins of North and South America and in the island arc systems of the western Pacific. Additional deposits are known from the Alpine mountain belt of southern Europe and southern Asia.

Porphyry-type deposits include approximately 70% of copper and 100% of molybdenum mined worldwide. A given ore deposit may contain 10 million to 1 billion tons of ore; commercial gold deposits are commonly associated with the primary ores. Although concentrations of metals are normally low, 0.3–1% Cu, they are typically distributed over large areas and lend themselves to relatively low-cost, open-pit mining methods. The widely distributed ores are locally concentrated in small fractures where acid-containing fluids leach and intensely corrode the country rock. The most common ore minerals are chalcopyrite and bornite (two Cu–Fe sulfides); pyrite is the most common accessory mineral. Molybdenite is mined as a byproduct. Intrusives delivering the ores have low $^{87}Sr/^{86}Sr$ ratios at island arcs (about 0.704) and higher ratios at the active continental margin of North America (up to 0.709).

Kuroko-type ore deposits (▣ Fig. 7.35) comprise another class of ore deposits. They are present in island arcs; the type locality is in Japan. Kuroko-type deposits belong to the group of VHMS (volcano hosted massive sulphide) ores and are characterized by polymetallic sulfidic ores with pyrite, chalcopyrite, sphalerite,

and galena, as well as gold. These ores are related to rhyolitic volcanos and their feeder dikes. Ore deposits can be associated with a subvolcanic network of veins in rhyolites (yielding copper, zinc, and gold) or occur as massive copper, zinc, and lead ore bodies that precipitate in sea water near volcanic centers.

7.16   The Backarc Basin

The island arc is continuous with the backarc basin located behind the magmatic arc (▣ Fig. 7.36). Japan, separated from the Asian continent by the Sea of Japan, is a good example with Japan constituting the arc and the Sea of Japan constituting the back-arc basin. In Japan, magmatism decreases towards the basin where shoshonitic volcanoes occur. The basin is underlain by oceanic crust that was formed when the Japanese Islands, containing continental crust, drifted away from the continent through some process of backarc spreading.

The island-arc garlands in the western Pacific were mostly formed by the same process. The effect of the suction caused by the roll-back of the subduction zones extended the lithosphere of the upper plate. The heat of the rising magma weakened the crust and initiated the break up of the area behind the magmatic belt. Consequently, new oceanic lithosphere formed in the backarc basin from the rising asthenospheric material. Formation of oceanic crust in this tectonic setting is generated by partial melting of rising asthenosphere and is related to processes that occur at a mid-ocean ridge. Magnetic stripes, however, are only partly evolved, probably because of irregular and closly spaced mantle currents that contribute to the formation of this crust. The Sea of Japan is sometimes referred to as a diffuse backarc basin because the new crust is formed over a broad area rather than focused on a distinct spreading ridge.

Active extension and seafloor spreading presently occurs in several backarc basins including the Marianas, the Tonga Islands (▣ Fig. 7.36), and the South Sandwich Islands (South Atlantic). In the backarc area of the Tonga Islands, the spreading velocity reaches 16 cm/yr and thus is comparable with that of the equatorial Pacific. In contrast, the backarc area of the Marianas has an extension rate of only 4 cm/yr. The processes and characteristics of fast and slow spreading mid-ocean ridge systems are comparable to spreading axes in backarc basins: smooth topography at fast spreading ridges and rough topography with a central graben at slow spreading systems. Even smokers can be observed in backarc basins. The asthenosphere of the mantle wedge is somewhat contaminated by fluids that contain elements derived from the subduction zone. Therefore, basalts of the backarc basin

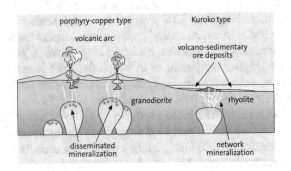

▣ **Fig. 7.35** Schematic diagram of ore deposits of the porphyry-copper type and Kuroko type. Both are related to acidic magmatites above subduction zones

Fig. 7.36 Distribution of backarc basins in the western Pacific (Karig 1971). In addition to the active basins that generate new oceanic crust, inactive basins exist, which according to their cooling stage, have differing amounts of heat flow

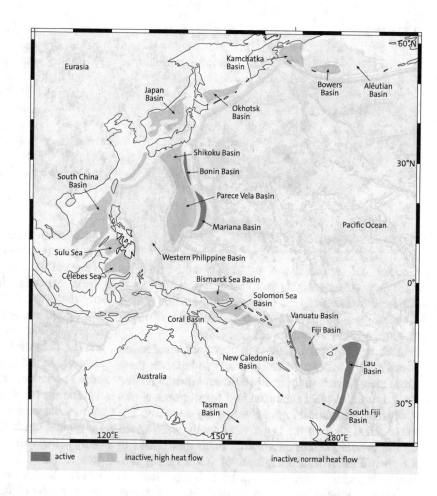

crust have, to some degree, the characteristics of subduction-related magmas. In most ways, they are very similar to MORB.

At active continental margins like the Andes, large-scale compression is transferred to the upper plate and consequently, the backarc area is characterized by compressional fold and thrust belts. Thrusts are directed towards the backside foreland, away from the volcanic belt (■ Fig. 7.5). The weight of the stacked wedges of the thrust belt depresses the adjacent crust and generates a retro-arc foreland basin.

Sediments deposited in backarc basins consist of arc-derived volcaniclastics and volcanic ashes, sand and mud eroded from the continent, and marine carbonates. The thickness of foreland basin sediments is highly variable and depends upon rate of basin subsidence and availability of eroded material. Both backarc and forearc basins contain large amounts of hydrocarbons.

Splitting of Intra-Oceanic Island Arcs

The intra-oceanic, ensimatic island arc of the Tonga Islands was split during the late Cenozoic (■ Fig. 7.37). The Tonga Islands represent the present active island arc and to the west, the Lau Basin represents a backarc basin (■ Fig. 7.36, high heat flow) and the Lau Ridge represents an inactive volcanic arc. The Lau Ridge is volcanically inactive because it is no longer located above the magmatic zone. The Lau Basin is composed of newly formed oceanic crust with backarc basin characteristics along with dispersed remnants of the inactive island arc. It has a very thin sedimentary cover due to its young age.

The Lau Ridge had an interesting Late Cenozoic history (Hawkins 1974). It began as the active island arc when the Pacific Plate subducted beneath it. However, slab roll-back (see above) caused the subduction zone to migrate eastward. The rapid eastward migration caused the arc to split and a backarc basin formed between the two pieces (■ Fig. 7.37). The split occurred because the arc consisted of hot, weak crust that was easily pulled apart. The Tonga Islands on the eastern margin continued to reside above active subduction and are presently the active island arc. The Lau Ridge, now removed from above the magmatic source, is inactive; the backarc Lau Basin separates the two although several fragments were imbedded in it as the older arc became dismembered.

The entirely oceanic Philippine Sea Plate contains an inactive, NW–SE trending spreading axis in its western and central part (◻ Fig. 7.36, area with normal heat flow). Spreading there was active in the Middle Cenozoic. Slab roll-back of the intra-oceanic Mariana subduction zone and backarc extension formed backarc basins along the eastern margin of the Philippine Sea Plate. The present backarc basin forms a narrow zone at the eastern margin of the plate, the Mariana Basin. The basin is characterized by young and hot ocean crust that evolved in similar fashion to the Lau Basin. The Parece Vela Basin to the west is an older backarc basin that became inactivated when the subduction zone migrated eastward and the Mariana Basin formed; it is presently cooling down.

7.17 Gravity and Heat Flow

Geophysical measurements of gravity and heat flow yield interesting insights into processes occurring at convergent plate margins. Gravity and heat flow values across a plate boundary indicate typical pairs of negative and positive anomalies (◻ Fig. 7.3).

Gravity in ocean basins in front of subduction zones and in backarc basins is nearly in balance; these areas are approximately in isostatic equilibrium related to the sublithospheric mantle. The forearc area shows a negative and adjacent positive gravity anomaly. The mass deficit causing the negative anomaly is explained by the bending of the subducting plate that results in a trench filled with a thick layer of water that is substantially lighter than rock material. Water-saturated sediments in the deep sea trench and in the accretionary wedge are also less dense and add to this anomaly. The positive anomaly results from the dense subducting plate plus the upper plate where it attains its full thickness including its dense lithospheric mantle (◻ Fig. 7.3). The positive anomaly is most characteristic at the magmatic front and decreases to normal gravity values towards the backarc basin. In the magmatic arc, the anomalously lighter asthenospheric wedge above the subduction zone partly compensates the positive gravity anomaly and contributes to the emergence of the volcanic belt.

As expected, heat flow abruptly changes at the magmatic front (◻ Fig. 7.3). Its value of ca. 50 mW per square meter (mW/m^2) in the open ocean is only slightly below the global average. This value significantly decreases in the forearc area and increases abruptly in the magmatic arc to 80–150 mW/m^2. These heat-flow values result from the downward transport of material in the forearc area and the rising of hot magmas in the magmatic arc. This situation also results in the paired meta-morphic belts. With decreasing volcanic activity the heat flow in the backarc area also decreases. However, in backarc basins with active spreading, the heat flow is significantly higher and can reach values approaching those of the magmatic arc.

7.18 Subduction and Collision

This chapter has presented the consequences of subduction activity. Only oceanic lithosphere is able to generate permanent subduction to great depths. When a crustal block with continental material that is con-

◻ **Fig. 7.37** Relations of the Lau Basin, Lau Ridge, and Tonga island arc above the Tonga subduction zone. The Lau Basin is an active backarc basin with a young spreading axis. Spreading in the backarc basin has split the original volcanic arc into active and inactive components

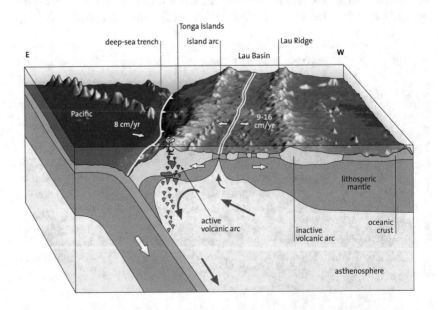

nected to oceanic crust reaches a subduction zone, it collides with the island arc or the active continental margin. During collision the continental block is partially subducted beneath the edge of the upper plate and then pushed or pulled also into deeper parts of the subduction zone. However, it is not transported to great depths because of its low density. The colliding continental margins are strongly deformed, sheared and dismembered into tectonic units (nappes) and are subjected to metamorphism characterized by an intermediate pressure gradient (Barrovian-type metamorphism; ◪ Fig. 7.24).

The eastern Sunda Arc provides an example of where collision has recently begun (◪ Fig. 7.9). The Australian continent and its marginal Sahul shelf are part of the Indo-Australian Plate that is drifting against the Sunda Arc. The collision has occurred adjacent to Timor and eastward. Underplating by the Sahul shelf has already caused the uplift of the outer ridge, which comprises Timor and the islands directly east of it. A trench exists but it is rather shallow, slightly more than 3000 m depth, because the underplated continental crust of the Sahul shelf is pushing upward. This condition documents the beginning orogenesis!

Transform Faults

Contents

© The Author(s), under exclusive license to Springer Nature Switzerland AG 2022
W. Frisch et al., *Plate Tectonics*,
Springer Textbooks in Earth Sciences, Geography and Environment,
https://doi.org/10.1007/978-3-030-88999-9_8

Transform faults represent one of the three types of plate boundaries. A peculiar aspect of their nature is that they are abruptly transformed into another kind of plate boundary at their termination (Wilson 1965). Plates glide along the fault and move past each other without destruction of or creation of new crust. Although crust is neither created or destroyed, the transform margin is commonly marked by topographic features like scarps, trenches or ridges.

Transform faults occur as several different geometries; they can connect two segments of growing plate boundaries (R–R transform fault), one growing and one subducting plate boundary (R–T transform fault) or two subducting plate boundaries (T–T transform fault); R stands for mid-ocean ridge, T for deep sea trench (subduction zone). R–R transform faults represent the most common type and are common along all mid-ocean ridges (■ Fig. 1.5).

The length of R–R transform faults remains constant; however, in contrast the length of R–T and T–T transform faults, with one exception, either grows or shrinks (■ Fig. 8.1). Transform faults that connect subducting plate boundaries typically cut through areas of continental crust. Examples include some of the most notorious transform faults, the San Andreas Fault in California (■ Fig. 8.7), the North Anatolian Fault in Turkey (■ Fig. 8.10), the Jordan Fault in the Middle East (■ Fig. 8.6), and the Alpine Fault on the southern island of New Zealand (■ Fig. 8.11).

8.1 Oceanic Transform Faults

Mid-ocean ridges are intercepted by R–R transform faults that normally are perpendicular to the ridge. Thus the ridges are subdivided into segments that

8

■ **Fig. 8.1** Sketch maps showing possible configurations of transform faults (**b**–**g**) compared to the geometry of a common strike-slip fault (**a**). All examples shown have right-lateral displacements. The movement velocity (v) is the relative movement between the two plates. Each respective lower image indicates the modification to fault geometry over time. R: mid-ocean ridge, T: deep sea trench (subduction zone)

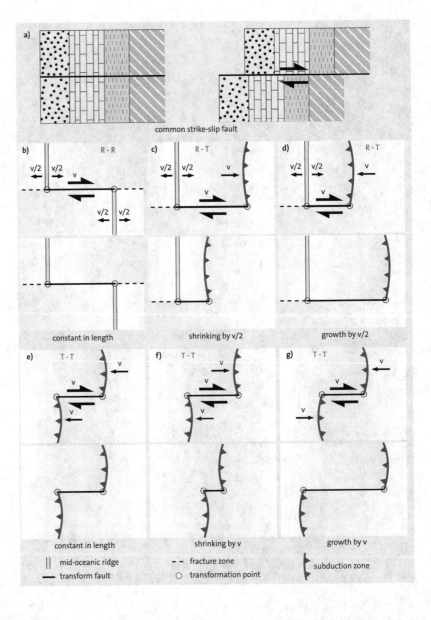

mostly have lengths of several tens of kilometers. At the East Pacific Rise west of Mexico, large transform faults occur at distances of several hundred kilometers and smaller ones occur at distances of 10 or 20 km. In each case the fault connects two segments that appear to be dislocated. However, the ridge segments maintain their distance from each other and are not dislocated by the fault; in fact, the fault forms a connecting link of constant length (◻ Fig. 8.1b). Although every transform fault is a strike-slip fault, not all large strike-slip faults represent a plate boundary (◻ Fig. 8.1a).

Because of the drift of the newly formed oceanic crust away from ridge segments, a relative movement along the faults is induced that corresponds to the spreading velocity on both sides of the ridge. The sense of displacement is contrary to the apparent displacement of the ridge segments (◻ Fig. 8.1b). In the example shown, the transform fault is a right-lateral strike-slip fault; if an observer straddles the fault, the right-hand side of the fault moves towards the observer, regardless of which way is faced. Transform faults end abruptly in a point, the transformation point, where the strike-slip movement is transformed into a diverging or converging movement. This property gives this fault its name. In the example of the R–R transform fault, the movement at both ends of the fault is transformed into the diverging plate movement of the mid-ocean ridge. Beyond the transformation point, the crust on both sides of the imaginary prolongation of the fault moves in the same direction with the same velocity. Therefore, no strike-slip movement occurs in the prolongation of the fault beyond the transformation point. The dashed lines in ◻ Fig. 8.1b mark this fissure where younger crust is welded against older crust beyond the transformation point.

The best confirmation for the mechanism of oceanic transform faults is the earthquake occurrence and the fault-plane solutions of the quakes (▶ Chap. 2).

Only the segments of the active transform fault between the ridge segments indicate continuous seismic activity; seismicity ends abruptly at the transformation points, the termination points of the faults. The fault-plane solutions always indicate horizontal displacement parallel to the fault with a sense of displacement which is consistent with the prediction of the theory.

8.2 Fracture Zones in the Ocean Floor

In spite of its abrupt end, the prolongation of a transform fault is topographically expressed on the ocean floor. This expression is caused by the different ages of crust across the prolongation. This age difference creates an isostatic imbalance so the crust lies at different depths. Across this prolongation, there is no plate boundary. With respective increasing age and increasing distance to the ridge, the ocean floor subsides (◻ Fig. 4.11). Younger crust is at the same time subsiding faster than older crust. Because the prolongation of the active part of the transform fault separates crustal segments of different ages, relative vertical movements occur along this fissure (◻ Fig. 8.2). Therefore, these fissures are planes of movement, mostly vertical in nature, and are called fracture zones. Beyond a transformation point at a subducting plate boundary, a fracture zone cannot continue because plate segments are not welded together at such a location but rather are split; crust on either side is of the same age (◻ Fig. 8.1c–g).

The fracture zones on the ocean floor are clearly visible in submarine digital relief maps. They represent zones of weakness within the oceanic plates and are thus easily reactivated. Prominent examples can be observed in the Eastern Pacific where several significant fracture zones occur. They are spaced at distances of approximately 1000 km and can be traced from the margin of the North American Plate westward into the

◻ **Fig. 8.2** Geometry of transform faults at a mid-ocean ridge. Although the strike-slip movement ends abruptly at the transformation point, a vertical movement occurs in its prolongation ("fracture zone") because the adjacent crustal segments of different ages subside with different rates

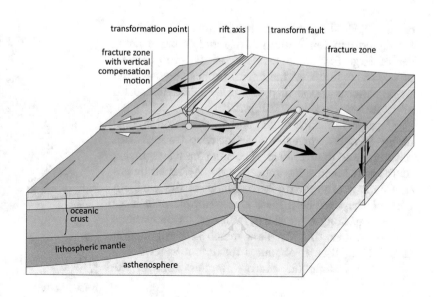

8

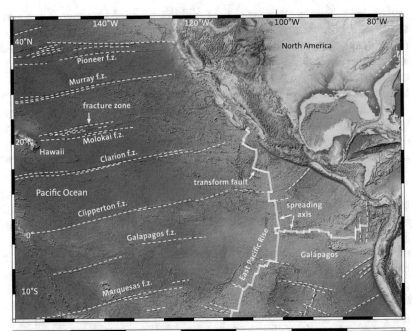

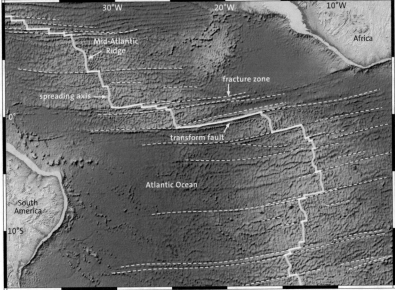

■ Fig. 8.3 Map showing major fracture zones and transform faults in the East Pacific and Central Atlantic

ocean (■ Fig. 8.3). Also in the Atlantic, especially in the region between Western Africa and Brazil, a number of large fracture zones occur as the prolongation of transform faults. At the fracture zones, the magnetic stripe patterns appear displaced in the same dimension as the mid-ocean ridges (■ Fig. 2.12).

Oceanic transform faults and fracture zones commonly form escarpments, ridges or deep troughs with relief greater than 2000 m. The intervening segments of the mid-ocean ridge appear as dome-shaped bulges (▶ Chap. 5). The Puerto-Rico Trough at the northern edge of the Caribbean Plate is an example of a large depression along a transform fault and its dimensions approach those of a deep-sea trench at a subduction zone (■ Fig. 8.4). It obtains a maximum water depth of 9219 m. Along the Murray fracture zone in the

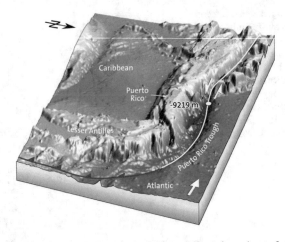

■ Fig. 8.4 The transform fault at the northern boundary of the Caribbean Plate forms the Puerto Rico Trough at more than 9000 m below sea level

northeastern Pacific (■ Fig. 8.3), a trough with a water depth of more than 6000 m is developed; in contrast, the abyssal plain is 1000 m higher. Also along this fracture zone are elongate ridges that protrude more than 1000 m above the ocean floor.

The occurrence of ridges and troughs along oceanic transform faults is a result of tensional and compressional processes. Tension or compression occurs along the fault when the plate motion direction slightly changes. Commonly, the narrow, elongate stretched plate segments between the transform faults and fracture zones may be locally rotated further complicating the fault geometry. Tension and compression may alternate at the same location or appear spatially laterally along a fault zone. Tension and compression generate distinct products on the ocean floor; tensional forces generate deep cracks in which sedimentary deposits form and occasionally produce intraplate volcanism whereas compressional forces generate reverse faults or folds, particularly within soft sedimentary deposits of the troughs, and ridges result. These movements along the fracture zones are capable of producing sporadic, small earthquakes.

The vertical and lateral movements along oceanic transform faults and fracture zones cause the exposure of various kinds of oceanic rocks; these rocks are commonly strongly deformed and metamorphosed. Sheared fragments derived from the deeper crust, including deformed and metamorphosed gabbros, have been found in various locations. At the Vema fracture zone in the Atlantic, the entire profile of the oceanic crust is exposed by the vertical movements (■ Fig. 5.12). Serpentinites converted from peridotites in the uppermost mantle, can ascend along the zones of weakness to the surface.

8.3 Continental Transform Faults

Continental transform faults are considerably more complex than their oceanic counterparts. Their geometry and complexity reflect the easy deform-ability of continental crust, its variable composition, and the presence of pre-existing structures developed during long periods of crustal evolution. Continental transform faults are generally accompanied by numerous other faults of variable size and type so that broad fault zones or fault systems develop. These fault zones range to several 100 km in width and comprise fragmented rhombic or lenticular blocks of variable dimensions. Some of the blocks are also moved in the vertical direction because of local compression or tension, even though the plate movement and main displacement occurs in the horizontal direction. Some continental transform faults are characterized by 1000 km or more of movement over millions of years. The San Andreas Fault system and its predecessors of California and western Mexico illustrate these points: the right-lateral system is approximately 2500 km long, 30 million years old, several hundred kms wide, has offset segments of continental crust more than 500 km, and consists of blocks that have experienced uplift or subsidence of 5–10 km. The southern half of the system has evolved into the Gulf of California, a transtensional, very recent (ca. 5 Ma) ocean basin.

Earthquake activity at continental transform faults is much stronger than that at oceanic transform faults because of the large thickness and complex structure of continental crust and the length of the faults, up to or more than 1000 km. Along with subduction zone faults, these fault zones produce the most devastating earthquakes. The 1906 San Francisco earthquake yielded displacements of more than 5 m along the San Andreas Fault (see ■ Fig. 8.9) and was one of the strongest earthquakes of the twentieth century. The North Anatolian Fault of Turkey also produced a dreaded sequence of earthquakes. The last devastating earthquake was that of Izmit in 1999, the strongest earthquake since 1939 along their fault (see ■ Fig. 8.10).

The complex structure of continental crust transform faults typically alters the geometry normally defined by the small circle to the pole of rotation (▶ Chap. 2), so consequently, the course of the fault consistently deviates from the theoretical track line. Thus zones of tension and compression develop in the crustal blocks gliding along each other. The complex and variable geometry of continental transform faults is illustrated in ■ Fig. 8.5. Transtension occurs where the strike-slip motion is also under tension, and transpression occurs where the motion is also under compression. Graben-like depressions and oblique

■ **Fig. 8.5** Block diagram showing geometry of a transform fault system with pull-apart basin, oblique pull-apart basin, transpressional structures (folding, reverse faulting) and fan-out of the fault. The insert indicates the principal directions of compression and tension in relation to the main fault

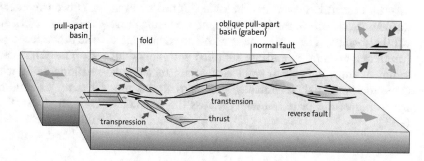

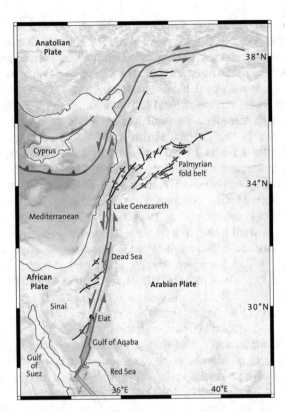

⬛ Fig. 8.6 Map of the Jordan structure, a large transform fault system that connects the spreading system of the Red Sea with the subduction zone of Cyprus. Pull-apart basins like the Dead Sea and Lake Genezareth line up along the transform fault. Transpression is responsible for the formation of the Palmyrian fold belt. The relative movement between African and Arabian plateYs is about 1 cm/yr

The Dead Sea, a modern pull-apart basin, exists in the Jordan Rift (⬛ Fig. 8.6). The system initiates in the Gulf of Aqaba, a graben-like transform fault with a left-lateral total displacement of 110 km that extends northward into southeastern Turkey. At several locations the main fault is offset at lefthanded stepovers that continue as a parallel fault. A left-lateral displacement with left-hand stepovers generates a pull-apart basin (⬛ Figs. 8.5 and 8.6). The Dead Sea has a basement that has subsided several thousand meters. Sedimentary input by rivers, however, is low due to the desert environment. Therefore, the 6 km-thick sequence of sedimentary material contains thick layers of evaporitic sediments like halite and gypsum. Much of this formed in response to repeated saltwater invasions from the Mediterranean that subsequently deposited evaporite sediment during desiccation. The present low precipitation and high evaporation rates in the Dead Sea mean that subsidence rates exceed sedimentation rates to form the deepest depression on Earth, nearly 400 m below sea level. North of Lake Genezareth, which is another pull-apart basin 200 m below the sea level, the transform fault exhibits a restraining bend with transpression. This bend is partly responsible for the compression that produces the Palmyrian fold belt.

8.4 San Andreas—The Dreaded Transform Fault of California

The 1100 km-long San Andreas Fault in California is probably the most studied fault on Earth. The San Andreas Fault, which forms the plate boundary between the Pacific and North American plates, has a displacement rate of 6 cm/yr (⬛ Fig. 8.7). The details of this fault system illustrate the complexity of transform faults in continental areas. The present plate boundary with its intense earthquake activity is one of the most dangerous faults in the world. The San Andreas Fault is part of a complicated, broad fault system accompanied by numerous other faults of various sizes and types. All of the parallel faults, some currently active, some apparently extinct, are right-lateral (dextral) faults. Although a total displacement of 1500 km can be documented on the system since its inception 25–30 million years ago, only 300 km of this displacement has occurred at the present San-Andreas Fault. The remaining 1200 km of displacement has been taken up by the other accompanying faults.

The Garlock Fault (⬛ Fig. 8.7), a WSW–ENE striking fault with left-lateral (sinistral) displacement, has an orientation perpendicular to the San Andreas system. Dextral and sinistral faults form a conjugate fault system because they belong to the same stress field and were alternately activated during the same pe-

tensional basins, termed pull-apart basins, are generated under transtension; surrounding uplands supply the rapidly subsiding basins with sediment. Transpression generates reverse faulting and folding in the adjacent crustal blocks. The normally rapid uplift results in high rates of erosion. The Ridge Basin along the San Andreas Fault is a well studied pull-apart basin and the mountains of the Transverse Ranges including the San Gabriel Mountains and San Bernadino Mountains are examples of transpressional uplift.

Pull-apart basins can also form where stepover geometry occurs along transform faults. Here two overlapping segments of the fault system generate a basin as wide as the distance between the segments and as long as the zone of fault overlap (⬛ Fig. 8.5). The continental crustal basement can become thinned or even dismembered, and in the latter case, new oceanic crust will be formed. The basins have high subsidence rates (up to several mm/yr) and are supplied with abundant sediment from the graben walls along with mixed volcanic rock material. Technically, pull-apart basins represent a growing plate boundary across the transform fault.

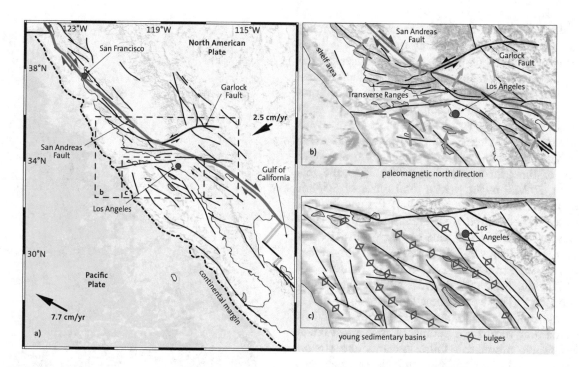

Fig. 8.7 The San Andreas Fault system in California (Christie-Blick and Biddle 1985) (**a**). Numerous faults approximately parallel the dextral San-Andreas Fault and have the same sense of movement; the oblique Garlock Fault moves in the opposite direction (sinistral). Green arrows (**b**) indicate the original north orientation of individual blocks of the Transverse Range. Many of them have rotated clockwise, some as much as 120°. Along the fault, bulges were caused by compression and basins were formed by extension (**c**)

riod. Because the Garlock Fault is nearly perpendicular to the San Andreas Fault and causes deviation in the course of the latter, a strong transpressive component develops along the plate boundary and the accompanying parallel faults of the San Andreas system.

The complexity of the San Andreas system is clearly documented by the interlocking fault pattern that has produced numerous small crustal blocks that interact with each other. The dextral sense of movement at the plate boundary has rotated some these blocks in ball-bearing fashion (**▣** Fig. 8.7b). The E–W-trending Transverse Ranges provide the best example. Their odd orientation produces the large E–W offset obvious along the Southern California coast. The up to 120° of rotation over the past 15 m. y. has been determined with the aid of paleomagnetic investigations. The magnetic polarity and direction prevailing at the time of formation is recorded in magmatic rocks during their cooling and in sedimentary rocks during their deposition and diagenesis. If a later rotation of a crustal block occurs the "frozen" magnetic direction does not coincide with the former position of the magnetic pole (magnetic and geographic poles coincide in average). From the difference of the pole directions, the rotation angle of the crustal block can be determined (**▣** Fig. 8.7b).

The geometry of the Transverse Ranges reflects the individual movements of the small blocks and their re-

sponse to space problems that originate because of their rotation. Collision, transform motion, or separation occurs along their boundaries and, in response, the rocks are deformed or compensated by compressional structures (folds, reverse and thrust faults) or tensional structures (tensional basins, normal faults). In the vicinity of Los Angeles, a number of compressional structures and sedimentary basins exist, each responding to local stress through time (**▣** Fig. 8.7c). Local relief can be extreme: NE of Los Angeles uplifted blocks tower 3000 m above adjacent basins slightly above sea level; several land-locked basins are actually below sea level. The style and intensity of deformation of the blocks bordered by the faults are highly dependent on their rock composition. Young sedimentary sequences are intensely deformed only where they are compressed between competent basement like granite and high-grade metamorphic rocks. Blocks composed of primarily low-grade metamorphic, especially schistose, rocks have been strongly compressed and folded throughout. Because of such differences, individual ranges of the Transverse Ranges can contrast strongly in their geology and morphology.

The San Andreas Fault and many accompanying faults are characterized by continuous earthquake activity; clearly the system is presently active. The oldest geologic units exposed at the surface show systematically increasing amounts of displacement. Displacement along transform faults is determined by piercing points, any geologic structure or unit that was once adjacent or continuous but is presently offset by the fault movement. Piercing points can include volcanoes, older faults offset by younger faults, folds, distinctive rock bodies, and geomorphic elements such as ridges or stream valleys. The latter two are useful in documenting youngest fault movement and rates. Older piercing points yield an average slip over longer periods of time; for example, an offset Oligocene volcano (ca. 30 Ma) might yield a total offset of 300 km since the formation of the volcano. If the volcano can be shown to be younger than the fault, then the average amount of offset per time can be determined, in this case, 10 km/1 m.y. A Middle Miocene lava flow (ca. 15 Ma) on the same fault segment might yield an offset of 200 km. Note that this rate is somewhat faster, 13.3 km/1 m.y. This would show that later fault movement was more rapid than earlier movement. When measured over very recent periods of time, most faults show episodic movement. Offset streams and stream terraces across the San Andreas Fault clearly document this as do the thousands of stations set up across the fault to measure fault activity. Surprisingly, total offset of the San Andreas Fault is difficult to measure. This is partly due to an absence of precise piercing points, but also due to the fact that fault motion on older faults, some as old as Jurassic, has caused previous offset to the present fault, some of this offset in a sinistral, rather than a dextral sense.

The arrangement of the earthquake epicenters indicate that the San Andreas Fault consists of a more continuous vertical fault plane at depth that fans-out into several shorter fault branches towards the top. Such a branching out towards the top is characteristic of many strike-slip fault systems. This pattern is called a "flower structure" because it resembles a bunch of flowers that fan upwards (◘ Fig. 8.8). In case of transpression, crustal wedges fan upwards resulting in a positive flower structure. In case of transtension, the wedges subside like a graben and a negative flower structure develops.

The earthquake epicenters are concentrated in the brittle upper crust that reacts by forming fractures with jerky, episodic movement. In 1978, 7500 earthquakes were registered in southern California alone. Some segments display a somewhat continuous creeping accompanied by very small but frequent earthquakes; this especially occurs where the fault is enclosed by sepentinites of the Franciscan Formation. Other segments, where movement is locked by competent rocks such as granite, gneiss, and quartzite, large stresses accumulate that generate strong earthquakes when the limiting value of rock strength is exceeded. Prehistoric earthquakes on the San Andreas Fault were identified by measuring displacement of Recent sediments that

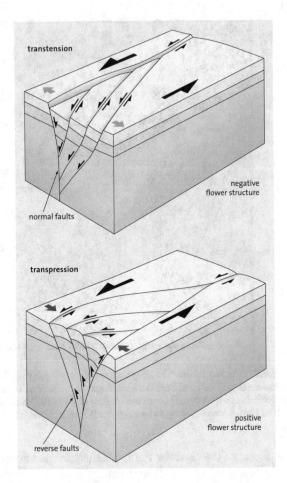

◘ **Fig. 8.8** Block diagrams showing negative and positive flower structures typical of large fault zones that are caused by transtension and transpression, respectively

have been dated by radiocarbon methods. In Southern California, during the last 1500 years, eight strong earthquakes have been identified occurring at intervals between 60 and 275 years, most of them with an estimated moment magnitude of 8 or more (Sieh 1978). This documents the periodicity of large-displacement events.

The San Andreas Fault has various topographic expressions. Erosion of crushed and highly fractured rocks along the fault zone creates long stretches of trenches and valleys. In places where competent rocks occur on either side of the fault, topographic expression is locally suppressed. In places where competent rocks abut young, poorly cemented sandstone and mudstone or older incompetent rocks, a sharp scarp is developed, locally with 3000 m of relief. Rivers running perpendicular across the fault are commonly offset a given distance by recent fault movement. In similar fashion, topographic ridges are offset by the fault. Some of the ridges are a result of transpression that led to the uplift of young, easily deformable sedimentary layers. There are places along the San Andreas Fault,

◻ Fig. 8.9 Dextral displacement of a fence along a discrete fracture formed during the 1906 earthquake in San Francisco (Photograph by M. Meschede)

where sedimentary rocks as young as Pliocene stand on end! The strong San Francisco earthquake of 1906 created displacements of several meters along discrete fracture lines (◻ Fig. 8.9).

Some basins along the San Andreas Fault yield large petroleum deposits. For example, the Ventura Basin 120 km NW of Los Angeles formed along the fault where the Transverse Range blocks were rotated. The thick Miocene Monterey Formation and related units originated as mostly deep marine deposits formed in waters rich in organics and later folded into complex anticlines that trapped the petroleum. The high organic content resulted from nutrient-rich upwelling of cold water along the Pacific Ocean. A high geothermal gradient (up to 55 °C/km) caused the formation of petroleum at a shallow depth.

8.5 The North Anatolian Fault in Asia Minor and the Alpine Fault in New Zealand

The 1200 km-long North Anatolian Fault forms the plate boundary between the Anatolian Plate and the Eurasian Plate (◻ Fig. 8.10). With a calculated dextral displacement of 2.5 cm/yr, it displays a clear northward convex bending which is best explained by the proximal location of the common pole of rotation of the two plates located near the north coast of Sinai (Stein et al. 1997). From 1939 to 1944 a series of four large earthquakes (moment magnitudes between 7.1 and 7.8) occurred along a 600 km-long fault segment; they migrated from east to west. Three more earthquakes (magnitudes between 7.0 and 7.4) occurred between 1957 and 1999, continued the pattern of westward migration, and moved an additional 300 km.

Time and location of the tremors in this earthquake series give insights to understanding the pattern and mechanisms of earthquakes. At each individual fault segment where an earthquake occurs, a spontaneous stress release is generated. Because the block motion occurs over a limited distance, new stress accumulates near the end of the active part of the fault. At this location of new stress, the next earthquake will form and so on. A large fault cannot have motion along its total length during any one event because the friction resistance is too strong. Therefore, the rupture always occurs in sections; for example, during the earthquake of Izmit in 1999, movement occurred over a length of 130 km. The progress of fault activity in stages is thus a normal process and another indication that movement in the upper brittle crust generally happens in an episodic, jerky way. Stages of quiescence at a given location between two large earthquakes may last for several decades or centuries. Along the North Anatolian Fault, evidence near Gerede (◻ Fig. 8.10) documented that eight large earthquakes occurred during the last 2000 years. This indicates a periodicity of 200–300 years. Near Erzincan, five large earthquakes during the last 1000 years yield a recurrence time of 200–250 years (Okomura et al. 1993).

The SW–NE striking sinistral East Anatolian Fault and the E–W striking dextral North Anatolian Fault form a conjugated system. Anatolia moves along these two faults towards the west relative to the neighboring plates. This westward movement is provoked by the Arabian Plate pushing from the south (◻ Fig. 8.10). The Anatolian Plate is pressed towards the west in similar fashion to one squeezing a stone from a plum between two fingers. Such a process is called "continental escape" or tectonic escape of blocks (see ▶ Chap. 13).

Fig. 8.10   The dextral North Anatolian Fault forms the plate boundary between the Anatolian and Eurasian plates; stars show epicenters of large earthquakes between 1939 and 1999 (Okomura et al. 1993). The numbers in brackets indicate the moment magnitudes of the earthquakes. The conjugated East Anatolian Fault has opposite sense of movement (sinistral). In the wedge between both faults the Anatolian Plate is pushed westwards ("tectonic escape")

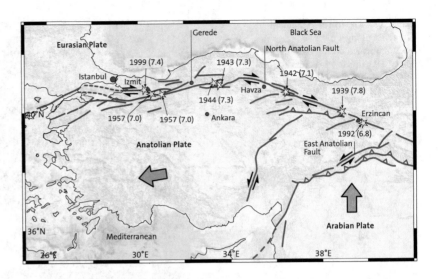

Fig. 8.11   The Alpine Fault in New Zealand is a dextral transform fault that connects two subducting plate boundaries (see insert for larger field of view). Along this fault the geological units (various colored symbols) are displaced by several hundred kilometers

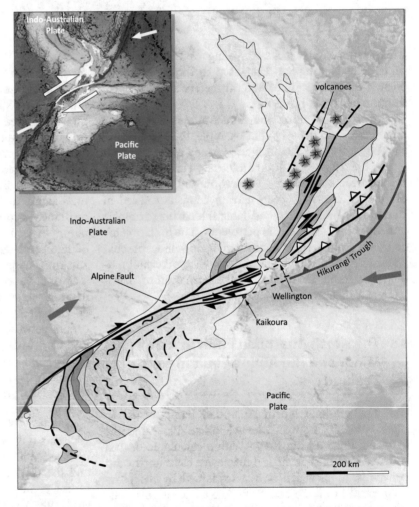

The Alpine Fault in the Southern Alps of New Zealand is a transform fault that connects two subduction zones, each with different polarity. North of the fault the Pacific Plate subducts beneath the Indo-Australian Plate; south of the fault subduction is the opposite (Fig. 8.11). Thus the fault lengthens over time

(Fig. 8.1g). The relative movement between the two plates is not exactly parallel to the fault, but rather oriented slightly oblique. This induces a transpressional component along the fault that is manifested by sharp local uplift along a complex bundle of faults.

At the end of the Oligocene, ca. 25 Ma, two directly opposed subduction zones developed. This unstable situation inevitably led to the formation of the dextral transform fault. Since that time, the fault has increased to a total length of 750 km, the length of the entire southern island of New Zealand. This yields an average growth of 3 cm/yr, the calculated average convergence rate between the two plates. At present, this rate has slowed down slightly. The transpression is responsible for the thrusting of the Pacific Plate onto the Indo-Australian Plate, which occurs at an angle of about 40° and generates uplift of 1 cm/yr. The movement at the Alpine Fault is also irregular. During the last millennium, four large earthquakes occurred with estimated magnitudes of approximately 8 and with displacements of up to 8 m. Three were located near the northern end of the fault. The last major earthquake occurred on November 13, 2016; it had a magnitude of 7.8 and the earthquake center was near the southern branch fault of the Alpine fault near Kaikoura (■ Fig. 8.11).

Terranes

Contents

© The Author(s), under exclusive license to Springer Nature Switzerland AG 2022
W. Frisch et al., *Plate Tectonics*,
Springer Textbooks in Earth Sciences, Geography and Environment,
https://doi.org/10.1007/978-3-030-88999-9_9

9

During the 1980s geological and geophysical investigation showed that large parts of the North American Cordillera are characterized by the welding of "allochthonous tectono-stratigraphic terranes" onto the North American continent. In the following years, a large number of terranes have been identified in other regions around the Pacific Ocean and elsewhere on Earth. It is now known that most mountain ranges including the Appalachians, Alps, and Himalayas are largely constructed of allochthonous terranes.

Tectonostratigraphic terranes, generally simply called "terranes" (the spelling is intentional in order to differentiate from the commonly used term terrain), consists of a block of geology, usually bounded by fault zones, that shows a geological evolution that contrasts with the geology of neighboring crustal blocks (Jones et al. 1982; Schermer et al. 1984). Therefore, it forms a tectonic and stratigraphic entity of its own. A characteristic of a terrane is its *allochthonous* nature (*Greek* other terrain). Allochthonous means that its place of formation was remote relative to the neighboring blocks. By inference, terranes have undergone plate drift as they wandered across ocean basins, until they collided or docked, and became welded to a larger continental block or another terrane.

Terranes are known to have several origins. A clue to one type of origin can be deduced by studying the components of modern, large ocean basins with long, complex histories, commonly several hundreds of millions of years in duration. The present Pacific Ocean contains numerous crustal pieces, commonly called oceanic plateaus, that differ from the surrounding normal ocean floor (◻ Fig. 9.1). These pieces generally possess greater crustal thickness and lower density than the normal ocean floor and are not easily subducted as

they enter a subduction zone. Rather, they collide with the upper plate, perhaps become partially subducted, and become welded onto the upper plate (◻ Fig. 9.2). The process of collision and welding is called accretion or docking. In principle, it is the same process that is responsible for a large orogeny when two continents collide (▶ Chap. 11). Also, the accretion of sediments in the accretionary wedge of a subduction zone is a similar phenomenon (▶ Chap. 7).

The distance between place of origin and the later accretion of a terrane can exceed several thousand kilometers. Crustal pieces that formed in the Pacific Ocean and later collided with one of the circum-Pacific continents are known to have traveled such distances. However, terranes are not defined by their travel distance, but rather by their own distinct geologic evolution. The size of terranes varies greatly; they range from small blocks, less than 100 km², to fragments of a small continent ("microcontinent") that may be 1000's of km². Terranes commonly break into several pieces that drift apart prior to or during accretion; such events makes terrane reconstruction difficult (see "dispersed terranes"; ◻ Fig. 9.2).

Terrane boundaries are generally faults or wide, complex fault zones. Commonly they are thrust faults along which the accreting terrane was pushed beneath or upon the margin of the continent. Many terrane boundaries are further complicated by overprinted strike-slip motion of considerable lateral displacement; after the terrane accretes, they can slide along the margin of the continent for several hundreds or thousands of kilometers. Like suture zones in major continental-collision orogens, terrane boundaries are commonly imbedded with ophiolites, splinters of the otherwise

◻ **Fig. 9.1** Various crustal fragments in the Pacific Ocean that could collide with, and be accreted to, one of the circum-Pacific continents as terranes (Howell 1985)

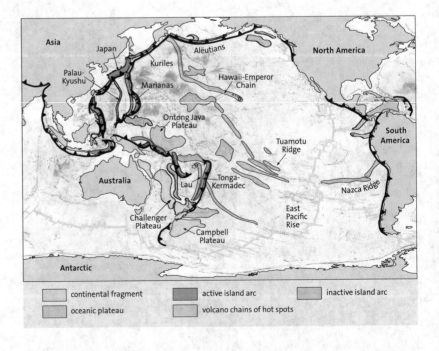

◨ Fig. 9.2 Map and cross-section views illustrating terrane accretion. A terrane approaches a continent by subduction of the intervening ocean floor. The eventual collision leads to underthrusting and accretion of the terrane. The subduction zone then jumps outboard of the terrane to the adjacent oceanic realm. During and after accretion, the terrane may be broken and dispersed along the continental margin

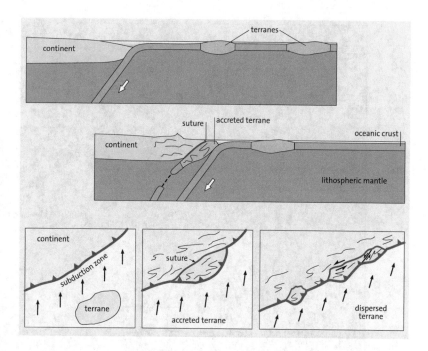

consumed oceanic lithosphere that was intermediate between the terrane and continent.

The crustal pieces that generate terranes can be constituted in various ways. Segments of thickened oceanic crust, oceanic islands, seamounts and volcanic chains above hot spots, active or inactive island arcs, oceanic plateaus, and continental fragments all exist presently in the modern oceans and each has been documented as the source of ancient terranes. Segments of oceanic crust that formed as either parts of mid-ocean ridges or backarc basins are especially common. Such hot and relatively light ocean floor complexes are accreted to the upper plate or obducted (thrust over its margin), rather than subducted. Examples of large obducted ocean floor segments are the Semail nappe in Oman (see ◨ Fig. 5.16) and the Troodos ophiolite complex in Cyprus.

Volcanic islands, seamounts, and volcanic chains above hot spots form obstructions during subduction and can be sheared off from the subducting oceanic basement and accreted to the upper plate. The northern end of the Emperor Seamount chain in the northwestern Pacific Ocean is presently colliding with the Asian continent in the Kamchatka subduction zone (◨ Fig. 9.1). Parts of this volcanic chain have become accreted as terranes. Accretion of a single volcanic edifice may create a microterrane with a size of only a few square kilometers. The western Pacific Ocean contains a number of active and inactive island arc systems, also candidates for becoming accreting terranes. Oceanic plateaus with their thickened oceanic crust (▶ Chap. 6) may also resist subduction and therefore become accreted.

Continental blocks may split off from larger continents and drift separately as so-called microcontinents. Commonly such fragments possess reduced crustal thickness because the splitting process occurred under extension and crustal thinning. Microcontinents with thinned crust may lie completely below sea level like the Challenger Plateau east of Australia (◨ Fig. 9.1).

Terranes may collide among themselves, usually in subduction zones or along transform faults. Presently such collisions are occurring in complex areas like the triangle between the Philippines, Java, and New Guinea, where several subduction zones, some with opposing subduction polarity, act simultaneously (see ◨ Fig. 13.1). The collision of terranes is called amalgamation; the outcome is a "composite terrane". Large composite terranes are called superterranes. Composite terranes usually consist of different types of crust and become accreted to a continent during a later event. The Insular Superterrane is several hundred kilometers wide and stretches over 2500 km from Vancouver Island, British Columbia to central Alaska. It is a composite of the Wrangell and Alexander terranes (◨ Fig. 9.3) and several smaller terranes. Geologic evidence shows that the two terranes were already amalgamated when they collided with western North America.

9.1 Documenting Terranes

Documenting a terrane's history, especially its travel distance and ultimate origin, can be a difficult and controversial process. Paleomagnetic research provides information concerning the drift history of a given ter-

9

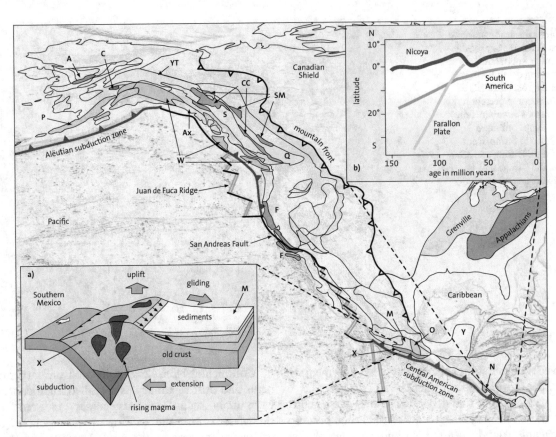

Fig. 9.3 Selected terranes along the west coast of North America and suspect terranes in Middle America (Howell 1985). Terranes: A: Angayucham, Ax: Alexander, C: Chulitna, CC: Cache Creek, F: Franciscan, M: Mixteca, N: Nicoya, O: Oaxaca, P: Peninsular, Q: Quesnel, S: Stikinia, SM: Slide Mountain, W. Wrangell, X: Xolapa, Y: Yucatan, YT: Yukon–Tanana. YT, S, CC, SM, and Q comprise the Intermontane Superterrane. P, Ax, and W comprise the Insular Superterrane. **a** Sketch showing the setting of the Xolapa complex in southern Mexico in Cretaceous time (Ratschbacher et al. 1991a). **b** The slow northward drift of the Nicoya complex in Costa Rica corresponds with the wander path of South America and precludes the drift as a part of the Farallon Plate from the Pacific region (pink wander path) (Frisch et al. 1992)

rane through geologic time. The magnetic orientation of iron-bearing minerals yields two pieces of information, declination (deviation from north) and inclination (deviation from the horizontal). This information is locked in the minerals during a past time when the mineral cooled through a certain temperature referred to as its Curie point. The inclination of the magnetic vector is a function of geographical latitude; orientation is horizontal at the equator and vertical at the poles. If the original horizontality of a rock can be determined, for example, sedimentary rocks that display bedding, then the inclination angle can be measured and the paleo-latitude can be reconstructed. A paleomagnetic survey can demonstrate the drift of a crustal block over several degrees of latitude and compare those measurements with measurements from adjacent tectonic blocks. The drift along a parallel cannot be measured by paleo-magnetic means.

Paleontological data also yields clues that can document past geographic positions of given terranes. Flora and fauna in rocks express past climate and environmental conditions. Sharp contrasts in flora and fauna

of the same age between adjacent tectonic blocks suggest that those blocks were widely separated during the time interval indicated by the fossils. Differences in paleo-latitude are generally easier to detect than distant places at similar latitude. However, some fossils are more provincial (of local extent) than others; in general, bottom-dwelling organisms that lack well-travelled larva forms are more provincial than swimming or floating organisms that are more readily widely distributed. This knowledge has been used to document the presence of exotic terranes in western North America. Certain tectonic units in the North American Cordillera contain Late Paleozoic strata that yield remnants of the East-Asian Cathaysia (Chinese) flora, whereas the remainder of North America shows a completely different, North American–European flora association during that period. The tectonic units containing the Cathaysia flora mainly drifted latitudinally over the area of the present Pacific Ocean. In this case, the distinct flora associations do not reflect strongly different climatic conditions but rather are explained by the large paleodistances that prevented mixing of floral elements.

Sedimentary facies, an expression of the sum of all features of a sedimentary rock, can provide strong evidence for past environmental conditions including climate, depositional environment, and the presence and composition of nearby mountains. Such information can then be used to speculate on the ancient geographic setting of terranes. Clay minerals form by weathering of magmatic and metamorphic rocks. They serve as paleogeographic indicators not only as to the source area of the clays, for example, igneous versus metamorphic, but also to the climatic conditions under which the clays formed. Carbonate rocks preferentially form in warm oceans and different limestone types are characteristic of various coastal, shallow water, or deep-water environments of formation. The composition and texture of sandstones and graywackes, sandstone with clay matrix as well as feldspar and lithic fragments, reflect whether their mineral components were derived from a large continental hinterland with metamorphic and granitic rocks or a nearby juvenile volcanic island arc complex with a steep relief. The study of heavy mineral spectra may also be of great value. Heavy minerals are extremely useful in terrane analysis. Heavy minerals occur in sandstones, have a specific gravity of 3 g/cm^3 or more, and are resistant to weathering; therefore, they are able to be transported and preserved over long distances. Their nature provides key information concerning age and composition of source areas of clastic sediment. One of the newest and most widely used techniques in terrane analysis involves the study of detrital zircons (ZrSiO$_4$; specific gravity 4.7). Zircons contain small amounts of U and that makes them easy to date using the latest U–Pb radiometric techniques. They form in acidic igneous rocks and are introduced into sediments either by direct ash fall that preserves delicate original crystal morphology or by weathering and erosion of igneous outcrops that produce rounded and abraded zircon crystals. The former are useful in directly dating the sediments in which they occur because they were directly deposited by the source volcano, and the latter are useful in dating the age of the source rocks of the sediment. An example is the Alexander Terrane of western Canada and southern Alaska. This terrane contains detrital zircons that yield an age that is incompatible with any rocks of western North America; this makes the Alexander Terrane "suspect" and suggests an exotic origin away from western North America. In fact, the age of the zircons along with carbonate facies information, fossil data, and age of included deformed rocks suggests that during the Precambrian, Alexander lay adjacent to Baltica or Siberia, both far removed from western North America at the time.

Each of the above helps to reconstruct the paleogeographic position of a terrane and to evaluate whether differences between the possible bordering terrane and adjacent continent can be explained by normal facies transitions over relatively short distances or rather require distant places of formation in their geologic past. To verify the existence of a terrane, all applicable methods should be used. Not all of the numerous terranes defined in the 1980's in the circum-Pacific region withstood subsequent examinations with more exact methods. Although some such as Alexander and Wrangellia have withstood later scrutiny and are believed to be truly exotic, others are now believed to be transported laterally along a given coast, to have formed as offshore island arcs, or to have formed as rifted terranes, later redocked to their continent of origin. Such terranes are generally referred to as periallochthonous or periautocthonous, depending on the speculated amount of displacement through geologic time.

9.2 Terranes in the North American Cordillera

In the Cordilleran Region of the United States and Canada, a large number of terranes have accreted since the Late Paleozoic. Over the past 300 Ma, the continental margin has accreted and shifted westward more than 800 km. Some of the terranes have travelled for more than 5000 km. During much of the Paleozoic era, western North America was a classic passive margin on which thousands of meters of sedimentary rock were deposited. This all changed in the Carboniferous and Permian as exotic island arcs and oceanic plateaus were accreted; from this time to the present, western North America has been characterized as active continental margin. Oceanic crust of the Pacific region was subducted beneath the North American continent and along with it came numerous terranes too buoyant to be subducted. Most of these terranes were subsequently accreted to the continental margin. To put in perspective the amount of potential terrane accretion and subduction that has occurred along the Cordilleran Region, consider that during the early Mesozoic, the Farallon Plate was the largest plate on Earth; all but a few small pieces of this once mighty plate have either been subducted or accreted to the west coast of the Americas!

One of the earliest and still most convincing arguments for terrane accretion is the presence of far-travelled fusulinids, large foraminifera (protozoans) of Late Paleozoic age, now present in the Cordillera. Many exotic crustal blocks contain these fusulinids that are known from the Tethys region, that oceanic realm from which later emerged the Alpine-Mediterranean mountain belt in southern Europe and Asia (◘ Fig. 4.8). Indigenous North American fusulinid faunas fundamentally differ from the Tethyan ones. They belong to another faunal province that had no direct connection to the Tethyan fauna.

The North American continent east of the Rocky Mountains consists of crust that was cratonized and stabilized during numerous Precambrian orogenies by crustal thickening and metamorphism; the Canadian Shield is an example (■ Fig. 9.3). On this stable platform terrestrial and shallow-marine sediments were deposited during the Paleozoic and Mesozoic. This region is in sharp contrast to that of the mobile terranes of the Cordillera. ■ Figure 9.3 is a greatly simplified map showing terranes accreted to North America (Jones et al. 1982; Howell 1985). The discussion below illustrates the diversity of some of the various terranes.

The *Angayucham Terrane (A)* in the Brooks Range of Alaska consists of late Paleozoic and Mesozoic oceanic basalts and siliceous and calcareous deep-water sediments. Accretion to Arctic Alaska occurred in the late Mesozoic. It represents obducted oceanic crust that carried individual volcanoes. The oceanic basalts and the mainly basaltic volcanoes, mostly formed by hot spot activity, can be discerned by their chemical characteristics, even though they have been disrupted and strongly deformed.

The *Chulitna Terrane (C)* is one a dozen small terranes in the Alaska Range. Each is composed of strongly contrasting geologic blocks surrounded by thick flysch successions. These microterranes represent pieces scraped off of larger units. The Chulitna Terrane consists of a sequence that is unknown elsewhere in all of North America. The sequence comprises a Paleozoic ophiolite complex with siliceous deep-water sediments overlain by Late Paleozoic to Triassic shallow-water sediments that were deposited on top of island-arc volcanics. This rock association probably represents a subduction-related volcanic chain in an intra-oceanic environment because it does not show any sedimentary influence from a continent. An Early Triassic ammonite fauna reveals that the strata were deposited in low latitudes. In contrast, the neighboring region of the Canadian Shield was positioned at about 40° northern latitude at that time. In the Late Triassic, the tectonic setting of the Chulitna Terrane changed dramatically. Influx of large volumes of quartz sand indicates the presence of a large, nearby continent with exposed metamorphic or granitic rocks. This is interpreted to reflect the accretion of the terrane onto the North American continent. In the neighboring Wrangell Terrane (see below) the Early Triassic sedimentary sequence shows a completely different setting as compared to the Chulitna Terrane.

One of the finest examples of terrane analysis comes from the study of the Intermontane Super-terrane (■ Figs. 9.3 and 9.4); the term Intermontane comes from the fact that the terrane lies between the folded and thrust Canadian Rockies to the east and the Coast Ranges to the west. The superterrane consists of four major components discussed below and many smaller fragments. The easternmost component consists of (1) the *Slide Mountain Terrane (SM),* mostly Triassic and Lower Jurassic ocean-floor basalt, deep ocean sediment and sediment derived from nearby continental uplands (■ Fig. 9.4b, c). Next, (2) the *Yukon–Tanana Terrane (YT)* and its southern extension, *Quesnel (Q)* consist of a continental nucleus composed of metamorphic rocks, volcanic rocks, and granites. Detrital zircon studies suggest this large, composite terrane was part of Precambrian North America (Laurentia) and was subsequently rifted some distance away during the late Paleozoic; how great this distance was is still a major controversy (■ Fig. 9.4a, b). Farther west is (3) the controversial *Cache Creek Terrane (CC)* that consists of upper Paleozoic through Lower Jurassic oceanic basalt, ophiolites, trench and mélange deposits, and meta-flysch deposits (■ Fig. 9.4d); much of this terrane is metamorphosed blueschist facies rocks. The westernmost terrane is (4) the *Stikine Terrane (S),* a complex of oceanic rocks, arc rocks, and deep marine sedimentary rocks, but like Yukon–Tanana and Quesnel, contains a nucleus of continental crust. Consensus among geologists concludes that the Intermontane Superterrane was amalgamated in the Middle Jurassic and accreted to North America by the end of the Jurassic (▶ Chap. 13); however, the details concerning the evolving paleogeography of the superterrane, especially its distance from North America, continue to be debated.

The Yukon–Tanana, Quesnel, and possibly Stikine terranes began as thinned and extended continental crust following the Late Precambrian rifting of western North America. After accretion of one or more island arc complexes in the Carboniferous (■ Fig. 9.4a), an active margin developed and an arc complex was built on the terranes; this was followed by Late Carboniferous to Permian backarc extension that separated the terranes, with their recently accreted arc material, from North America. The Slide Mountain backarc basin formed during this event. Meanwhile, subducted under the west flank of the terranes was the Cache Creek Ocean, part of the huge, late Paleozoic-early Mesozoic Panthalassa Ocean. The Cache Creek Terrane formed at the margin of the subducted plate as trench, obducted ophiolite, and accretionary prism deposits. Within the accretionary prism were accreted Tethyan fusulinid-bearing limestones as described above (■ Fig. 9.4d). The tectonic setting during most of the Triassic was, from east to west, the western margin of North America, the Slide Mountain backarc basin, the Stikine(?)–Yukon–Tanana–Quesnel island arc, and the Cache Creek subduction zone (■ Fig. 9.4d). By the end of the Triassic, the Slide

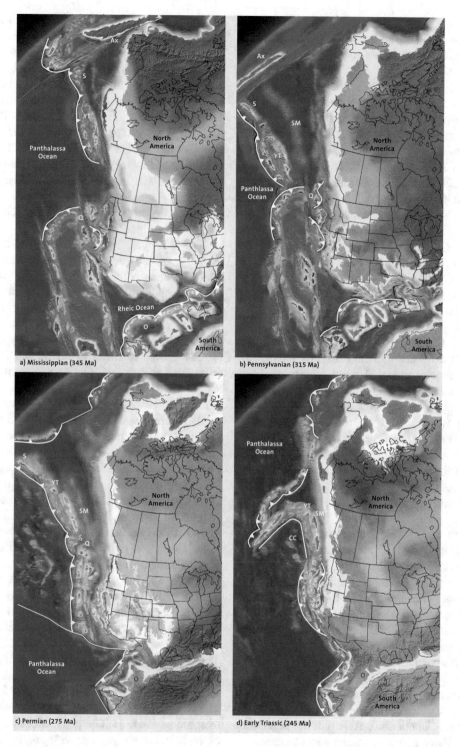

Fig. 9.4 A series of paleo-geographic maps of part of North America and surrounding regions showing the complexities of terrane accretion. Abbreviations are the same as in ▪ Fig. 9.3 with several additions explained below. **a** Mississippian, Early Carboniferous and **b** Pennsylvanian, Late Carboniferous; the elements of the Intermontane Terrane are widely scattered along the Cordilleran margin. Oaxaca, Yucatan, and adjacent unlabeled terranes accreted to southern North America with the closing of the Rheic Ocean. Some models show these terranes as part of South America rather than as a ribbon continent as shown here. Alexander may have rifted from Baltica or Siberia and subsequently translated northward into the Panthalassa Ocean. **c** Permian; following a complex amalgamation involving one or more arcs, Stikine, Yukon–Tanana, and Quesnel separate from North America by backarc spreading and the formation of the Slide Mountain backarc basin. Pangaea, which lay mostly to the east of the map, was assembled by the Permian and Oaxaca and Yucatan were accreted to North America. Terranes with Permian Tethyan fusulinids drift into the Cordilleran region. **d** Early Triassic; Stikine folded oroclinally towards Quesnel closing the Cache Creek Ocean between. Some models show this as a transform fault system rather than oroclinal fold. The Slide Mountain backarc basin was being closed

Mountain backarc basin was closed; detrital remnants of Yukon–Tanana were deposited on western North America signifying arc collapse against the continent and subsequent mountain building. Following the closure of the Slide Mountain Basin, the tectonic history becomes controversial. Stikine accreted to Yukon–Tanana–Quesnel from the west closing the Cache Creek Ocean and forming the Cache Creek Terrane between them. There are several tectonic possibilities that can explain this: (1) Stikine was exotic to North America and collided from the west, (2) Stikine was marginal to North America and along strike of Yukon–Tanana–Quesnel; it was subsequently transposed along transform faults and then collapsed against the continent to close the Cache Creek Ocean, or (3) Stikine formed an oroclinal fold, a fold that bends back on itself by 180°, that closed the Cache Creek Ocean (□ Fig. 9.4d). Regardless of tectonic detail, the event that closed the Cache Creek Ocean was completed by the Middle Jurassic. The amalgamated superterrane then became the leading edge of North America and a large Late Jurassic and Cretaceous arc was built on it (▶ Chap. 13). Deformed and metamorphosed flysch occurs on both margins of Yukon–Tanana–Quesnel. Flysch is mainly formed along steep and tectonically active slopes along active continental margins. It is commonly deposited in deep-sea trenches and therefore is a strong indicator of converging plate margins. Because subduction precedes continental collision, flysch sequences are commonly precursors of collision. The flysch on the margins of the Yukon–Tanana Terrane marks terrane collision and backarc closure.

Although currently located from northern Washington to SE Alaska, the events of the Intermontane Superterrane just described occurred as much as 800 km farther south, relative to North America, than their current location. During much of the Cretaceous, transform faults moved the superterrane northward (▶ Chap. 13). Also during the Cretaceous, the Insular Superterrane, described below, was accreted onto the Intermontane Su-perterrane. Late Cretaceous sedimentary rocks were deposited on the folded flysch deposits and overlap terrane boudaries indicating that the Yukon–Tanana Terrane arrived in its present position relative to the North American continent at the beginning of the Tertiary. Paleomagnetic studies confirm this.

The Insular Superterrane consists of three major components (□ Fig. 9.3). (1) the *Wrangell Terrane (W)* consists of a Late Paleozoic volcanic island arc topped by thick Permian and Triassic sedimentary and volcanic rocks deposited in a regime of crustal extension. Permian limestones contain fusulinids that are clearly different from the Tethyan forms in the neighboring Intermontane Superterrane. The Wrangell Terrane probably originated from a distant place in the vast Panthalassa Ocean far away from the Tethys Ocean (□ Fig. 4.8). Late Jurassic and Cretaceous sedimentary sequences are common with the neighboring (2) *Peninsular Terrane (P)* and (3) *Alexander Terrane (Ax)*. The exotic history of the Alexander Terrane was briefly outlined above. Overlapping Late Jurassic sedimentary rocks and cross-cutting Jurassic plutons indicate amalgamation of these three terranes in the Middle Jurassic. Common also are Cretaceous folding, thrusting, and metamorphism. These events reflect the collision of the large composite terrane with North America. As a consequence of this collisional event, parts of the terrane became dispersed and therefore appear disconnected today.

The Wrangell Terrane is involved in a major controversy of western North American tectonic history. Paleomagnetic studies and some supporting geologic evidence indicate that after its initial accretion to western North America, the terrane captured parts of western North America and moved southward, as far as Baja California, Mexico in some hypotheses. It then rapidly moved northward during the Late Cretaceous to its current location. This rapidly moving terrane has been coined "Baja BC", BC standing for British Columbia where major portions of Wrangellia are currently located (▶ Chapter 13). Subsequent studies have modified the extreme southern limit somewhat and suggest 800–1500 km of lateral translation rather than the 3000 km as required in earlier hypotheses (Wyld et al. 2006; Umhoefer and Blakey 2006). Analysis of Farallon Plate motions during the Jurassic and Cretaceous suggest alternating dextral and sinistral transform motions with respect to North America and add credence to the south and north transform motions of Baja BC.

A terrane of completely different nature is the *Franciscan Terrane (F)*, which is exposed along the Californian coast on both sides of the San Andreas Fault (□ Fig. 9.3). Rocks range in age from Upper Jurassic to Lower Tertiary and include ophiolites, representing ocean floor fragments, chaotic blocks, and sediments deposited in a deep-sea trench. Paleomagnetic and paleontologic studies demonstrate that many blocks present in the Franciscan are exotic to North America. The chaotic internal structure of the terrane is that of a tectonic mélange (▶ Chap. 7): blocks of various types of resistant rocks, present in sizes from several meters to kilometers, "float" in a strongly sheared matrix of easily deformable, originally argillaceous rocks. Metamorphism of some Franciscan rocks to blueschist facies shows that the mélange formed during a subduction process.

The Franciscan is not without controversy. Original interpretations suggested that the Franciscan was the trench-accretionary prism portion of a classic forearc trench setting and the adjacent sandstone and mudstone of the Great Valley Group formed in the forearc basin. More recent studies, still in accord with the earlier general interpretations, suggest more complex tectonic history with sedimentation in various forearc and interarc settings but with considerable lateral juxtaposition by coeval and subsequent transform processes. Noting that the Franciscan is never in stratigraphic contact with the supposedly adjacent Great Valley Group and using paleomagnetic and paleontologic data, some workers have suggested that the entire terrane is exotic to North America and that most of it was originally deposited off southern Mexico and Central America and later translated northward during the Late Cretaceous and Tertiary.

9.3 Suspect Terranes in Mexico and Middle America

As mentioned above, many terranes have been defined around the Pacific Ocean without conclusive proof. As long as unequivocal proof is missing, the term "suspect terrane" should be used. Studies in Mexico and Middle America refute the terrane nature, the derivation from a distant place, for crustal blocks that were proposed to be terranes. It is possible that other suspect terranes represent blocks that did not travel afar but rather were displaced along fault zones from neighboring blocks.

The suspect *Xolapa Terrane (X)* in southern Mexico forms a block that is elongate parallel to the coast (■ Fig. 9.3). The boundaries with the suspect *Mixteca Terrane (M)* and suspect *Oaxaca Ter rane (O)* are not thrusts or transcurrent faults, but rather low-angle detachment zones with normal displacement (■ Fig. 9.3a). Normal faults indicate crustal extension and not terrane accretion and thrusting. There is no indication that the extensional structures overprint older structures formed by compressional or lateral motions. The Xolapa Terrane was an active magmatic belt above the Middle American subduction zone that experienced uplift as the magmatic melts penetrated the crust. Accompanying heating weakened the rocks until they became plastic and easily deformable. Finally, this led to gravity gliding of part of the upper crust away from the uplifting magmatic belt. This culminated in the formation of the normal detachment zone between the uplifted Xolapa complex and the units north of it.

This interpretation is backed by further data. Paleomagnetic studies show consistent directions between the magnetic vectors of several suspect terranes in southern Mexico. Moreover, some of these terranes contain a metamorphic basement that is typical of the North American continent; they contain rocks gener-

ated during the Middle Proterozoic Grenville orogeny and are common to rocks in the Canadian Shield. The Grenville orogenic belt is known to extend from eastern Canada to southern Mexico, although in places it is covered by younger sediments (■ Fig. 9.3).

There are documented exotic terranes in Mexico. The *Yucatan Terrane (Y)* contains crustal elements, so-called Cadomian basement, only found in terranes that broke off Gondwana during the early and middle Paleozoic (■ Fig. 9.4b). Today these peri-Gondwanan terranes extend from Mexico, to the Appalachians, Southern Europe, and Central Asia eastward to China. They represent one or more ribbon-shaped continents that rifted from Gondwana. But again there is controversy with these peri-Gondwanan terranes. Did they rift to form a separate ribbon continent (■ Fig. 9.4a) or did they collide with North America and other continents as the leading edge of Gondwana and were later stranded as the Gulf of Mexico, Atlantic Ocean, and Neotethys Ocean opened in the Mesozoic?

Another suspect terrane is the *Nicoya Terrane (N)* in Costa Rica and Panama. It represents a complex of oceanic crust (ophiolites) and Jurassic to Cretaceous deep-sea sediments. Following tectonic events with nappe thrusting in Late Cretaceous time, the deformed sequence was covered with shallow-water sediments. Most investigators of the region have proposed that the ophiolites are remnants of ocean floor from the Farallon Plate in the Pacific region that travelled into the present position in Late Cretaceous time. However, paleomagnetic data demonstrate that the Jurassic ocean floor and the overlying deep-sea sediments all formed near the equator (■ Fig. 9.3b). The drift of the oceanic Farallon Plate was northeasterly through these long periods of time and with a velocity of more than 5 cm/yr. Therefore, if the sediments of the Nicoya complex were formed on Farallon crust, they would have been deposited at southern latitudes far from the equator. Such an interpretation contradicts the paleomagnetic data. Rather, these data indicate paleolatitudes for the Nicoya complex that are very similar to those of the neighboring parts of South America and reflect only very slow northward movement. It is therefore likely that the ophiolites of Middle America formed on the Caribbean crust that was attached to South America during this interval. The Caribbean crust was displaced relative to South America only since the Late Cretaceous. It has subsequently been translated more than 1000 km to the east–northeast with a resulting northward shift of only few hundred kilometers. This raises an interesting question; if this newer interpretation is correct, should the Nicoya Terrane be considered exotic?

Early Precambrian Plate Tectonics

Contents

© The Author(s), under exclusive license to Springer Nature Switzerland AG 2022
W. Frisch et al., *Plate Tectonics*,
Springer Textbooks in Earth Sciences, Geography and Environment,
https://doi.org/10.1007/978-3-030-88999-9_10

The Precambrian comprises geologic time before 540 Ma, the time before the Cambrian. The term Precambrian has been used by geologists for a long time and although it is no longer recognized as a formal term in the most recent timescales, it is still a useful term and used accordingly herein. The Precambrian consists of the Archean Eon (4000–2500 Ma; *archaios, Greek* very old) and the Proterozoic Eon (2500–540 Ma; *proteros, zoon, Greek* earlier [than] animals). Proterozoic plate tectonic events are widely documented in many places including the Wopmay orogen of NW Canada and the Yavapai and Mazatzal orogens of Arizona. Although there is some debate on the rates at which Proterozoic plate tectonics occurred, there is little question that plate tectonics, as we know it today, has been around for over 2 billion years. What remains uncertain is the existence and nature of Archean plate tectonics; this and a discussion of the oldest rocks on Earth are the subjects of this chapter.

The oldest preserved rocks are known from the Canadian Shield—the Slave Province in the Northwest Territories, the Nuvvuagittuq greenstone belt in northern Quebec, and the area around Isua in western Greenland. However, zircons from Australia yield even older ages (see box below). The granitic protoliths of gneisses in the Slave Province crystallized approximately four billion years ago according to radiometric age dating. The sequence in Isua revealed an age of more than 3800 Ma. Remarkably, the Isua rocks indicate surface temperatures that are similar to current ones: they comprise sedimentary rocks that were deposited in water. According to studies on carbon isotopes ($^{13}C/^{12}C$ ratio) it was also assumed that primitive, bacterial life existed (Schidlowksi 1983); however, this conclusion is still being debated.

The time before the age of the oldest rocks and following the age of the origin of the Earth, generally believed to be 4550 Ma, represents the nascent time of our planetary system and therefore also Earth. It is called the dawn of Earth or Hadean Eon *(Hades, Greek* the invisible, God of Orcus). The crust of the Earth that formed in the Hadean was initially ultrabasic but then became increasingly basic in composition. The proto-crust was completely destroyed by subduction or other processes and recycled into the mantle. Therefore, we have no direct information from rocks during this time except from small zircon crystals contained in younger rocks (see box)—thus the name Hadean. In contrast, Archean rocks are known from all continents (◻ Fig. 10.1). Archean rocks were incorporated into many younger orogens and overprinted by deformation and metamorphism.

Archean rocks are difficult to comprehend—in general, the more ancient the rock, the more difficult its history is to decipher. The same can be said of the study of human history! One of the most discussed questions in geology is whether plate tectonic processes, as we presently understand them, can be traced back into the Archean. This raises the question about the characteristic features necessary to document plate tectonic processes in earlier periods. Important characteristics include: (1) the drift of plates and imbedded continents, and (2) the constant generation of oceanic crust and its re-integration into the mantle by subduction. Subduction generates a specific style of magmatism and paired metamorphic belts. Continent collision is the consequence of subduction; the sutures between colliding continental blocks are generally impregnated with ophiolites. Archean rocks have not yielded definite ophiolites nor high-pressure metamorphic rocks; hence, no ophiolitic sutures or paired metamorphic belts. However, Archean eclogites (3200–2500 Ma) occur in much younger magmatic rocks as inclusions formed in and carried from the mantle during later magma ascent (Foley et al. 2003). They are of great importance for the reconstruction of Archean plate tectonic scenarios (see below).

◻ **Fig. 10.1** Distribution of the oldest rocks on Earth (Condie 1997). The Archean cratons (stable continental blocks that escaped later overprint) consist of rocks with radiometric ages between 4000 and 2500 Ma. Several of the rock complexes are labeled

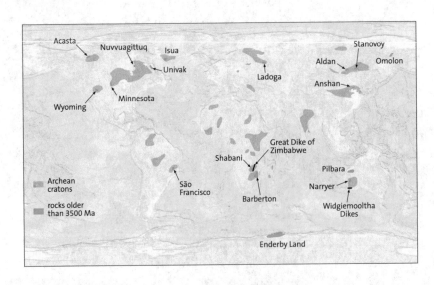

The Oldest Rocks and Minerals

Oceanic crust is generally recycled into the mantle and destroyed. However, basic magmatic rocks that have been transformed to amphibolites by metatmorphism and that could represent a sort of early oceanic crust, have been dated in the Nuvvuagittuq greenstone belt in northern Quebec, Canada. An unusual dating method using neodymium isotopes suggests that the protolith material was extracted from the mantle some 4200 Ma ago (O'Neil et al. 2008). This age is still under discussion.

In contrast to basic rocks, the granitic and sedimentary rocks of continental crust resist subduction so the chance of finding ancient rocks in continental crust is much higher. To find such ancient crust, radiometric dating of zircon crystals with the uranium-lead method is suited. The mineral zircon is zirconium silicate ($ZrSiO_4$) that incorporates a number of external elements in its crystal lattice, including rare earth elements, uranium, and thorium. Uranium and thorium decay into lead and from the ratios of mother and daughter isotopes, the time elapsed since crystallization of the mineral can be calculated. Zircon is present in minor amounts in many granitic rocks, and granitic rocks are common in continental crust. Zircon is very resistant and largely survives metamorphism and deformation without change. Therefore element exchange between zircon and neighboring crystals is generally insignificant so that the U, Th, and Pb contents remain unaltered. The mineral thus keeps its "memory" from the time of initial crystallization.

With the ion microprobe, a 20 µm-thick ion beam is focused onto the polished surface of a crystal and is able to measure the concentration of isotopes. In this way, the age of a spot within the crystal can be determined. Because minerals can grow over long periods of time, different spots may give different ages. Zircon that grows in a magmatic melt can later acquire a rim formed by overgrowth during a subsequent metamorphic event. Therefore, outer shells of the crystal will then yield lower ages than the core.

The oldest known continental rocks that were dated by the U–Pb method using zircons were found in the arctic region of Canada, in the Acasta Gneiss of the Slave Province (◘ Fig. 10.1). The gneiss was originally a magmatic rock, tonalite and granodiorite, that was transformed by later metamorphism. Portions of zircon in these rocks gave ages between 4002 and 4065 Ma (Bowring and Williams 1999). These rocks therefore mark the last period of the Hadean.

The chances of finding even older rocks seem small; repeated meteorite bombardment between 4550 and 3900 Ma destroyed early Earth crust. Never the less, older crystals have been found. In the Narryer Gneiss in Western Australia, zircon crystals were found that were eroded from their magmatic protolith by weathering and then deposited in younger sandstone. These zircons are much older than those from the Acasta gneiss. They endured weathering and erosion of the original magmatic rock, as well as the transport by water. They kept their magmatic crystallization age. The Narryer Gneiss was formed by the metamorphism of a sedimentary rock sequence. Gneisses are metamorphic, feldspar-rich rocks that can be derived from either magmatic or sedimentary protoliths. Because the zircons in question are detrital, the gneiss must have had a sedimentary precursor. The sequence of events that formed the Narryer Gneiss was complicated. The original zircon formed in granite that was part of an ancient continental crust. The granite was eroded and the zircon was transported in water to form sandstone. The sandstone and other sedimentary rocks in the sequence were then metamorphosed to form gneiss.

Detrital zircons from the Narryer Gneiss were analyzed by U–Pb radiometric dating methods. In different spots of one zircon, grain ages between 4300 and 4400 Ma were obtained; the oldest spot yielded an age of 4404 ± 8 Ma (Wilde et al. 2001). From the exact chemical and isotope composition of the zircon it was deduced that the protolith was a granitic rock that formed by melting of older continental crust in the presence of water. The presence of water was deduced from the oxygen isotope ratio ($^{18}O/^{16}O$). This evidence demonstrated the presence of water on the surface and thus some sort of ocean at 4400 Ma. Ocean water was transported into deeper crustal levels along with sediments that subsequently experienced metamorphism and were then melted to form granitic magma. Consequently, only some 150 m. y. after its accretion, the Earth exhibited a surface that had already cooled down to temperatures between 100 and 0 °C; These events occurred in an aqueous environment more than half a billion of years before deposition of the rocks of Isua in Greenland. These results are surprising. They suggest that the geologic setting on Earth relatively shortly after its creation was in certain respects, not unlike to the present-day world. Of course, subsequent meteorite bombardment would have changed that in the following several hundred m. y.

In Phanerozoic continents, blocks that were originally distal but are now welded together can be documented through paleontological and paleomagnetic studies. Fossil associations are not useful during the Archean because lifeforms were not diverse enough. Paleomagnetic measurements are difficult to interpret in Precambrian rocks because of the many orogenic events that have affected them, especially for Archean rocks. Paleomagnetic data suggest that at the end of the Archean Eon there may have been only one or two large continents. Later in the Proterozoic Eon there exists more differentiated information about continental drift and thus plate tectonics, and there is conclusive evidence for the existence of a single supercontinent, Rodinia, between 1000 and 750 Ma (◘ Fig. 6.7).

Does the absence of clear evidence mean that Archean plate tectonics were null? Were the processes of subduction and continental suturing absent? Perhaps the original evidence has been destroyed by later tectonic events. Or perhaps Archean plate tectonics existed but in a different form from the familiar style we know today. The Archean mantle was likely 100 to 300 °C hotter than today (Richter 1985) and must have undergone convection because heat dissipation by conduction is much less efficient than by convection. In order to generate heat dissipation up to the Earth's surface, spreading systems driven by convection must have existed and new crust must have formed. The crustal growth must have been compensated elsewhere by some sort of subduction. There are only two alternatives to convection and subduction: (1) Earth expanded in size; there is much direct and indirect evidence that this was not an important Archean process. Comparing Earth to other rocky planets in the Solar System, accretion and expansion by meteorite impact occurred mostly in the Hadean or very earliest Archean. (2) Earth was completely molten and lacked a solid crust. We have seen above that the earliest Archean rocks disprove this hypothesis and that water was present on the surface 4 billion years ago.

During the Hadean, the temperature was initially high enough so that mantle material was partly in a molten state up to or very near the surface. Probably an ultrabasic crust (<45 wt% SiO_2) formed. The higher the portion of melt, the more the melt approximates the peridotitic (ultrabasic) composition of the protolith. Heat production was higher because of the nature of radioactive decay. There were large amounts of short half-life unstable isotopes. Presently there are no traces of these isotopes as all have decayed. There were larger amounts of long half-life isotopes as well. Consequently, heat flow 4500 Ma ago was about 4.5 times higher than today and 3800 Ma ago 2.5 times higher (Kroner 1981). Kinetic energy released during the formation of the Earth's core and by meteorite impacts also contributed to these early high temperatures.

The higher temperatures in the mantle could only be released to the surface by faster convection. Therefore, spreading rates were either higher or spreading zones were much longer, or, likely, both. During the Early Archean, a six-fold rate of production of oceanic lithosphere compared to the present has been proposed (Bickle 1978). Also implied is a six-fold increase in subduction rates as compared to today. More lithosphere must have disappeared in subduction zones. Therefore, the difference between Early Archean and today is of quantitative rather than of qualitative nature.

The discussions above strongly suggest that plate tectonics was an important process in the Hadean and Archean eons and that it dominated the dynamics of the outer solid shell of the Earth. Production and subduction of lithosphere as well as plate motions occurred at substantially higher rates than today. The distances between spreading axes and subduction zones must have been considerably shorter and the plates must have been smaller in size and more numerous. This also implies that plates were generally much younger when they entered a subduction zone. Hot spots may have contributed to plate dynamics, but there is no direct evidence for this suggestion.

The young age of subducting plates coupled with a higher temperature regime raises the question about the driving forces behind plate drift in the Archean. Today, slab pull, the down-pulling force of dense oceanic lithosphere in the subduction zones, plays the dominant role. Few if any eclogites were present in subduction zones more than 3200 Ma ago, implying that slab pull was much less efficient than today. Presumably ridge push, induced by the upward-directed mantle flow below the spreading centers, was much more efficient than today due to the faster flow regime. Therefore ridge push is considered to have been the main driving force of plate drift during the Archean (Keary and Vine 1990).

Archean rock sequences are found on all continents (◘ Fig. 10.1). They comprise two different rock associations: the weakly metamorphosed greenstone-granite belts and the highly metamorphosed granulite-gneiss belts. Canadian geologists describe the Archean terrains north of Lake Superior as "islands of greenstone in a sea of gneiss".

10.1 Greenstone-Granite Belts

The greenstone-granite belts mainly occur in North America, South Africa, and Western Australia. They contain thick sequences of volcanic rocks intercalated with detrital and chemical sedimentary rocks. Low-grade, greenschist-facies metamorphism transformed the abundant basic and intermediate volcanic rocks into greenstones that are dominated by green minerals like chlorite, epidote, and amphibole. Among the greenstones are magnesium-rich peridotitic and basaltic lavas, the so-called komatiites, unusual rocks that are nearly completely missing in post-Archean terrains (see box). Komatiites document that temperatures in the mantle source region were higher than they are today.

The Barberton greenstone belt in South Africa contains the type locality for komatiites and is one of the

◘ Fig. 10.2 Geological sketch of the Barberton greenstone belt in southern Africa (Visser et al. 1984)

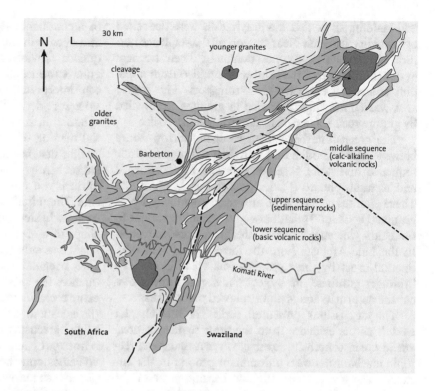

oldest of its kind (◘ Fig. 10.2). The oldest parts of the 130 km-long belt were dated between 3600 and 3500 Ma. The sequence is divided into three groups. In ascending order they consist of ultrabasic to basic volcanic rocks, calcalkaline volcanic rocks, and sedimentary rocks (◘ Fig. 10.3). The komatiites and tholeiitic basalts in the lower group have been interpreted as oceanic crust that formed the basement of a marginal basin above a subduction zone (Anhaeusser 1978). The andesites, dacites, and rhyolites in the middle group belong to the calcalkaline magmatic clan and are typical of island arcs, at least compared to similar rocks in younger Earth history.

Cherty sedimentary rocks that occur in certain horizons of the Barberton greenstone sequence consist of quartz and represent chemical sediments precipitated from seawater possibly assisted by bacteria. The enhanced dissolution of silica (SiO_2) was driven by excessive concentrations of volcanic exhalations in the primordial sea. Commonly these sediments contain large amounts of iron oxide, also precipitated from the seawater. Together, this assemblage forms the famous banded iron ores (◘ Fig. 10.3); the banding is thought to reflect sedimentary bedding. Rocks of this nature are restricted to the Archean because an oxygen-free atmosphere is necessary so that dissolved iron can be transported into the sea without being precipitated on the way. Detrital sedimentary rocks including conglomerates, graywackes and other immature sandstones, and claystones are common in the higher parts of the sequence and represent the erosional products of the thick volcanic rocks. Preserved sedimentary structures

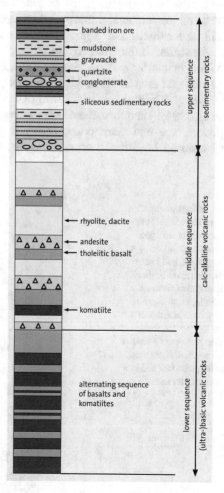

◘ Fig. 10.3 Schematic rock column of the Barberton greenstone belt (Drury 1981)

and bedding show that the graywackes were deposited by turbidity currents. Near the top of the sequence mature, quartz-rich sandstones dominate. Their richness in quartz shows that the sand was supplied from a distant source by extensive river transport. The source area was continental crust rich in granites as revealed by granite rock fragments in the sedimentary rocks.

The greenstone belts form elongate narrow zones between granite bodies that intrude into the greenstone sequence and were together, subsequently deformed and metamorphosed ("older granites" in ◻ Fig. 10.2). These granites occupy large areas and are mostly I-type granites (▶ Chap. 7) that have a composition similar to the acidic volcanic rocks of the greenstone sequence. In the Late Archean Eon, the formation of new continental crust by these granites was voluminous. The "younger granites" are largely undeformed and cross-cut the deformed and metamorphosed older rocks.

The discussions presented earlier in this chapter as well as the evidence from the three sequences present in greenstone belts strongly suggest that plate tectonic mechanisms were a dominant process in the Archean. The lower, komatiitic and basaltic parts of the greenstone belts were suggested to represent oceanic crust by some researchers; however, this interpretation is still under discussion. Some greenstone belts in Canada, Australia, India, and Granulite-gneiss belts South Africa primarily rest on continental crust and yield evidence that the basalts and komatiites were contaminated by continental material during their ascent (Bickle et al. 1994). Therefore, these volcanic rocks cannot represent true midocean crust; rather it appears that they erupted on thinned crust along the margin of a continent (◻ Fig. 10.4). The presence of andesites and dacites in the middle portion of the greenstone sequence provides additional documentation of a marginal continental setting. A comprehensive, more recent interpretation suggests that the komatiites and basalts were derived from ascending mantle flows above a subduction zone by relatively high proportions of partial melting of mantle material. These melts ascended during distension of the continental crust above and subsequently penetrated the thinned crust. This scenario is not unlike the setting of modern backarc basins. The initial back-arc basins may have evolved into oceanic basins with "true" oceanic crust. The original shape of a greenstone belt was probably rather circular but it was subsequently deformed into its present elongate, irregular shape. During the evolution of the sequence, the subduction-related component of the volcanic rocks increased, thus leading to the generation of the calcalkaline volcanics. In contrast, an interpretation of the greenstone belts as flood basalt sequence above a hot spot cannot explain the presence of the younger volcanics and the associated immature sediments that indicate subaerial relief. Viewed as a whole, greenstone belts suggest backarc basins and island arc complexes rather than deep, isolated ocean basins (◻ Fig. 10.4).

During the advanced stages of greenstone belt evolution, granites intruded the greenstone sequence followed by metamorphism and deformation. This compressional event possibly reflects continental collision with another continent that arrived from the oceanic side of the arc complex. However, the youngest granites are not affected by this tectonometamorphic (mountain building) event indicating that they are post-orogenic.

◻ **Fig. 10.4** Diagrammatic cross sections showing settings of major Archean rock types. Upper: possible evolution of a greenstone-granite belt in the position of a backarc basin above a subduction zone (Tarney et al. 1976). Lower: granulite-gneiss belts as deep crustal levels of a subduction related magmatic belt and collision zone with a passive continental margin. Note that the granulite-gneiss belt forms at much deeper crustal levels

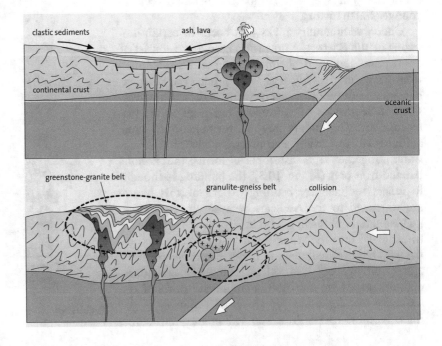

This model is in concert with evidence that indicates that the rocks were not buried to depths greater than 20 or 25 km as revealed by their greenschist-facies metamorphic grade.

10.2 Granulite-Gneiss Belts

The granulite-gneiss belts mainly consist of highly metamorphosed gneisses derived from tonalites, trondhjemites, and granodiorites. These are magmatic rocks related to granites that contain substantial amounts of plagioclase and quartz. In contrast to normal granites they contain little or no alkali (K–Na) feldspar. Minor metamorphosed sedimentary and volcanic rocks include quartzites and marbles whose sedimentary protoliths are sandstones and limestones respectively. The sediments were deposited in shallow water shelf environment. The pure quartz sandstones reflect a distant continental hinterland source area.

The tonalites, trondhjemites, and granodiorites probably formed by the partial melting of metamorphosed basaltic rocks and represent juvenile continental crust. They are strongly deformed and were subjected to a metamorphic overprint at temperatures exceeding 900 °C. However, these rocks have an acidic composition and in the presence of aqueous fluids should begin to melt at about 650 °C. Therefore, water was removed—possibly by CO_2 degassing from the underlying mantle, as evidenced by gas inclusions rich in CO_2 that are contained in the minerals of these rocks (Drury 1981); in this fashion, the gneisses were largely "dried out" and thus were prevented from remelting. Gneisses that were metamorphosed at temperatures above 650 °C under dry conditions, and therefore remained in a solid state, are called granulites (granulite facies of metamorphism; see ◻ Fig. 7.24). In contrast, in areas where aqueous fluids remained in the rocks, migmatites were formed. These ductile-deformed rocks consist of bodies that formed as "mixtures" of solid gneiss and molten granite and typically display diffuse rather than sharp boundaries with surrounding bodies of rock. It is also possible to produce the granulites in deep continental crust following partial melting; the aqueous granitic melts separated and rose to higher crustal levels leaving behind the dry granulites. This would be in accord with the fact that the granulites tend to be somewhat more basic in their composition as compared to the granitic gneisses (Tarney 1976). It is assumed that from Archean to present, the lower continental crust is generally rich in granulites.

The Archean granulites and migmatites exposed in shield areas today are underlain by ca. 40 km of continental crust. During metamorphism they had been buried to depths of more than 50 km. Consequently,

the thickness of the crust during metamorphism must have been ca. 90 km. Either it became thickened by underplating of layers of acidic magmatic bodies, or it thickened by collision of continental masses that were stacked by large-scale thrusting. The presence of associated sedimentary rocks that formed on a stable shelf clearly favors the second hypothesis; during collision a passive continental margin is pulled or pushed down the subduction zone (◻ Fig. 10.4). Further confirmation of crustal stacking comes from Canada where deep seismic profiles document imbricate stacks of thick crustal sheets. This is some of the strongest evidence for Archean continental growth by collision and subduction; each incoming crustal block was partially subducted under the adjacent continent much like present India is being subducted under central Asia. Crustal thickening, regardless of mode of origin, leads to isostatic uplift, erosion and exhumation of deeper crustal levels. Although gneiss-migmatite-granulite complexes have formed throughout the geologic record, they are especially concentrated in the Late Archean—a result of the high level of orogenic activity and mobility of the crust.

In summary, the granulite-gneiss belts probably represent the deeper levels of an active continental margin or ensimatic island arc after collision with a passive continental margin. The greenstone belts formed in backarc basins. During compressional orogenesis, each was deformed and metamorphosed. Following collision, both belts were exhumed by isostatic uplift and erosion. Although both belts formed in the same proximity and at present are side by side, the granulite-gneiss belts represent much deeper crustal levels than do the greenstone belts (◻ Fig. 10.4).

Komatiites

Komatiites are peridotitic, ultrabasic (<45% per weight SiO_2) or basaltic, basic (45–53% SiO_2) volcanic rocks that form from mantle peridotite through a high degree of partial melting. They are characterized by high contents in magnesia (MgO > 18%). They transition to normal tholeiitic basalts (MgO < 12%). Komatiites are found in the lower parts of greenstone belts.

Komatiites are abundant in Archean greenstone belts but almost completely absent in post-Archean sequences; an exception is the Upper Cretaceous komatiites from Gorgona Island, Columbia. Their post-Archean absence reflects the lowering of the geothermal gradient in the upper mantle. Komatiites require melting temperatures above 1600 °C. Such temperatures were achieved at relatively shallow mantle

Fig. 10.5 Photomicrograph of komatiite showing the typical elongate olivine crystals ("spinifex texture")

1mm

depths, in which the magmas formed, only in the Hadean and Archean eons. To generate komatiitic melts, 40–70% of lherzolitic peridotite (▶ Chap. 5) must be melted. In contrast, tholeiitic basalts require 15–25% of melting of the same source at temperatures of 1100–1300 °C.

The term "komatiite" is derived from the Komati River that crosses the Barberton greenstone belt, South Africa and Swaziland, one of the oldest and best studied greenstone belts (□ Fig. 10.2). The volcanic nature of the komatiites is documented by the presence of pillow lavas that formed when lavas were extruded underwater (▶ Chap. 5). So-called spinifex textures are a characteristic feature of komatiites. They show that the hot lavas were chilled extremely fast. Spinifex textures (□ Fig. 10.5) are characterized by long, skeletal prisms of olivine and pyroxene, minerals that usually grow as short prisms or equigranular grains. The lengthy prisms typically grow at high temperatures in the melt. When the lava is quenched during extrusion, their shapes become "frozen" and thus preserved. In contrast, when a melt cools at slower rates, mineral growth continues so that the crystals exhibit a more compact habit. Spinifex is a spiny-leaved grass in the Australian steppe, where also greenstone belts occur. Because the lengthy crystals in the komatiites resemble the spiny leaves of the grass, these textures were named after it.

10.3   Towards an Archean Plate Tectonic Model

Both Archean greenstone belts and granulite-gneiss belts can be related to plate-tectonic processes, therefore demonstrating that plate tectonics have operated through most of geologic time. The occurrence of komatiites shows that mantle material was at higher temperatures and that melting was more widespread than it is today. Neither conclusion is surprising given higher rates of radioactive decay in the Archean. Such conditions presented much higher temperatures than those found beneath present-day middle oceanic ridges, backarc basins, or hot spots.

Applying these conditions and the actualistic principles that accompany them, an Archean plate-tectonic model can be hypothesized. The andesites, dacites, and rhyolites of the greenstone belts must have erupted above a subduction zone. The associated detrital sedimentary rocks represent the erosional products of cal-calkaline volcanic rocks in areas with steep relief. Their low maturity reflected by abundant feldspar and dark, ferromagnesian minerals, and low amounts of quartz, as well as their origin as turbidites suggest deposition in forearc, interarc, and backarc basins. Increasing maturity upwards in a given sequence, especially as reflected in quartzites, demonstrates ensuing uplift and erosion and a long transport history from a large continental hinterland. What is unclear is the amount of time represented by such a cycle, but almost certainly it was shorter in duration than Phanerozoic plate-tectonic cycles.

Although no true oceanic crust is preserved from Archean times, just as no Paleozoic or early Mesozoic ocean crust is preserved today, one perplexing piece of the Archean plate-tectonic puzzle that is missing is high-pressure metamorphic rocks such as eclogites. For some years, this was used as evidence that there were no processes corresponding to present-day subduction. However, it can be argued that the lack of Archean eclogites can readily be explained by the hotter temperature regime in the subduction zones. Also, Archean eclogites have been found as xenolith (*xenos*, *lithos*, *Greek* foreign rock) inclusions in much younger kimberlites. Kimberlites are peridotitic rocks from the mantle that are found in subvolcanic vents related to hot-spot magmatism. Many peridotites are Cretaceous age (▶ Chap. 6). The eclogite xenoliths in the Cretaceous peridotites yield ages between 3200 and 2500 Ma. Their protoliths were mostly gabbros but also partly volcanic rocks with the composition of basaltic komatiites. In fact, the composition of oceanic crust is likely to have been mainly that of basaltic komatiites in the Archean Eon. The magmas differentiated into basalts and gabbros as well as olivine-pyroxene cumulates as complementary rocks at the base of the crust (Foley et al. 2003).

Based on the Archean xenoliths, it can be speculated that by 3200 Ma, conditions were reached that permitted the formation of eclogites. Due to the slowly decreasing mantle temperatures the oceanic crust became thinner and approached the composition of tholeiitic basalts. This implies that the temperature regime in the subduction zone also became cooler. The uppermost portions of the basalts were infiltrated by ocean water and thus altered by hydrothermal solutions (▶ Chap. 5). Basalts that contain hydrous minerals are metamorphosed to garnet amphibolite, a garnet bearing hornblende-plagioclase rock, in the subduction zone. Hornblende fixes water in its crystal lattice. When the subduction zone enters the asthenosphere and attains temperatures in excess of 1000 °C, the garnet amphibolite becomes partially melted. The solid part of the rock is transformed into water-free eclogite that consists of pyroxene and garnet. The extracted melts have tonalitic to granodioritic composition and intrude the crust of the plate above the subduction zone. Therefore the tonalite-trondhjemite-granodiorite associations of the granulite-gneiss belts are likely to represent the partial melts rising from subduction zones and as such build the lower parts of island arcs or adjacent active continental margins (Foley et al. 2003). The production of these rocks rapidly increased at ca. 3200 Ma.

The formation of eclogites in the subduction zones at 3200 Ma suggests that by that time slab pull became the main driving mechanism of plate drift. Because continental crust existed before 3200 Ma, processes that produced tonalites, trondhjemites, and granodiorites must have operated back to Early Archean and possibly Hadean times, as shown by the 4400 Ma old detrital zircons from western Australia. Probably hydrothermally altered basaltic rocks from the uppermost crust were pulled or pushed into a subduction zone; there they were partially melted and subsequently ascended to form the acidic magmatic rocks that constituted the primordial continents.

From the various lines of evidence concerning the characteristics of early oceanic crust, it may be deduced that it evolved from an overall ultrabasic (peridotitic-komatiitic) to a basic (theoleiite-basal-tic) crust in Hadean and Archean times over the first two billion years of Earth history. Trough-out the Archean Eon, the rates of creation of new lithosphere at the spreading centers as well as the rates of subduction decreased, the overall length of spreading axes and subduction zones became shorter, the oceanic crust became thinner, and the plates became larger. Due to the increased distance between the spreading axis and subduction zone, oceanic lithosphere would cool more effectively. At ca. 3200 Ma, eclogite formation in subduction zones was possible and the plate tectonic scenario very much resembled the present regime.

The interpretation of the greenstone-granite belts and the granulite-gneiss belts as regions of island arcs and active continental margins supports the fact that these are the places where new continental crust is formed in the plate-tectonic model. Through the processes of erosion of quartz-bearing rocks and their recycling during subduction, collision and re-melting, increasing amounts of acidic crust were formed. The resulting granites consist of quartz, alkali feldspar, and plagioclase in sub-equal amounts. Towards the end of the Archean, these granites appeared in voluminous amounts. Many researchers agree that the Late Archean was the period during which the most abundant formation of new continental crust occurred (see below).

The Archean plate-tectonic model dovetails with the modern model and suggests that plate tectonic processes have existed throughout most of Earth history. However, it must be emphasized that early plate tectonics featured certain peculiarities and that it took two billion years, nearly half of Earth's history, until the creation and destruction of plates operated according to the present-day scheme. Towards the end of the Archean Eon a setting existed that was characterized by few, stable continents and large oceanic basins. The number of plates and the overall length of spreading axes and subduction zones were probably similar to those of the last two billion years.

10.4 The Growth of Continents

Archean terrains form the cores of continents and are called "shields" because in map view they resemble the shapes of shields carried by warriors. It has long been observed that during later orogenies younger mountain belts were attached to these central shield continental masses, thus systematically enlarging the continents. It has also been hypothesized that the formation

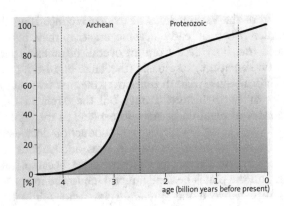

◻ Fig. 10.6 Graph showing possible growth of continental crust with time (present volume: 100%)

of new continental crust was increased during orogeny and that the rate of continental growth was approximately linear when averaged over long periods of geologic time. However, radiometric dating has challenged this last assumption over the last four decades. We now know that in many younger orogens, Archean crustal material was reincorporated and again deformed, metamorphosed, or melted. Despite this later overprint, it is possible to date the original rock by the uranium-lead method applied to zircon crystals. As a result of these age dates, it can be asserted that at the end of the Archean Eon, much larger areas of continental crust existed than would be suggested from the actual distribution of existing Archean rocks (◻ Fig. 10.1).

Early continental crust has repeatedly been destroyed by meteorite bombardment, especially in the period before 3900 Ma. Enormous crustal growth occurred in the second half of the Archean era, between 3200 and 2500 Ma (◻ Fig. 10.6). tis is due to the considerably more efficient production of rocks constituting continental crust above subduction zones since ca. 3200 Ma; meanwhile, plate mobility and subduction activity were still high. In this time span, most of the greenstone-granite and granulite-gneiss belts were formed. Before 3200 Ma there were only small, unstable continents. Supported by continent collision, they became considerably enlarged and stabilized between 3200 and 2500 Ma. These 700 million years probably reflect the most important change in the history of the Earth.

Radiometric dating incorporated with the ideas above suggest that by the end of the Archean Era, between two thirds and three quarters of the total volume of continental crust had been generated (◻ Fig. 10.6). From chemical patterns in magmatic rocks, it is con-

cluded that the crust of North America, except for that under Phanerozoic mountain ranges, had nearly reached its present-day normal thickness of ca. 30 km at that time (Condie 1973). In the Early Archean, shallow-water shelves existed but they were few and narrow; this probably reflects the dynamics of the times. The size and amount of shelves changed ca. 3000 Ma when thick, extensive sediment sequences were deposited in shelf environments. At the end of the Archean Eon, large and stable continents with abundant granites dominated. Passive continental margins featured broad shelf areas. The resulting stable continents are called cratons (*kratos*, *Greek* strength). In general, large Archean-Proterozoic cratons such as the one that comprises much of North America have escaped extensive subsequent orogenic activity, although there are several important exceptions. Rather, the Phanerozoic record of cratons is mainly recorded in the mostly non-deformed layers of sedimentary rocks that overlie the Precambrian basement.

10.5 Possible Younger Equivalents of Greenstone-Granite Belts

Although greenstone belts containing komatiites are a typical feature of Archean terrains, there are similar rock complexes in later Earth history that have been compared with the Archean belts. Due to the more recent lower temperatures in the upper mantle, they do not contain komatiites.

The *Rocas Verdes* (*Spanish* greenstones) complex extends over 800 km in southern Chile between 51° S latitude to Cape Horn at 56° S. It represents a 150–120 Ma old Late Jurassic to Early Cretaceous backarc basin (Tarney et al. 1976). The basin, positioned between the South American continent to the east and an island arc to the west, was floored by oceanic crust. Cherty and clayey sediments were deposited in the basin. Close to the island arc and extending into the basin center, these sediments are overlain by calcalkaline volcanic rocks, mainly andesites, and immature detrital sediments, mainly graywackes supplied with the volcanic material. The island arc is underlain by older continental crust. During the middle Cretaceous, the basin was tectonically squeezed into an elongate zone between the island arc and the continent. It experienced strong deformation and metamorphism to greenschist facies. After this event, large granitic and granodioritic magmatic bodies intruded into the sequence.

The Great Dike of Zimbabwe

In Zimbabwe, formerly Rhodesia, a 500 km-long and 6–8 km-wide dike of basic and ultrabasic magmatic rocks is called the "Great Dike of Zimbabwe" (◧ Fig. 10.1). Its age is about 2575 Ma (Armstrong and Wilson 2000). This largest "dike" on Earth, which has vertical walls, crystallized from a magnesium-rich tholeiite-basaltic magma that differentiated into grabbros and ultramafic cumulates that generated a layered internal structure.

The Great Dike of Zimbabwe is generally considered to be a deep-level exposure of a continental rift exhumed by erosion. The continental crust became separated by the basaltic magma but no further opening ensued. The Zimbabwe craton, which is dissected by the dike, formed by orogenic processes betweeen 2900 and 2600 Ma and was a stable continent or craton at the time of the intrusion of the dike. This documents that towards the end of the Archean era, large continents existed. The Great Dike is the oldest of its kind.

The Widgiemooltha Complex in western Australia is a similar geological feature but consists of several dikes, the oldest of which intruded around 2400 Ma. The largest of these dikes is 320 km long and 3.2 km wide. The dikes dissect a greenstone-granite belt.

The Penninic units of the Eastern Alps, exposed in the large Tauern Window (▶ Chap. 13) also contain a sequence that is reminiscent of greenstone-granite belts (Frisch et al. 1990). Developed on at least partly oceanic crust, an island-arc system formed above a subduction zone that underwent a long-lasting evolution in the latest Precambrian and Early Paleozoic. This so-called Habach Terrane (see ◧ Fig. 12.5) was jammed between continental blocks during the Variscan orogeny ca. 320 Ma and thus became part of the emerging supercontinent Pangaea. During this event, the former ocean floor and island-arc sequence was intruded by a large number of tonalite, granodiorite, and granite bodies that later, during the Alpine orogeny, were transformed into gneisses. The association of basic (basaltic) and more acidic (andesitic-dacitic) volcanics, detrital sediments containing large amounts of volcanic material, and large volumes of granitic rocks resembles the greenstone-granite belts in many respects.

Plate Tectonics and Mountain Building

Contents

© The Author(s), under exclusive license to Springer Nature Switzerland AG 2022
W. Frisch et al., *Plate Tectonics*,
Springer Textbooks in Earth Sciences, Geography and Environment,
https://doi.org/10.1007/978-3-030-88999-9_11

11

One of the greatest strengths of the modern plate tectonics theory is its ability to explain the origin of virtually all of the present and most ancient mountain belts on Earth. In other words, mountain building (orogeny) stands in strong causal interrelation with the global plate drift pattern. The motor of orogeny is subduction. To enable subduction, a basin floored by oceanic crust must be present. The process of orogeny becomes initiated by subduction of ocean floor and finds its climax in the collision of continents and island arcs. Mountain belts are elongate zones characterized by crust thickened to more than 70 km, in comparison to normal continental crust that is 30–40 km thick. The most classical *style* of orogeny involves continental collision that follows a lengthy period of subduction and completes a Wilson cycle. Ensuing orogeny leads to crustal thickening, deformation, metamorphism, and uplift. This style is called the Alpine style of orogeny. In contrast, orogenic belts that face Pacific-style oceans do not culminate with continent–continent collision, but rather involve long periods of ocean-slab subduction beneath continental margins with repeated episodes of collision that involve island arcs, oceanic plateaus, and microcontinents. This orogenic style is exceptionally rich in volcanic/plutonic production and is called the Cordilleran style of orogeny.

The following discussion overviews the above orogenic styles; however, it is necessary to realize that transitions exist between each style and that many mountain belts have been generated by combinations of both of these. But first we will examine some of the variations in the types of subduction zones and active continental margins that lead to these different styles of orogeny.

11.1 Types of Active Continental Margins Within Orogenic Styles

The Alpine and Cordilleran orogenic styles describe long-lasting orogenic cycles that commonly involve numerous phases of orogeny. Each style consists of smaller events that are related to specific geometries of the active continental margin within the larger orogenic cycle. We refer to these as *types* of orogeny. ◘ Figure 11.1 illustrates these types, although it must be emphasized that other types are also possible.

The *island-arc type* of orogeny (◘ Fig. 11.1a–c) forms following lengthy periods of subduction activity within an ocean or along its margins; the subduction generates island arcs and backarc basins such as those presently found in the western Pacific region. Subduction in these cases is generally initiated by the high density of old oceanic lithosphere ("spontaneous subduction"; see ◘ Fig. 7.5). The persistent subduction that

accompanies island arc evolution eventually results in closure of the ocean when the subduction velocity exceeds the unilateral spreading rate at the mid-ocean ridge for a long time. During closure, the island arc collides with the approaching passive continental margin and is thrust over it. However, the convergence process is not terminated because the motion of the plates is driven by the global plate drift pattern and oceanic realms are still present to be subducted. The subduction zone typically jumps across the accreted arc into the adjacent ocean realm and a new island arc is built (◘ Fig. 11.1a). Following arc accretion and renewal of subduction, the polarity of the subduction zone may change; a new volcanic arc is then built on the collided terrain (◘ Fig. 11.1b). When the new oceanic lithosphere is still hot, it may be obducted or overthrust by the plate convergence because it is too buoyant to sink by its own density ("forced subduction"). There may be several intra-oceanic subduction zones in a given region such as it is presently the case around the Philippine Sea Plate and the Molucca Islands (◘ Figs. 11.1c and 13.1).

The final closure of all oceanic realms results in collision of two or more continents. The two continent margins and the intermediate island arc systems become overthrust and underthrust, folded, and metamorphosed where they were dragged or pushed into depth. During an orogenic cycle of the island-arc type, new continental crust is created from the long lasting subduction by lengthy magmatic activity. An example for this type of orogeny occurred during the Late Precambrian Panafrican orogen in northeast Africa and the Arabian Peninsula (▶ Chap. 12). However, as was stated above, not all oceans close.

The *Andean type* of orogeny (◘ Fig. 11.1d) is represented by and named after the Andes Mountains. Along this active continental margin, mountain belts were generated by long-lasting magmatic activity, subduction, and terrane accretion, processes common in the geologic history of the North American Cordillera (▶ Chap. 9). Andean orogenic style is characterized by subduction directly under the continent rather than beneath a fringing island arc system.

The *Alpine style* of orogeny (◘ Fig. 11.1e, f) is described by a "Wilson cycle". Such a cycle, named after J. Tuzo Wilson, the discoverer of the transform faults, starts with the break-up of a continent and growth of an ocean. Such oceans may remain limited in size or attain the dimensions of the Atlantic Ocean. Eventually, the ocean closes during continent–continent collision ending the cycle. The types of active margins present adjacent to the closing ocean basin can be either island arc type, Andean type, or both. Active margins may be present on either or both of the closing continental margins. The size of the ocean between the continents will determine whether crustal growth by magma-

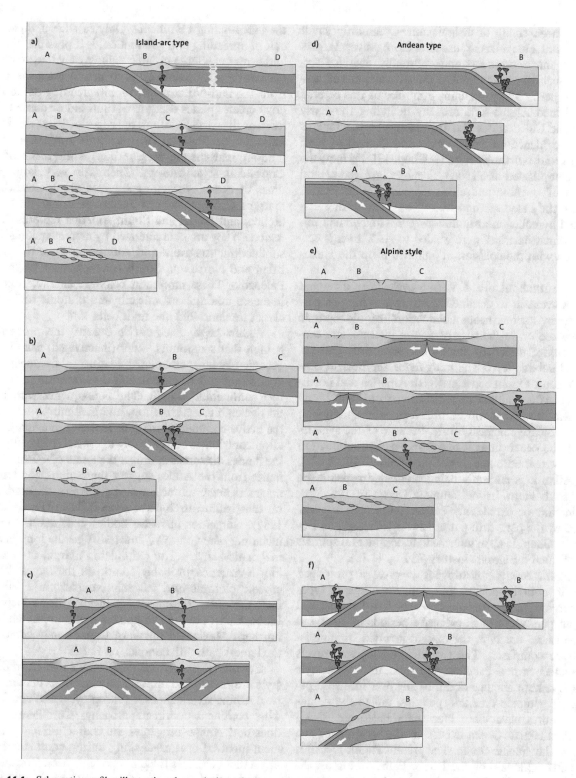

◘ Fig. 11.1 Schematic profiles illustrating the evolution of orogens as a consequence of subduction and collision. The island-arc type **a–c** creates complex scenarios with opposing or flipping subduction zones—three possible scenarios are shown. The Andean type **d** is characterized by abundant magmatic growth (like the island-arc type), sporadic terranes may contribute to crustal thickening. The Cordilleran style is a combination of the Andean and island arc types but without subsequent continental collision. The Alpine style **e, f** represents the "normal" collisional orogen, where two continent masses collide and intermediate continent splinters (example e: Alps) may be involved. Example e shows a complete Wilson cycle, typical of Alpine-style orogens

tism is large, small, or insignificant. Commonly, small continental blocks separate from the passive margins to form microcontinents and eventually, accreted terranes. Such blocks complicate the subduction and collision process. There are many examples of this type of complicated Alpine-style orogeny including the Early Paleozoic Caledonides and the closing of the Iapetus Ocean; evidence for this event is now present in both Europe and North America (▶ Chap. 12). Perhaps the most complicated of all such orogenies involved several phases of opening and closing of the Tethys Ocean. During the Paleozoic and Mesozoic, each closing was marked by collision as numerous peri-Gondwanan microcontinent terranes were welded to Asia. Events culminated with the collision of India to form the Himalayas.

The turnabout in the Wilson cycle from divergent plate movement to continent convergence may be performed by "spontaneous subduction" of old oceanic lithosphere or, in the case of young intermediate oceanic realms, forced by the global plate drift pattern, "forced subduction". The margins of the present central Atlantic Ocean between North America and northwestern Africa consist of old, Lower to Middle Jurassic ocean floor that will be transformed in subduction zones in near geological future. Because of the symmetry of the ocean floor it may happen that subduction zones form at both margins of the ocean (◻ Fig. 11.1f). The Atlantic is currently being subducted beneath the Caribbean Plate. In the Penninic Ocean of the Alps, subduction occurred only on one side and was enforced by the plate drift pattern as Africa closed on Europe (▶ Chap. 13). An intermediate continental splinter complicated the closure history (◻ Fig. 11.1e).

Cordilleran-style of orogeny involves a prolonged case of both island arc and Andean orogenic types of margins. In the Andean type, an active continental margin persists through extensive periods of time and in the island arc type, the margin involves fringing island arc complexes. The Cordilleran-style orogeny is associated with a Pacific-style of ocean. Pacific-style oceans remain as huge ocean basins over immense periods of geologic time and perhaps, unlike the Atlantic or Tethys, never close. During the Paleozoic, the giant Panthalassa Ocean existed at the location of the present-day Pacific Ocean. The present Pacific Ocean is currently underlain by the Pacific Plate, an almost exclusively oceanic plate; during the Mesozoic, it was underlain by the Farallon Plate, an equally large oceanic plate. However, oceanic crust older than Jurassic is not found in the Pacific Ocean today. Long-lasting and mostly rapid subduction has been compensated by accordingly rapid spreading. Through the subduction of old, dense lithosphere, rapid subduction was stimulated that, in turn, brought about rapid spreading along the mid-ocean ridge thus preventing collisions of the surrounding continents. Only in such a rapid circuit of spreading and subduction, is it possible to form and leave open large oceanic realms; only these circumstances permit oceanic crust to persist great distances from the mid-ocean ridge; rapid spreading of oceanic crust creates plates that are not too old or dense to remain at the surface. Along the edges of large, semi-permanent (Pacific-style) oceans, both Andean and island-arc margins are present. This is the realm of the Cordilleran-style orogeny. Therefore, very large volumes of ocean crust are subducted along Cordilleran margins. Orogenic belts in Eastern Australia, East Antarctica, and North and South America represent such regions. They are characterized by extremely long-lived subduction, Australia and Antarctica since the Cambrian and North and South America since the middle Paleozoic. These mountain belts are characterized by immense amounts of volcanic and plutonic rocks and classic foreland fold and fault belts.

Within large, Pacific-style oceans, the probability is high that seamounts, older (inactivated) island arcs, oceanic plateaus, or isolated continental fragments will drift along and become accreted as terranes to the active continental margin. The consequences are crustal thickening, deformation, and metamorphism along the active continental margin. However, magma generation contributes considerably to crustal thickening—in the Andes, the crust attains a thickness of 70 km. Estimates from the Andes suggest that subduction-related magmatism caused oceanward growth of the continental crust of up to 200 km since 200 Ma (Drake et al. 1982). Numerous intrusive bodies penetrate the crust, including older plutons, and testify to the long lasting magmatic activity. The Cordilleran margin of western North America probably comprises the largest Phanerozoic plutonic and volcanic complex on Earth. In fact, the once continuous Peninsular–Sierra Nevada–Idaho–Coast plutonic complexes that now stretch from northern Mexico to northern Canada may constitute the largest batholith complex on Earth.

Comparisons between Alpine style and Cordilleran style orogenies yield important distinctions between the two. Alpine style defines a Wilson cycle and concludes with continent–continent orogeny; Cordilleran style does not. Alpine orogenies are characterized by basement-involved crustal stacking during continental collision that generates thick nappe sequences; Cordilleran orogenies tend to involve thick sedimentary sequences and sedimentary thrust sheets. Alpine style generates melting within the thickened crust including melting of sedimentary rocks to produce S-type granites; Cordilleran style generates vast primary melts and produces new (juvenile) crust and I-type granites. Alpine style produces bilateral foreland fold and thrust belts and foreland basins that form the sites for thick molasse deposits; Cordilleran style produces a single foreland fold and thrust belt behind the active arc with thrusting directed towards the continent; thrusting generates a single foreland basin, commonly called a retroarc foreland basin. Alpine orogenies tend to destroy older foreland basins by structural, metamorphic, and plutonic processes; Cordilleran orogenies commonly preserve foreland basins, sometimes with little deformation. Both styles commonly have accreted terranes, although Cordilleran orogenies tend to have more oceanic-type terranes and Alpine orogenies tend to have accreted microcontinents. Both styles can contain numerous ophiolite sequences. Classic Alpine-style orogenies include the Caledonian-Acadian (Silurian-Devonian of Europe and eastern North America that closed the Iapetus Ocean), Variscan-Alleghenian (Carboniferous-Permian of Europe and North America that closed the Rheic and related oceans), Urals (Carboniferous-Permian that sutured Europe and Asia and closed the Ural Ocean), and the greater Alpine-Caucasus-Himalayan (late Mesozoic and Cenozoic that sutured Eurasia and parts of Gondwana and closed the Tethys Ocean). Cordilleran-style orogenies include the greater Nevadan-Sevier (late Mesozoic and early Cenozoic of Western North America), Andean (Mesozoic and Cenozoic of South America), New England (Permo-Triassic of Eastern Australia), and numerous orogenic events that affected parts of Antarctica, South America, South Africa, and SE Australia throughout the Phanerozoic.

11.2 Continent–Continent Collision

Alpine-style orogeny culminates in continent–continent collision. The collision of continents causes considerable deformation and large-scale overthrusting on the order of a 100 km or more. The condition of the subduction process persists during the early stages of collision. In front of the overriding plate, a deep trench or trough exists that becomes filled with large volumes of turbidites (flysch sediments) derived from the raising nappe pile in the hinterland (▪ Figs. 11.2 and 11.3). During this time, the trough is underlain by continental crust of the subducting plate and the flysch sequence becomes deposited on top of the shelf sequence of the passive continental margin of that plate. Flysch generally forms the youngest sedimentary deposits of such a sequence because the sediment pile becomes overthrust beneath the margin of the upper plate in the following stage. Similar but smaller-scale conditions exist in front of individual nappes within the plates (▪ Fig. 11.2).

During the stages of plate collision, the axis of the sedimentary trough migrates towards the subducted plate as the nappes advance (▪ Fig. 11.2). Initially the top of the sedimentary prism resides below sea level, even at near-abyssal depths. During this time, the trough is filled with deep-water flysch, commonly turbidites, that is derived from the uplands on the upper plate (▪ Figs. 11.2a and 11.3a). The flysch accumulates in the main trough between the two plates and also in

▪ **Fig. 11.2** Collision of two continents and evolution from **a** the flysch stage (flysch sedimentation in deep troughs in front of the upper plate or of individual nappes) to **b** the molasse stage (filling up of the foreland trough with the debris derived from the ascending mountain range)

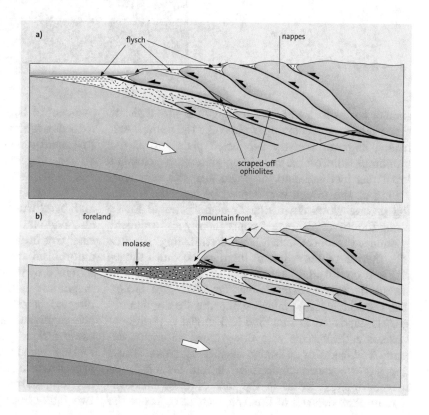

■ **Fig. 11.3**   Flysch sequence in the Eastern Alps (Rhenodanubian flysch; photograph by M. Meschede) and molasse sequence in Tibet (Qiuwu molasse, see ■ Fig. 7.18; photograph by W. Frisch). In the deep-marine flysch beds, layers of sandstone, and mudstone alternate; each layer is deposited by a turbidity current and shows graded bedding (becoming finer grained towards its top). The folded terrestrial molasse beds in Tibet show alternating conglomerate-sandstone (light layers) and mudstone (dark layers)

smaller basins formed between individual thrust sheets along the front of the upper plate. As the main flysch trough migrates over the subducted plate and as the mountain front on the overriding plate rises, sediment input rate increases and the basin fills with sediment. Filling of the trough is enhanced because the thrusting process slows down due to high frictional forces along the thrust planes and consequently less sediment is removed from the trench by underthrusting. Eventually, the top of the sediment interface becomes subaerial and continental depositional systems replace and overlie the marine deposits (■ Figs. 11.2b and 11.3b). The resulting coarse-grained sediments are dominantly gravels deposited in shoreline and fluvial depositional systems and alternate with fine-grained sediments deposited in lakes or stillwater reaches of rivers. These deposits are called molasse. In humid settings, coal may be intercalated as in the case of the Pennsylvanian molasse deposits of the Appalachian basins of the

east-central United States. In arid settings, continental redbeds are formed as in the case of the Devonian Old Red Sandstone of northwest Europe during the Caledonian orogeny. For etymology of the terms "flysch" and "molasse", see p. 104.

The burial of continental domains during collision and overthrusting not only causes deformation, but also regional metamorphism across large parts of the emerging orogen. The most important factors of metamorphism are (1) the lithostatic pressure, which is determined by the depth of burial and acts with the same magnitude in all directions, (2) the temperature, (3) the directed pressure, which creates a tectonic stress that acts at higher magnitude in the direction of tectonic compression, and (4) the fluid phase, mainly water. During metamorphism the mineral assemblage (paragenesis) of a rock is adapted to the prevailing pressure and temperature conditions by mineral reactions. These two factors are mainly responsible for the paragenesis of the metamorphic rock. Stress, in contrast, causes deformation, rotation, and recrystallization of minerals under preferred orientation—it is therefore responsible for the cleavage, a typical texture of most metamorphic rocks. The fluid phase considerably accelerates mineral reactions and recrystallization—in a "dry" environment that lacks a fluid phase, reactions remain largely incomplete.

During continent collision, large volumes of cool rock from the upper part of the subducting continental margin are brought to increasing depth. Therefore, rocks involved in an orogeny usually initiate metamorphism with conditions of a high pressure/temperature (P/T) ratio as is typical of high-pressure or Barrow-type metamorphism (see ◘ Fig. 7.24). In an advanced stage of burial, the P/T ratio gradually lowers due to thermal adjustment: the rocks are heated to the ambient temperatures of the given depth after burial has been completed and exhumation started. Therefore, the P/T conditions usually describe a loop (◘ Fig. 11.4). During burial, the temperature increase lags considerably behind the normal gradient of 30° C/km because rocks exhibit low thermal conduction; the extreme case is that of high-pressure metamorphism in subduction zones. During exhumation of rocks the temperature gradient increases—the heated rocks are lifted to shallower depth and subsequently cool under

delayed conditions. Therefore, as compared to burial, the P/T ratio is considerably lower.

Continent collision commonly leads to anatexis, the partial melting of crustal rocks. The presence of "wet" metamorphosed mudstone and sandstone favors the formation of melts that initiate around 650 °C. Anatexis generates large bodies of magmatic rock, chiefly granites to granodiorites, that are called batholiths; they are surrounded by zones of high-grade metamorphism. The temperatures of the melts usually only slightly surpass the melting temperature of the rock; therefore, the magma bodies cannot ascend larger distances but rather solidify near their place of formation. These rocks typically classify as S-type granites, formed from sedimentary protoliths (▶ Chap. 7).

Alpine-type continent–continent collisions are characterized by distinct features (see also above). These include (a) large nappes that express large-scale stacking of continental crust; (b) broad belts of deformation and regional metamorphism in which the metamorphic history traces a typical pressure–temperature loop; (c) the occurrence of ophiolitic sutures that mark the seam of collision and represent remnants of the ocean floor, commonly transformed by high-pressure metamorphic conditions; (d) island-arc magmatism. The latter may be of minor importance due to limited duration of subduction of small ocean basins, in which cases the growth of continental crust is also rather limited.

11.3 Uplift, Erosion, and Elevation of Mountains

Crustal stacking during continent collision typically generates crustal thicknesses between 50 and 70 km. As a consequence, the newly formed orogen experiences isostatic uplift that is responsible for the morphological feature called a mountain range. However, in many orogens, including the Alps and the Himalayas, uplift of the surface to high elevations was not the immediate consequence of crustal thickening but occurred after a delay of several million years.

During and following the collision process, the subducting continental margin is pulled downward by the attached oceanic slab. The dense oceanic slab forms a counterweight against the thickened and buoyant continental crust. Continued slab pull causes ongoing thrusting of the colliding continent margins and leads to the creation of new nappes and greater thrusting distances; however, new, high mountains do not develop yet. With time and continued subduction, the resistance against compression in the collision zone becomes strong enough so that the dense, heavy oceanic lithosphere in the subduction zone breaks off; this results because rocks possess low tensile strength. The de-

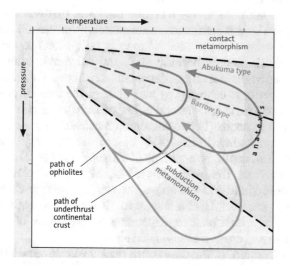

◘ **Fig. 11.4** Typical pressure–temperature loops from rocks in a collisional orogen. Oceanic crust or continent splinters can be deeply subducted and experience subduction metamorphism; during their ascent they will be overprinted in amphibolite or greenschist facies (green paths). Other parts of continental crust experience pressure-emphasized regional metamorphism (Barrow-type) or even anatexis (partial melting) during burial and reach fields of the Abukuma-type regional metamorphism during ascent (brown paths). During their ascent the rocks are much hotter than during descent at the same depth, because both heating and cooling are slow processes. Compare ◘ Fig. 7.24

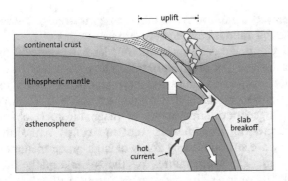

◻ **Fig. 11.5** Cross section showing effects of slab breakoff. Break-off of the subducted, dense oceanic lithosphere after collision is triggered by the buoyancy of the lower-plate continental crust and the decelerating plate convergence (caused by increasing frictional forces). The slab breakoff allows hot asthenospheric mantle to ascend into the newly created space and to cause partial melting. The melts can rise and penetrate the crust of the upper plate. In addition, the loss of the heavy counterweight enables rapid uplift of the mountain range

11

tached slab freely sinks into the mantle (◻ Fig. 11.5). This process is called slab breakoff. Slab breakoff causes the thickened continental crust to lose its counterweight and commence isostatic uplift. By analogy, a small boat being pulled down by a heavy anchor will bob up if the anchor is cut free. Isostatic uplift allows mountain ranges to reach great heights.

Climate plays a key role for shaping the mountain range. In arid regions like the Altiplano of the Andes or the Tibetan Plateau of the Himalayan region, denudation by erosion is very slow. Therefore, uplift of the crustal stack generally equals uplift of the surface. Elevated surfaces in arid climate commonly show surprisingly low relief (◻ Fig. 11.8, upper). In humid regions erosion rates are accelerated and relief is increased. This is especially apparent where glaciation is present, as ice is a very efficient agent of erosion. Glacial erosion creates deeply incised valleys with strong relief. Although erosion reduces the mean elevation of a mountain range, summit areas, which generally exhibit lower than average erosion rates, commonly increase in elevation due to uplift. The effect of climate applies well to the Himalayas where the southern monsoon generates enormous amounts of precipitation. Here, the erosion rates are considerably higher than in the much dryer Tibetan Plateau that lies in a rain shadow behind the mountain front. As a result, the Himalayas show lower mean elevation but higher summits than the Tibetan Plateau (◻ Fig. 11.8, lower; see also below). Mountain topography is complex and reflects many factors including rock type, structure, isostatic uplift, and climate.

According to the principle of isostasy, a body floating in a denser medium will stand higher in proportion to its thickness. For example, a thick iceberg will attain higher elevation than a thin one. With respect to continental crust, this means that thick crust is characterized by high mountains or plateaus and thin crust is char-

acterized by lowlands or continental shelves below sea level. If isostatic equilibrium is attained, the weight of any rock column of a given area that rests on a theoretical horizontal plane at depth is equal (◻ Fig. 11.6). Because the configuration of the Earth below the lithosphere is rather uniform, a plane of equilibrium can be defined at or slightly below the base of the lithosphere. Examples, as illustrated in ◻ Fig. 11.6, show the isostatic influence of crustal thickening during an orogenic cycle and changes in the thickness of the lithospheric mantle due to syn-orogenic thickening or post-orogenic delamination.

The models presented in ◻ Fig. 11.6 assume the following: (1) Continental crust has an average density of 2.8 g/cm^3; the rocks of the upper and middle crust generally show slightly lower density but the more basic rocks in the deep crust have higher specific weight. (2) The peridotites of the lithospheric mantle have a density of 3.3 g/cm^3. (3) Peridotites of the asthenosphere, which is hotter and contains minor amounts of melt, have a density of 3.25 g/cm^3. Although the actual densities may slightly deviate from these values, this would not change the results significantly.

Column 'a' assumes an initial continental lithosphere with 30 km of continental crust and 70 km of lithospheric mantle. Therefore, a weight of 64,000 kg per square centimeter would rest on a theoretical plane at 200 km depth (◻ Fig. 11.6a). Such a rock column would be at sea level, actually slightly above (which is neglected here for simplicity).

Column 'b' shows a crustal thickness that was doubled during continent–continent collision to 60 km; the weight of the rock column is only 62,650 kg/cm^2 at 200 km depth because thick and buoyant continental crust replaces more dense asthenoshpere. This mass deficiency is balanced by buoyancy of the rock column that is provided by lateral inflow of asthenospheric material at depth. The rock column regains isostatic equilibrium, when it rises 4154 m because the weight at 200 km depth is again 64,000 kg/cm^2 (◻ Fig. 11.6b). Therefore, the now 204.154 km-high rock column would stand 4154 m above the original surface elevation, sea level. Over a large area, this scenario could result in a high plateau, depending on the nature of climate and erosion as discussed above.

Column 'c' illustrates both thickened crust and lithospheric mantle, each doubled from column 'a'. Such a geometry can be generated during an orogenic cycle—although in many orogens the thickness of the lithospheric mantle is not well constrained; it may attain a thickness of 140 km or more. In this case, uplift of the rock column to 3077 m would be sufficient to establish isostatic equilibrium (◻ Fig. 11.6c). The lithospheric mantle, slightly denser than the asthenospheric mantle replaced by it, compensates for a smaller part of the mass deficiency caused by the thickened crust.

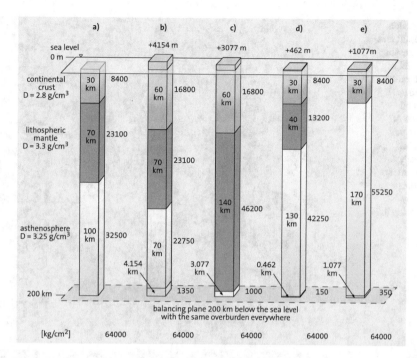

Fig. 11.6 Effects of changes in thicknesses of continental crust and lithospheric mantle with respect to surface elevation. A balancing plane is assumed at 200 km depth on which the pressure per unit of area (here taken as kg/cm²) must be equal, when equilibrium is attained. Thickening of the relatively light crust and thinning of the dense lithospheric mantle both cause uplift of the rock column. **a** Initial conditions with a 100 km-thick lithospheric plate containing continental crust. **b** Doubling of the crust during orogeny. **c** Doubling of both crust and lithospheric mantle (possible under mountain ranges). **d** Delamination of a part of the lithospheric mantle as compared to the initial condition. **e** Complete delamination of the lithospheric mantle

Columns 'd' and 'e' demonstrate the effects of delamination of the lithosphere and thermal transformation of lithosphere to asthenosphere. Both processes will cause uplift. Either part of or the whole lithospheric mantle may delaminate and sink freely into the deeper upper mantle. Equally plausible is a temperature increase that transforms the whole or the lower part of the lithospheric mantle into asthenospheric mantle as both consist of peridotite. In these cases, dense lower lithosphere becomes replaced by slightly less dense asthenospheric material. Compared to the original scenario of column 'a', in 'd', a 30 km-thick crust is maintained but lithosphere delamination or thermal transformation results in the lower 30 km of the lithosphere being replaced by asthenosphere and an uplift of the rock column of 462 m results (Fig. 11.6d). The uplift effect is only one ninth of that in the case of doubling crustal thickness (column 'b'). If the whole lithospheric mantle is replaced by asthenosphere, as shown in column 'e', uplift will be 1077 m (Fig. 11.6e).

Each case shown in Fig. 11.6 results in mass deficiency that causes surficial uplift. However, uplift is a slow process and will partly be compensated by erosion. Due to the very low erosion rates in arid climates, surface uplift will nearly equal the uplift of the rock column (rock uplift). This is valid for the Altiplano or the Tibetan Plateau. In the Tibetan Plateau, there is some precipitation and erosion, but drainage is internal as no river breaks through the margins of the plateau. Therefore, the mass balance remains constant. The mean elevation of 5000 m corresponds to a crustal thickness of 60–70 km and a relatively thin lithospheric mantle, as has been demonstrated beneath parts of the plateau. The thin lithospheric mantle probably formed by thermal transformation of lithospheric mantle to asthenospheric mantle due to hot convection streams.

When isostatic equilibrium is attained, uplift will cease. Erosion reduces the thickness of the relatively light crust by removal of material at the surface, again leading to mass deficiency in the rock column. To balance this loss of mass, the rock column will again be uplifted, but the new surface elevation, or mean elevation in case of a relief, will be lower than before because the eroded crust is compensated by dense asthenosphere at depth. Erosion of 1000 m of crust results in a decrease in surface elevation of 138 m and rock uplift of 862 m as shown in Fig. 11.7a–c. The formula is

$$\text{rock uplift} (862\,\text{m}) = \text{erosion}(1000\,\text{m}) + \text{surface uplift}(-138\,\text{m}),$$

where by surface uplift is negative. The relation between 138 m decrease in surface elevation and 1000 m erosion (i.e., 13.8%) remains constant, if the same densities for the rock units are assumed (see example of Himalayas below, Fig. 11.9).

■ **Fig. 11.7** Effects of erosion and relief on mountain uplift and topography. Erosion reduces the mean elevation but the uplift of peaks follows the pace of valley incision. **a** Conditions shown in ■ Fig. 11.6b. **b, c** Erosion of 1000 m causes reduction of surface elevation by 138 m und rock uplift of 862 m. **d** When erosion concentrates in valleys and spares the peaks, the latter will rise 862 m, whereas the mean elevation will be reduced by 138 m. The volume of erosion in (**c**) and (**d**) is the same

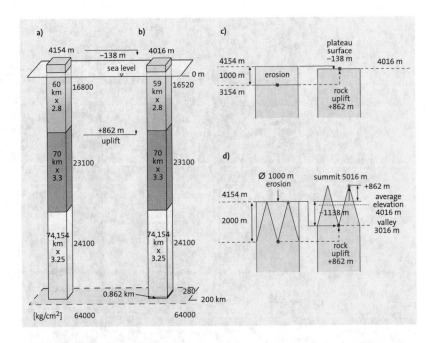

11

■ **Fig. 11.8** Photographs contrasting Tibetan Plateau with Himalayas. The Tibetan Plateau (above) forms a flat depression with limited relief, only interrupted by individual mountain ranges (NW Tibet, photo taken on ground at 5100 m, photograph by W. Frisch). The Himalayas (below, aerial photograph by W. Frisch) achieved their high relief due to the intense monsoonal precipitation (Cho Oyu, 8200 m, to the right; Tibetan Plateau in the background)

However, erosion does not result in a planar surface, but rather causes valley incision and creates a relief that can be considerable, depending on uplift and elevation. This means that valleys, places where erosion is greatest, become deeply incised whereas summit areas, where erosion is low, will rise relative to their initial position. When 2000 m-deep valleys form, the summits do not experience erosion, and erosion averages 1000 m as in the above example; rock uplift will be 862 m accompanied by a decrease in *mean* elevation of 138 m. To summarize, the valley floors will eventually become 1138 m deeper than the surface in its original position, whereas the peaks will rise 862 m (■ Fig. 11.7d). For simplicity, idealized V-shape valleys have been illustrated, which although not realistic, leads to an acceptable conclusion. When the valleys are more U-shaped, the valley floors become less deeply incised. Also, it was assumed that erosion is absent on the peaks, also not completely true in most cases. Therefore the actual resulting relief will be less than in the example shown. Nevertheless, the example illustrates how erosion and relief exert decisive influence on the evolution of the geomorphology of a mountain range. As erosion lowers mean elevation, summits rise and considerable relief is generated.

In the case in the Himalayas, the heavy monsoon rains, increased during the rise of the Himalayan mountain range, can explain how a landscape typical of the Tibetan Plateau was transformed into the present High Himalayas along its southern margin (■ Fig. 11.8). Analysis of a digital elevation model, a digitized topographic map, shows that the mean elevation along the shown profile (■ Fig. 11.9) is 5000 m in the Tibetan Plateau and 4570 m in the High Himalayas; these values will be slightly different in other profiles.

To explain the lower mean elevation of the High Himalayas (−430 m) relative to the Tibetan Pla-

teau, an average of 3115 m must have been eroded (■ Fig. 11.9). Isostatic balancing generates a rock uplift of 2685 m (2685 = 3115 − 430; the relation of these figures is the same as in the example of ■ Fig. 11.7d, each multiplied by 3.115). Assuming idealized V-shaped valleys and no erosion at the peaks, the peaks would attain an elevation of 7685 m, the valley floors 1455 m. The relief would amount to 6230 m. According to the simplified assumptions, the actual relief is expected to be slightly less. Because the Tibetan Plateau also bears some mountain ranges exceeding 6000 m in elevation, even higher peaks have to be expected in the Himalayas.

Despite the simplification of this example, the predicted results in the High Himalayas are perfectly met. The 7000–8000 m-high peaks are located close to valleys that are incised to less than 2000 m above sea level. Therefore, widely accepted hypothesis that the morphological difference between Tibetan Plateau and High Himalayas is largely due to the erosive processes caused by the powerful monsoon rains, is strongly supported by the above calculations.

Of course, crustal thickening, uplift and erosion are processes that overlap and act simultaneously. The results are rising mountain ranges with high elevation and relief. Only in places where erosion is limited, will high plateaus form.

11.4 Collapse and Crustal Escape

Rapid exhumation of deeply buried rocks is not necessarily coupled with strong surface uplift as is the case in the Himalayas. Burial during collision causes heating and softening of rocks. Rocks rich in quartz, one of the most common minerals in continental crust, be-

■ **Fig. 11.9** Topographic profile across the Tibetan Plateau and the Himalayas, and theoretical elevation profile illustrating the relief formation in the Himalayas by monsoonal precipitation. The theoretical profile follows the scheme shown in ■ Fig. 11.7d. The evaluation of the digital elevation model (map) shows that the mean elevation in the High Himalayas is 430 m lower than in the Tibetan Plateau. Evaluation by B. Székely

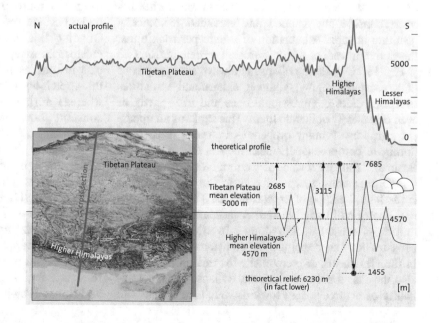

■ **Fig. 11.10** Escape of crustal blocks in southeastern Asia (Tapponnier et al. 1986). The escape motion towards the Pacific region is caused by continued northward migration of India. It is accompanied by crustal extension as revealed by graben structures and metamorphic domes (shown in green). The SE motion of Southeast Asia is taken up in the adjacent subduction zones. Similar structures to the NE are greatly reduced in size because there is no room for "escape". The box shows an analog experiment with plasticine where the pushing indenter and the escaping wedges duplicate the structures in southeastern Asia

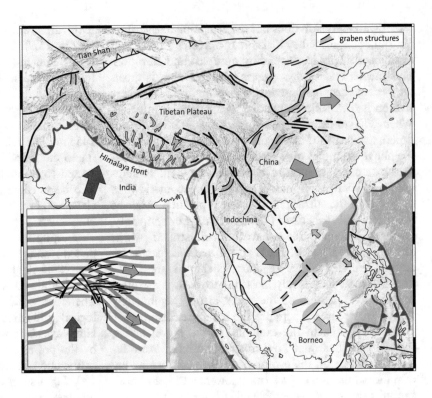

come ductile at ca. 300 °C and 10 km depth; therefore, they will react to stress by plastic deformation, not by fracturing. They will lose much of their strength and undergo plastic flow at depths of more than 10 km without becoming melted. This process is called thermal weakening. Through this process, the heated crustal stack will respond to the gravitational instability caused by orogenic thickening and react by lateral flow that is driven by gravity forces. The result is horizontal crustal extension and is called "gravitational collapse". It enables deeply buried crustal material to approach a near-surface position very rapidly; such is the case in metamorphic domes (see ■ Fig. 3.19). The overlying rocks of the upper crust are not removed by erosion, but rather become considerably thinned and extended tectonically resulting in rapid exhumation of the deeper rock units; this is the process of tectonic erosion or denudation (see ■ Fig. 13.9). Because collapse is always a consequence of crustal thickening, younger extensional structures overprint older compressional ones and may partly or even completely obliterate them. This explains an apparent paradox of many orogens—extensional structures dominate compressional ones. Such is the case in the Basin and Range Province of western North America.

Another consequence of collision in many orogens is the lateral escape of crustal blocks. This feature was first described in southeastern Asia. Here, the eastward motion of crustal wedges along subvertical faults compensates for the ongoing post-collisional compression in the Himalayas. A prerequisite for this process—as is also the case for the gravitational collapse—is the availability of a "free space" towards which the blocks can

move. East of the Himalayas and the Tibetan Plateau, crustal blocks greater than one million square kilometers in area escape towards the east where they are able to move freely against the convergent plate margins in the western Pacific region (■ Fig. 11.10). Similar blocks are also known from Asia Minor (■ Fig. 8.10) and the Eastern Alps (■ Fig. 13.9). In the Eastern Alps, the combination of gravitational collapse and lateral escape led to a complex process described as "lateral tectonic extrusion" (▶ Chap. 13; Ratschbacher et al. 1991b).

Thick continental crust becomes gravitationally unstable because of the heating and softening of the rocks as described above. Therefore, crustal thickening to greater than 70 or 80 km cannot occur because the rocks become too weak to support the crustal stack. Ongoing thickening by compression is prevented by collapse or escape. Collapse is the more efficient process insofar as it leads to crustal thinning (horizontal stretching compensates for vertical shortening), whereas during escape the crustal thickness remains constant (horizontal stretching compensates for perpendicular horizontal compression). Incidentally, an initial stage of gravitational collapse is also occurring in Tibet; this is in addition to the escape motions of the crustal blocks. Numerous young, active graben structures (■ Fig. 11.10) and several metamorphic domes indicate west–east extension under vertical shortening. Due to the size and elevation of the Tibetan Plateau, a dramatic change can be expected in near geologic future; it would not be unexpected if the region evolved to a state similar to that of the North American Basin and Range Province.

Old Orogens

Contents

© The Author(s), under exclusive license to Springer Nature Switzerland AG 2022
W. Frisch et al., *Plate Tectonics*,
Springer Textbooks in Earth Sciences, Geography and Environment,
https://doi.org/10.1007/978-3-030-88999-9_12

Since Early Proterozoic times, Wilson cycles have operated, each beginning with the breakup of continents, evolving into the stage with a mature ocean framed by passive continental margins, and concluding with subduction and collision (▶ Chap. 11). Remnants of oceans, present as ophiolite complexes in orogens, are the best indication of plate tectonic processes and mountain building. However, ophiolites older than 800 Ma are uncommon. Older ophiolite complexes do exist that resemble those in young Alpine orogens. They were derived from oceans formed 2500–2000 Ma ago by the break-up of Archean cratons and 1600–600 Ma by the break-up of the supercontinents Panotia and Rodinia. Examples are the Zunhua–Wutaishan ophiolite in northeast China, the Purtuniq ophiolite in northeast Canada, and the Jormua ophiolite in Finland (◻ Fig. 12.1). These ophiolites suggest that oceanic crust at that time was at least as thick as in modern oceans (ca. 6 km). Probably oceanic crust was generally even thicker than today and thus reflects a hotter mantle with a higher percentage of basaltic melts beneath the mid-ocean ridges. In the mentioned examples, the ocean became subducted; this led to collision of the opposing continents and mountain building, similar to the case in younger Earth history. In fact, the Early Proterozoic Wopmay orogen in Canada shows an evolutionary history (including the sedimentary record) that resembles modern orogens in much detail; therefore, it is generally accepted that the plate tectonic processes at that time closely mimicked those of the present.

Most old orogens are no longer topographically prominent and are strongly eroded; therefore, only originally deeply buried rock units are generally preserved. The geologist can only reconstruct the evolution of such an orogen from the record in the metamorphic rocks and from structural features.

12.1   2500–2000 Million Years Old Ophiolites

The Archean North China craton encompasses large parts of northeast China. West and east of Beijing, the fragmented *Zunhua–Wutaishan ophiolite belt* stretches over several hundred kilometers (◻ Fig. 12.1; Polat et al. 2006). It contains harzburgite and dunite with chromite as well as gabbro and basalt. Despite their tectonic and metamorphic overprint, some basalts reveal their original pillow structure. Sulfide mineralization is associated with the basalts and is interpreted as products from black smokers. The great interest of this ophiolite belt is its age: chromites and volcanic rocks yielded ages between 2550 and 2500 Ma thus establishing the oldest known ophiolite—the ocean, from which it derived, opened and closed before the Archean/Proterozoic boundary (2500 Ma). During the ensuing continent–continent collision, parts of the ocean floor were sandwiched between continental blocks, tectonically dismembered, and imbricated with continental material.

The oldest North American crust, the Canadian Shield, contains a complex network of Early Proterozoic orogens that formed between ca. 2000 and 1800 Ma by crustal growth and the continent–continent collision of Archean cratonic blocks. The metamorphosed *Purtuniq ophiolite complex* is part of the 500 km-wide Trans-Hudson orogen that crosses Hudson Bay (◻ Fig. 12.1) and consists of several tectonostratigraphic terranes. The 2000 Ma-old ophiolite encompasses 5 km-thick tholeiitic basalts that contain preserved pillow structures despite their deformation, a dike complex, and gabbros (Scott et al., 1992). The dikes are partly split into half-dikes (see ◻ Fig. 5.3). A younger magmatic sequence is also tholeiitic but differs from the older one by its trace element and isotope

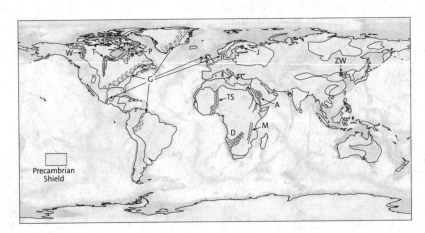

◻ **Fig. 12.1**   Present-day distribution of Precambrian shield areas. Orogens and localities described in the text are shown. *Early Proterozoic:* ZW: Zunhua–Wutaishan ophiolite. J: Jormua ophiolite. P: Purtuniq ophiolite. T: Trans–Hudson orogen. W: Wopmay orogen. *Middle Proterozoic:* G: Grenville orogen. *Panafrican orogenic belts (Late Proterozoic):* A: Arabian–Nubian Shield. D: Damara-Katanga orogen. M: Mozambique belt. TS: Trans-Sahara belt

chemistry; it represents the plutonic level of a volcanic edifice that formed from a hot spot similar to that of Hawaii.

The *Jormua ophiolite complex* is the northernmost exposure of a chain of ophiolite fragments in the Baltic Shield of Finland. Radiometric dating shows that by 1970 Ma, an Archean craton broke apart and formed an ocean that subsequently closed at 1900 Ma by continent–continent collision. The resulting Svekokarelian orogeny embraced large parts of Sweden and Finland (Kontinen, 1987). The Archean rocks across the broad flanks of the ophiolites are penetrated by numerous basaltic dikes oriented subparallel (NNW–SSE) to the ophiolite chain. They indicate the extensional rifting stage in the continental crust before its break-up. The Outokumpu complex at the southern end of the Finnish ophiolite chain contains important ore deposits including copper, cobalt, and zinc that were formed by volcanic exhalations on the sea floor. They closely resemble modern deposits of black smokers along mid-ocean ridges (▶ Chap. 5).

12.2 The Wopmay Orogen in Canada

The Wopmay orogen in the Canadian Northwest Territories (◘ Fig. 12.1) developed between ca. 2100 and 1800 Ma. It describes a complete Wilson cycle and represents a collision orogen between two Archean cratons (Hoffman, 1980). The evolution of the eastern passive continental margin after continental break-up in N–S direction is well documented by the sedimentary sequences that are overprinted by metamorphism. It is the earliest detailed example of a passive continental margin that corresponds to modern patterns.

Through the formation of a rift system, a larger Archean continent split up. The main rift system developed two triple junctions of the RRR type. The failed third arm of each became inactive and filled with thick sediments. These are the earliest documented aulacogens (▶ Chap. 4). The rifts are associated with strongly alkaline intrusions including carbonatites, rocks that are unknown from the Archean era but occur in the Early Proterozoic. Rifting, accompanied by alkaline magmatic suites, is characteristic of stable continental crust that breaks apart such as occurred along Pangaea later in the Phanerozoic.

The sedimentary sequences of the Wopmay orogen were stacked in several nappes. Reconstruction across the nappes reveals several zones, from east to west, inner shelf, outer shelf, and continent slope. During nappe thrusting, turbidites were deposited that were derived from the advancing nappes to form thick flysch sequences in front of the thrust units. The flysch sedimentation eventually graded into molasse sedimenta-tion that indicates the uplift of a mountain range in the hinterland. Interestingly, this molasse includes the oldest eolian deposits on Earth.

12.3 The Grenville Orogenic Cycle and the Formation of the Supercontinent Rodinia

The Grenville orogen, named after a settlement near Montreal, Canada, generated a long mountain belt that stretched from southern Scandinavia through a strip in Scotland, eastern Greenland, and large parts of eastern North America to South America (◘ Fig. 12.1). When North America is fit against Europe and South America in the Pangaean reconstruction, the trend of the orogen is perfect. This orogeny also describes a Wilson cycle that initiated with continent break-up around 1300 Ma. In the area of the Great Lakes, eastern Canada, and southern Greenland, the activity of hot spots and related rifting reflect the break-up process. An expression of this activity is alkaline intrusions, flood basalts, and dike swarms trending parallel to the margins of the later Grenville orogen (rift related) or radial with reference to the center of a hot spot. Numerous aulacogens branch off from the orogen with a high angle.

In the United States and Canada, the orogen consists of metamorphosed sequences that contain several basic-ultrabasic rock complexes interpreted as ophiolites. The ophiolitic rocks are partly transformed into glaucophane schists and eclogites that are among the oldest-known high-pressure metamorphic rocks (see below). In the area of the Great Lakes, the ophiolites are overlain by tholeiites, typical of primitive island arcs, and then by andesites. This island-arc sequence is intruded by large bodies of granodiorite and granite, which together with the andesites, indicate a mature stage of island arc magmatism. Thus, an evolution from primitive to mature island arc can be reconstructed. Two large fault zones separate the island arc sequence from gneisses that represent two different continents. Collision of these continents marks the orogenic climax and started around 1100 Ma. The Wilson cycle therefore lasted for approximately 200 m. y. The Grenville orogeny and related orogens assembled the supercontinent Rodinia (see ◘ Fig. 6.7), which by 750 Ma started to disintegrate.

12.4 The Panafrican Orogeny and the Formation of Gondwana

The break-up of Rodinia initiated an episode of high crustal mobility in the region of the continents of the present southern hemisphere. This occurred because of the rapid generation of new oceanic crust and its subsequent subduction. The welding together of dispersed continental pieces led to the formation of a complex network of mountain belts near the end of the Precambrian at approximately 550 Ma. The result of these events was the creation of the giant continent Gondwana, a large landmass that included the present southern continents plus India, as well as parts of present North America and Europe. Although it was not a single, unified orogenic event, this orogenic period is called Panafrican orogeny. Actually, a number of continental blocks and island arc systems collided over a period of 200–250 m. y. In the region of the Arabian–Nubian Shield, the area encompassing the Arabian Peninsula and large parts of Egypt and Sudan, a complicated system of island arcs evolved that resemble the present setting in the western Pacific region. The term Panafrican orogeny originated because the event welded together the different parts of present-day Africa, as well as other regions.

In the area of the Arabian–Nubian Shield (◪ Fig. 12.1) on either side of the Red Sea, several Late Proterozoic island arc complexes separated by ophiolite belts can be discerned. The ophiolites and associated large volumes of calcalkaline magmatic rocks correspond in detail to events documented in rock complexes of the Phanerozoic (Frisch and Al Shanti 1977). Some ophiolites contain a nearly complete profile through oceanic crust including sheeted dike complexes. Some ophiolites have been interpreted as back-arc oceanic crust. The island arcs evolved from a primitive stage with basalts and andesites to a mature stage with andesites, dacites, and rhyolites intruded by diorites and granodiorites. Sedimentary rocks, though of minor volume, consist of mainly sandstones and graywackes that contain the debris from the eroded volcanics. Widespread Kuroko-type mineral deposits are also present (▶ Chap. 7).

The Arabian–Nubian Shield was situated along the margin of Gondwana, whereas other Panafrican mountain belts formed by collision of continental masses and therefore cross interior parts of Gondwana. Panafrican collisional orogens that underwent a Wilson cycle in Late Proterozoic times are the Mauritanian and the Trans-Sahara mountain belt in northwestern Africa, the Damara orogen in southwestern Africa, and the Mozambique belt in eastern Africa. In the Trans-Sahara orogen the oldest-known paired metamorphic belt has been reported: a high-pressure belt with eclogite facies and a high temperature belt with cordierite-bearing gneisses.

Some eclogites of the *Trans-Sahara mountain belt* (◪ Fig. 12.1) contain coesite and hence experienced burial to at least 80 km depth (Jahn et al. 2001). Their age of 620 Ma makes them the oldest ultrahigh-pressure metamorphic rocks (▶ Chap. 7). Eclogites and glaucophane schists are rare in the Precambrian. Archean eclogites are known only indirectly (▶ Chap. 10), with one exception: the 2600 Ma old eclogites of the Churchill province in Canada. Approximately 2000 Ma-old eclogites are reported from Tanzania, and 1070 Ma-old eclogites occur in the Grenville orogen. Additionally, 900 Ma- and 700 Ma-old glaucophane schists were found in China. Probably a higher temperature gradient in the upper mantle in early Earth history prevented the exhumation of high-pressure metamorphic rocks without destroying the high-pressure minerals. The rare or missing occurrence of such rocks is therefore not a stringent argument against their former existence.

In Europe the Panafrican orogeny is expressed in the *Cadomian mountain belt* (named after the French town Caen, *Latin Cadomus*). The rocks were originally positioned at the northern margin of Africa (Gondwana) and formed between 700 and 550 Ma. Cadomian rocks are found in the Armorican Massif (Bretagne and Normandie), in the Bohemian Massif, on the Iberian Peninsula, and in basement complexes of the Alps. In fact, the complete continental basement of the Variscan orogen of Europe was bound to Africa (see below) where it was affected by the Panafrican–Cadomian orogeny. Cadomian rocks have also been found in many parts of eastern North America and parts of eastern Mexico, and more recently in the Cordilleran areas of Canada and the USA. These rocks were rifted from Gondwana in the Paleozoic and Mesozoic and are referred to as peri-Gondwanan terranes. Because all these terranes became overprinted by later orogenies, they are not shown as "Precambrian shields" in ◪ Fig. 12.1.

Gondwana was welded together during the Panafrican orogenic cycle in the late Precambrian and early Paleozoic. Subsequently, pieces were rifted from its northern and western margins and incorporated into North America and Europe in the middle Paleozoic. In the late Paleozoic, all the pieces were integrated during the Variscan and Alleghenian orogenies to form the supercontinent Pangaea. Collectively, these events occurred within a timespan of less than 300 m. y.

12.5 The Caledonides—A Wilson Cycle Around the Iapetus Ocean

The Caledonian orogen of northern Europe and coeval Acadian orogen of northeastern North America describe a Wilson cycle that began ~600 Ma in the Late Proterozoic and culminated in the Silurian and Devonian ~400 Ma. The Iapetus Ocean opened in a similar position as the present North Atlantic Ocean, between Laurentia (North America), Baltica (northern Europe), and Gondwana (including Africa as well as southern, western, and central Europe) (◪ Fig. 12.2). During the early Paleozoic, Avalonia rifted from Gondwana as a minor continent named after the Avalon Peninsula in Newfoundland. Avalonia collided with Baltica in the Late Ordovician (time not well constrained), and Avalonia/Baltica with Laurentia at the Silurian–Devonian boundary, thus terminating the Caledonian cycle. The resulting Alpine-style mountains may have been similar in height to the highest mountains on Earth today.

Presently the Caledonian orogen is found on both sides of the Atlantic Ocean: in East Greenland, Spitsbergen, along the western coast of Scandinavia, on the British isles, and along the eastern coast of North America well into the Arctic. It continues southward into the Appalachians where it became overprinted by the collision between Laurentia and Gondwana in Late Paleozoic times.

The Significance of Scotland and the Greek Mythology

The definition of the Caledonides as an independent orogen is based on the work of Eduard Suess (1885). The name derives from the Latin word for Scotland: Caledonia. The term "Iapetus", however, derives from Greek mythology: Iapetos one of the titans, was the son of Uranos (Heaven) and Gaia (Earth). Two of his sisters, Tethys and Rhea, also lent their names to oceans in Earth history (see below). Okeanos ("ocean") was one of his brothers. At the time of the Iapetus Ocean, northwestern Scotland was part of Laurentia, whereas today Avalonia is split be-

Fig. 12.2 Plate tectonic evolution of the Caledonian, Appalachian, and Variscan mountain belts in the Paleozoic era (modified after Tait et al., 1997). The sketch maps show the settings in the Early and Late Ordovician (485, 445 Ma), the Early Devonian (400 Ma), the Late Carboniferous (310 Ma), and today

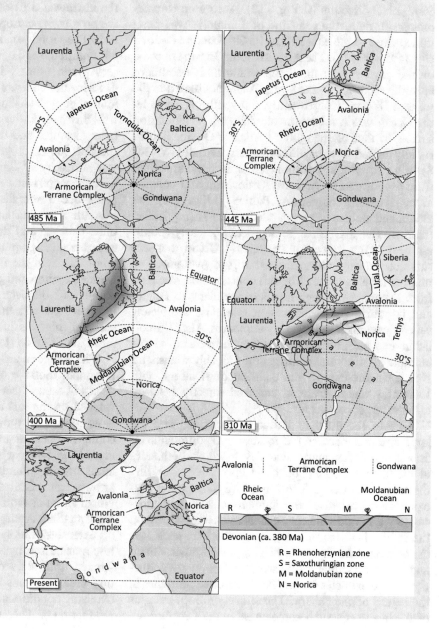

tween North America and Europe (◻ Fig. 12.2). Incidentally, when plate tectonics mentions an "ocean", generally the areas underlain by oceanic crust are meant. The term "continent", in turn, includes the shelf areas under the sea, because they are underlain by continental crust.

In the Scottish Caledonides George Barrow developed his concept of index minerals in the late nineteenth century, delineating metamorphic zones (Barrow, 1893). He discerned zones with characteristic metamorphic minerals in schists of the Dalradian (Late Precambrian to Cambrian) and demonstrated that rock bodies that experienced different metamorphic conditions became exposed at the surface one beside the other by later folding and erosion. The pressure-dominated regional metamorphism was named after him (see ◻ Fig. 7.24), characterized, amongst others, by the index mineral kyanite.

It can be said that the birth of modern geology in the late 18th century originated from the early studies of the Caledonian orogeny. On a precipitous headland of Scotland, Hutton described his famous unconformity where Devonian arkose of the Old Red Sandstone (see below) overlies folded Silurian flysch and graywacke (Hutton, 1795). Hutton correctly surmised that complicated events necessary to generate such an outcrop must have taken vast amounts of geologic time and that the Earth was, therefore, very old.

The Iapetus Ocean had its trend in approximately a SSW–ENE direction (◻ Fig. 12.2). According to paleomagnetic data, the coastline of Laurentia was positioned close to the equator, ca. 10–20° S, in the Ordovician (495–440 Ma). The African–South European coast on the other side of the ocean was at a high southern latitude of 60–70° S. The South Pole was situated in the western Sahara. Consequently, the width of the Iapetus Ocean was a minimum of ~5000 km, a dimension similar to that of the modern Atlantic Ocean. Baltica rapidly drifted from medium southern latitudes (60–30° S) to an equatorial position. Avalonia probably rifted from Gondwana in the Early Ordovician and collided along its eastern margin with Baltica (◻ Fig. 12.2). Avalonia and Baltica were separated by the Tornquist Sea (named after a German geologist of the early twentieth century), a branch of the Iapetus Ocean that was subducted during the northward drift of Avalonia. The mountain belt generated by this collision now stretches from Denmark, through northern Germany to Poland and is poorly known because it is almost completely veiled by younger sedimentary sequences. Also, the collision was strongly oblique, nearly transform, so the amount of structural deformation probably was slight. Not so in the case of the collision of Avalonia with Laurentia where Avalonia smacked directly head-on into the larger continent.

Due to the different latitudinal positions and the large distances between the coasts, clearly differentiated faunas and sedimentary facies developed on both sides of the Iapetus Ocean (Cocks and Fortey 1982). During the Early Ordovician, limestones with warm water faunas were deposited on the Laurentian shelf, and limestones with faunas indicating moderate temperatures on the Baltic shelf. In contrast, the Gondwanan shelf was characterized by faunas indicating a cool environment and by clastic sediments, because the carbonate production in cool climates is low. Glaciers formed moraines in the region of the Sahara. The faunas on both margins of the Iapetus Ocean became mixed in the Late Ordovician and indistinguishable in the Silurian. This reflects the northwad drift of Avalonia in the Ordovician and the progressive closure of the Iapetus Ocean in the Silurian. The Gondwanan margin remained in latitudes ca. 60° S until the end of the Silurian.

Around the Silurian–Devonian boundary, ca. 415 Ma, the collision of Laurentia with Baltica led to intense nappe stacking in Scandinavia. The nappe stack was thrust over the foreland of the Baltic Shield for great distances and the event marks the termination of the Caledonian orogeny. The Baltic continental margin was buried by a W-dipping subduction zone and shortened by approximately 400 km. Ophiolites of the Iapetus Ocean were also subducted but eventually thrust over the Scandinavian nappe stack. Parts of the ophiolilte package and the rocks of the continental margin experienced high-pressure and ultrahigh-pressure metamorphism. In the fjords and islands south of Trondheim, eclogites and gneisses contain coesite and diamonds with diameters to 50 μm (0.05 mm) as inclusions in other high-pressure minerals. The diamonds and associated minerals indicate that the rocks were buried more than 130 km (Dobrzhinetskaya et al., 1995).

The Caledonian–Acadian orogen contains paired Devonian molasse belts, the Old Red Sandstone and "Catskill facies" (several geologic formations) that developed on the east and west flanks respectively of the large continent–continent collision. Both of these units contain abundant arkose, sandstone and conglomerate rich in feldspar that was derived from the hinterlands of the mountain belt. These deposits are now known from Central Europe to Spitsbergen and in North America from Tennessee and North Carolina to the Canadian Arctic and Alaska. The molasse formed on extensive river systems, coastal plains, and in shallow marine settings within large foreland basins that flanked both sides of the Caledonides. The sedimentary rocks are mostly red because they formed under oxidizing conditions and contain iron derived from the weathering of igneous and metamorphic rocks.

12.6 The Variscides—A Broad Mountain Belt in Central Europe

The Variscan orogeny (named after the tribe of the Variscans in Franconia, Germany; the term Hercynian orogeny that is used as a synonym, refers to the Harz (Hercynian) Mountains in Germany, this term, however, is largely abandoned) took place from the Devonian through the Carboniferous and, based on its timing, marks the direct continuation of the Caledonian orogeny. In Caledonian times the eastern margin of Laurentia collided with part of Avalonia, whereas the margin of the large continent Gondwana was still remote. However, in Variscan time the two large continents collided and the Caledonian (Acadian) mountain belt along the eastern coast of North America became overprinted by later orogenic movements that persisted into the Permian. In Europe, both orogens, Caledonian and Varsican, are spatially separated; the collision of Laurentia and Baltica and the closure of the Iapetus Ocean terminated the orogenic events along the Caledonian collision zone. The Variscan orogeny followed farther south and formed an orogenic belt as much as 1000 km wide, the southernmost part of which was later overprinted by the Alpine orogeny. Large parts of Spain, France, and Germany were involved in the Variscan orogen and its basement almost everywhere was overprinted on the older Cadomian orogeny.

The classical subdivision of the Middle European Variscides contains three main tectonic units, from north to south, the Rhenohercynian, the Saxothuringian, and the Moldanubian zones (■ Fig. 12.3; Kossmat 1927). Ophiolite and subduction-related magmatic complexes were once thought to be rare. This led to the proposal by German geologists in the 1970's that the Variscides were an intra-continental orogen, not formed by continent collision but rather by compressive deformation of a broad strip of continental crust.

More recent studies have provided strong support for a plate tectonic scenario concerning the Variscan orogeny as both ophiolite bodies indicating oceanic realms and calcalkaline magmatic suites and high-pressure metamorphic rocks indicating subduction are now well known. A portion of these rocks are contained in the basement of the Alps and were neglected for a long time in the reconstructions of the Variscides. Moreover, the concept of Kossmat has to be revised because evidence for a suture zone—indicating continent collision—has been found within the Moldanubian Zone.

The northward drift of Avalonia caused the opening of an oceanic realm in its wake, the Rheic Ocean (Rhea, sister of Iapetos) (■ Fig. 12.2). The Rhenohercynian Zone (■ Fig. 12.3) was part of Avalonia and therefore part of Baltica/Avalonia during the Devonian. A remnant of the Rheic Ocean, which was subducted towards the south, is the Lizard ophiolite complex along the southern coast of Cornwall in SW England which is part of the Variscides and belongs to the Rhenohercynian Zone. The Rhenohercynian zone was stretched during the opening of the Rheic Ocean, creating a basin with thinned continental crust in the area of Germany and southern England. At the Silurian/Devonian boundary, a volcanic arc formed on the northern edge of the Saxothuringian Zone, which is part of the Armorican Terran Complex. The subduction zone of this convergent plate boundary plunged towards the south (■ Fig. 12.2: cross section). Deep stockwerks of this subduction zone are now exposed in the Mid-German Crystalline Zone (Kroner et al. 2008). The closure of the Rheic Ocean occurred during the Devonian and Carboniferous and is marked by the formation of a north-vergent nappe stack.

The Armorican Massif, the French Central Massif, the Vosges, Black Forest, and the Bohemian Massif are all part of the Moldanubian Zone that contains the second important suture zone of the Variscides, the Moldanubian suture. The Armorican Terrane or Ar-

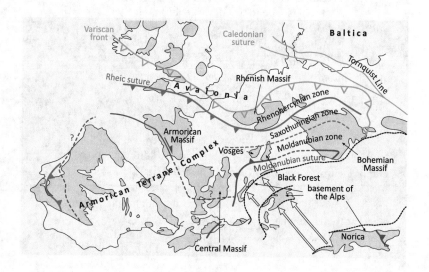

■ **Fig. 12.3** Geological sketch map of the European Variscides (modified after Franke 2002). The Rheic suture and the Moldanubian suture show opposite polarity of thrusting. Iberia is rotated back to its Late Paleozoic position

morica is located between the Rheic and the Moldanubian suture. It represents an agglomeration of various terranes forming the Armorican Terrane Complex that, like Avalonia, split off from Gondwana and drifted ahead of it towards the north (■ Fig. 12.2). The Armorican Terrane Complex probably separated from Gondwana in the Silurian and reached tropical latitudes in the Early Devonian as indicated by coral reefs. In its wake the Moldanubian Ocean opened (Franke 2002). The closure of this ocean led to the collision of Gondwana, now drifting rapidly northward, with the northern continent containing the welded terranes. The Moldanubian subduction dipped towards the north and thus created a southvergent nappe stack. Above the subduction zone on the Armorican Terrane, a subduction-related magmatic belt formed, the Ligerian Cordillera (■ Fig. 12.2: profile).

The Moldanubian suture stretches from Bretagne via the southern Black Forest (see box) to the Eastern Alps (■ Fig. 12.3). In its eastern part the collision occurred in the Early Carboniferous, in its western part it already started in the Devonian. This scissor-like closure is the result of oblique collision of irregular continental margins. Both the Moldanubian suture zone and the suture between the Armorican Terrane Complex and the Modanubian Zone (■ Fig. 12.3) contain high-pressure to ultrahigh-pressure metamorphic rocks, the former with coesite bearing eclogites (Central Massif), the latter with diamond bearing gneisses (Saxonian Erzgebirge).

By welding together Laurentia/Baltica, commonly called Laurussia, with Gondwana and incorporating the continental fragments and island arcs in between, most continents worldwide were unified in the incipient Pangaea supercontinent by the Late Carboniferous. During and after formation of the European Variscides, westward movement of Gondwana led to compression in the Appalachians, an event that persisted into Permian and completed the orogenic process there. Also in Permian times the Urals were formed by the collision of Siberia with Baltica (■ Fig. 12.2). The birth of Pangaea was finalized.

A Variscan Suture in the Southern Black Forest

Two highly metamorphosed gneiss complexes in the central and the southern Black Forest (Schwarzwald) are separated by the Badenweiler–Lenzkirch Zone, an area comprised of a weakly metamorphosed sequence of detrital sedimentary rocks ranging in age from Ordovician to Early Carboniferous. The Central Black Forest Gneiss Complex contains splinters of ophiolites with eclogites. Radiometric dating yielded Ordovician ages for the formation of the ocean floor rocks. The eclogites, formed during subduction of the ocean floor, yielded Early Carboniferous ages (Chen et al. 2003). The entire area of the southern Black Forest experienced south-vergent nappe stacking in the Early Carboniferous. A tectonic unit at the southern margin of the Central Black Forest Gneiss Complex was interpreted to be the remnant of the plutonic level of subduction-related magmatism. To the south of this zone was a deep-sea trench along a convergent continental margin in which argillitic and sandy sediments including graywackes were deposited and subsequently imbricated in an accretionary wedge, now the Badenweiler–Lenzkirch Zone (■ Fig. 12.4).

Recent research revealed a suture zone in the southern Black Forest, along which the collision between two continent masses occurred (Hann et al. 2003). This suture marks the boundary between Armorica to the north and Gondwana to the south. Although the suture itself lacks ophiolites, several important characteristics were discovered that enabled the reconstruction of a subduction zone adjacent to an oceanic realm (■ Fig. 12.4). Dating of the magmatic rocks and the eclogites proved the persistence

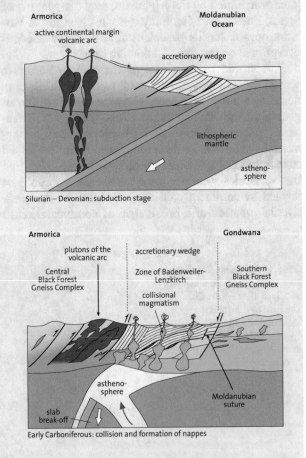

■ Fig. 12.4 Geological evolution of the Moldanubian suture zone in the southern Black Forest (Hann et al. 2003). South is to the right

of subduction activity for at least 100 m. y. (Silurian to Early Carboniferous), hence the subducted ocean must have had a considerable width, far in excess of 1000 km. This scenario orchestrates well with the drift of Gondwana from high to low southern latitudes during the same interval of time (◻ Fig. 12.2).

Following collision in the Early Carboniferous, the subducted oceanic part of the plate broke off and enabled the asthenosphere to ascend to the base of the crust (◻ Fig. 12.4). This process caused heating of the crust and the formation of large volumes of granitic melts. The granites penetrated the crust as a number of plutonic bodies.

12.7   The Variscan Orogen in the Alps

Portions of the Variscan orogen were later involved in the Alpine orogeny and therefore form most of the basement of the Alps. The rocks exposed in the Alps, the southern Variscides, experienced southvergent nappe stacking. The Alpine overprint included strong deformation and metamorphism that veils the older structures and makes reconstruction of the Variscan processes difficult. However, modern methods enable astonishingly good insights into this older episode of mountain building.

The basement of the Western Alps is the direct continuation of the southern Central Massif and the southern Black Forest and was thus part of Gondwana until its collision during the Variscan orogeny (◻ Fig. 12.3). In the Eastern Alps, the restoration of the original arrangement of crustal blocks is much more difficult because of the Austroalpine mega-unit (see ◻ Fig. 13.6). During the Alpine orogeny, this complex was thrust for long distances over those units exposed in the Western Alps. However, the Austroalpine basement was also part of Gondwana although it was partially split off from this large continent before the Variscan collision. The Austroalpine unit consists of several tectonic units, each defined by different characteristics. This led to a geodynamic interpretation in the frame of the terrane concept (▶ Chap. 9).

The northeastern part of the Austroalpine basement comprises the *Pannonian Terrane* (◻ Fig. 12.5) that experienced deformation and metamorphism during the Devonian. Present is a subduction-related magmatic complex that can be correlated with the Ligerian Cordillera in the western Armorican continent (◻ Fig. 12.2: profile). Northward subduction of the Moldanubian ocean floor enabled the formation of the Koriden unit, an accretionary wedge with probable flysch deposits that formed in a deep-sea trench (◻ Fig. 12.5). These deposits were welded along the southern margin of Armorica. Ophiolite bodies of the Plankogel unit, a tectonic mélange, became also accreted. The Plankogel unit includes ocean floor remnants, basalts of an intra-oceanic seamount, and limestones and sandstones that float in an argillitic matrix; these components were all metamorphosed to amphibolite facies conditions in both Variscan and Alpine times. The Plankogel unit represents the Moldanubian

suture zone and is therefore considered to be the continuation of the suture exposed in the southern Black Forest (see box). In contrast to the suture in the Black Forest, the rocks in the Eastern Alps are highly metamorphosed and numerous ophiolites are present. Ophiolites were also found in the Central Massif along this suture zone (Matte, 2001).

During subduction of the Moldanubian Ocean, a continental fragment rifted from Gondwana and opened an oceanic area in a scissor-like manner, wider towards the east. This continental fragment is named the Noric Terrane or Norica (◻ Figs. 12.2 and 12.5). The *Noric Terrane* is a composite terrane that forms the greatest portion of the Austroalpine basement. It consists of the older Cadomian basement, a magmatic arc of Late Proterozoic and Early Cambrian age. Large volumes of volcanic and plutonic rocks were added to the crust during Cadomian times. The Noric Terrane consists of two smaller terrane units. The magmatic arc is represented by the *Keltic Terrane* and the back-arc basin is represented by the *Speik Terrane*. The arc collided with Gondwana in the early Paleozoic and the backarc ocean floor, the Speik Terrane, was thrust over the arc, the Keltic Terrane. The resulting amalgamation formed the composite basement of the Noric Terrane (◻ Fig. 12.5).

The basement of the Noric Terrane was covered by a sedimentary sequence that spanned the Ordovician to Early Carboniferous. The sedimentary rocks contain some intraplate volcanic horizons that represent crustal distension that culminated in the scissor-like separation of Norica. Collision between the Noric and Pannonian terranes occurred in concert with the closure of the Moldanubian Ocean in the southern Black Forest and the Bohemian Massif (◻ Fig. 12.3). The oeanic wedge between Norica and Gondwana became part of another large oceanic realm, the Paleotethys Ocean (Tethys, another sister of Iapetos) that widened considerably towards the east and merged with the giant Panthalassa Ocean (see ◻ Fig. 4.8: Permian). Additional continental fragments rifted from Gondwana to open the Neotethys Ocean in the Triassic.

The *Habach Terrane*, exposed in the Tauern Window in the Eastern Alps, consists of a completely different nature. It is not part of the Austroalpine but rather is part of the structurally lower Penninic mega-unit of the Alpine nappe edifice (see ◻ Fig. 13.6). It probably

■ **Fig. 12.5** Paleozoic terranes in the Eastern Alps that were assembled during the Variscan orogen (Frisch and Neubauer, 1989)

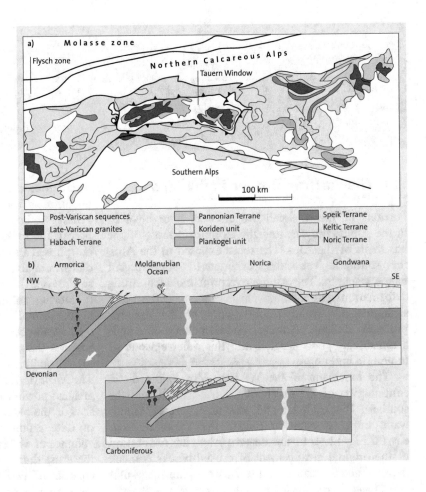

formed somewhere in the oceanic widths to the east where its position relative to the other terranes remains unknown. It developed as an island arc on oceanic crust from the Late Proterozoic for more than 300 m. y. until terminated by the Variscan orogeny. During the Variscan orogeny, the Habach Terrane became sandwiched between colliding continental blocks and was amalgamated as part of supercontinent Pangaea. Partial melting during collision and persistent subduction activity along the Tethys continental margin generated large volumes of granitic melts that intruded into the ocean floor and thick island arc sequences in the Carboniferous. The granites were transformed to gneisses during the Alpine orogeny. The association of island arc volcanics, largely metamorphosed to greenschists and related rocks, and granites resembles Archean granite greenstone belts (▶ Chap. 10).

12.8 Paleozoic Mountain Building in Eastern and Southern North America

The Appalachian Mountains today form a low mountain range that stretches from the Maritime Provinces of eastern Canada to northern Georgia. The Appala-

chians were not overprinted by later orogeny, hence part of their original form is still intact. However, parts are missing including much of the ancient hinterland that now lies below the Atlantic Coastal Plain or was destroyed by the opening of the Atlantic Ocean. The basement of the Appalachians ranges from crystalline rocks formed during the Grenville orogeny to various crystalline blocks of ancient terranes that were deformed, intruded, and metamorphosed during the various phases of Appalachian orogenies. In general, three main orogenic events have been traditionally recognized although modern research has complicated this interpretation somewhat. Most Appalachian events have general correlatives to the events discussed above in Europe.

The main structural trends of the Appalachians plunge under the Atlantic and Gulf Coastal Plains but deep seismic reflections trace the orogen under the Mississippi River Embayment westward into Arkansas and Oklahoma where it reappears as the Ouachita Mountains. From there, correlation proceeds southwestward into the Marathon Mountains of Texas and from there southward into northern Mexico. The complexity and number of orogenic events of this great mountain chain generally increases from SW to NE—the Marathons and Ouachitas have one major event each, the

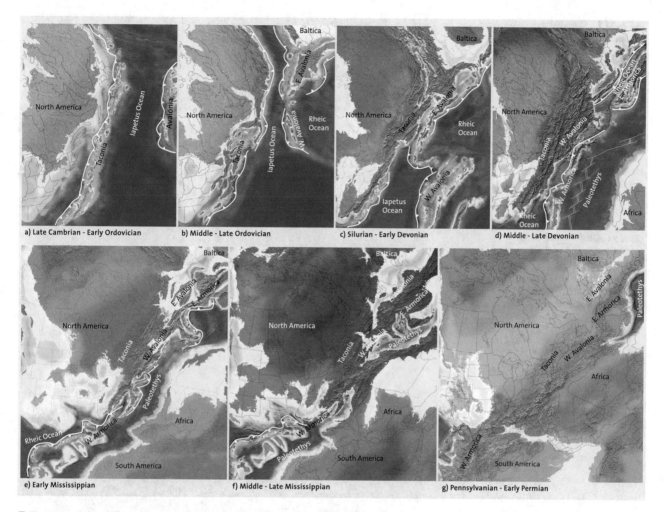

Fig. 12.6   Paleogeo-graphic maps that show evolution of Appalachian Mountains during Paleozoic. Note that Paleotethys corresponds to Moldanubian Ocean in Fig. 12.2

Southern Appalachians several events, and the Central and parts of the Northern Appalachians may have five or more major events.

The oldest orogenic event widely recognized in eastern North America is the *Taconic orogeny*, named for the Taconic Mountains between New York and Vermont, and coeval *Grampian orogeny* of Scotland and *Humberian orogeny* of Maritime Canada. Remember that Scotland was part of Laurentia during the Paleozoic. The Taconic orogeny has been dated as Middle Ordovician to Early Silurian, older in Scotland and Canada and younger in New England. The orogeny was originally defined by thrust sheets in the Taconic Mountains in which deep-water flysch was thrust over marine shelf deposits. Subsequent work has documented Taconic intrusives, metamorphic sequences, and ophiolites. A well-developed foreland basin sequence of Late Ordovician and Silurian age is present across much of the western Appalachians, especially in New York State. The Taconic and related orogens have been interpreted as the result of collisions between is-

land arc complexes and the microcontinents on which they were built with the eastern margin of Laurentia. Laurentia was the lower plate of the subduction zone so the arcs were thrust over the larger continent (Figs. 12.6a, b and 12.7a, b). The microcontinents may have originated as rifted slivers from North America during the opening of the Iapetus Ocean. Arcs were built on the microcontinents and then obducted onto North America. The Taconic and related orogenies may partially correlate in time and orogenic style with the *Finnmarkian orogeny* of west Baltica, generally considered the earliest phase of the Caledonian orogeny. Because the arcs and microcontinents associated with these early orogenies were intra-oceanic, their collisions did not close the Iapetus Ocean, only parts of it.

As Avalonia joined Baltica, the two bore down on NE Laurentia and the ensuing collision generated the Salinian and Main phases of the *Acadian orogeny*. The collision was slightly oblique so that Greenland, then part of Laurentia, and Baltica collided along their re-

◻ Fig. 12.7 Schematic cross sections that show events in evolution of Appalachian Mountains during five time segments shown in ◻ Fig. 12.6. **a** Iapetus Ocean prior to Taconic orogeny (see ◻ Fig. 12.6a); **b** Taconic orogeny (see ◻ Fig. 12.6b); **c** Avalonia collides with Laurentia during Acadian orogeny and closure of Iapetus Ocean as Hun Terrane (Western Armorican Terrane) rifts from Gondwana (see ◻ Fig. 12.6d); **d** accretion of Hun Terrane; late phase of Acadian (north) to early phase of Alleghenian (south) orogenies (see ◻ Fig. 12.6f); **e** Alleghenian orogeny marks final collision of Gondwana and Laurentia to form Appalachian Mountains and Pangaea (see ◻ Fig. 12.6g)

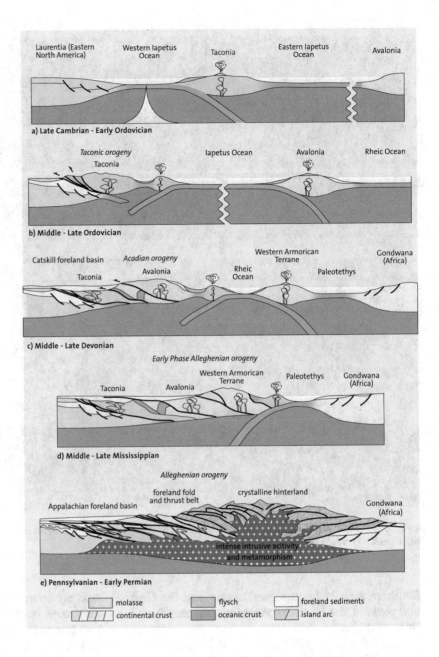

spective northern coasts in the Middle Silurian and scissored southward so that in New England, the collision with West Avalonia was Early and Middle Devonian (◻ Figs. 12.6c,d and 12.7c). The Baltica–Greenland collision was described above. Farther south from the Maritimes of Canada to the central Appalachians, the orogeny generated major thrusting, metamorphism, and intrusion of major granite batholiths. Substantial portions of the Atlantic Coast states were accreted to eastern Laurentia (compare ◻ Fig. 12.6c, d). The well-developed foreland basin to the west received clastic sediment from the high mountains; variously called the Catskill or Acadian foreland basin, it is the North American equivalent to the Caledonian foreland basin of Europe in which the famous Old Red Sandstone was deposited.

As described above, the earliest phases of the Variscan orogeny involved arc tectonics and ophiolite obduction, Lizard ophiolite and Ligerian orogeny. Similar events occurred at various places in the Appalachians, especially New England, as various arcs and backarc basins accreted onto the continent. The accreted Meguma Terrane, now part of the continental shelf along Nova Scotia, may have been a straggler of West Avalonia and later collided with Laurentia or was driven into the continent by younger, incoming terranes. The next succession of incoming terranes, recently named the West Hun Terrane (named for the Huns, a marauding band of peoples that hassled the Roman Empire at its fall), was part of a larger succession of ribbon continents that rifted off Gondwana and accreted to Europe and North America closing the Rheic Ocean (◘ Figs. 12.6d,e and 12.7c). Armoricia (from the Celtic–Roman word Ar(e)morica for Bretagne and Normandy) formed the large island to the NE of ◘ Fig. 12.6d and described above in the Variscan orogeny was part of this chain as were parts of what are now eastern North America from New England to Mexico. We prefer the older term, Armorica, for the whole complex. In North America, the accretion of these terranes was very complicated and strongly time-transgressive, older to the northeast, younger to the southwest. They generated the late phases of the Acadian orogeny and early phases of the Alleghenian orogeny as well as the *Ouachita and Marathon orogenies* (◘ Figs. 12.6e–g and 12.7d, e), the latter of which culminated in the Permian. Some of these terranes may have "escaped" around the southern margin of North America and are now incorporated in Cordilleran terranes in California!

How Many Orogenies?

As the discussions of the orogenies of Europe and North America imply, many orogenic events affected southern Laurussia, the name for combined Laurentia, Baltica, and Western Europe, during the Paleozoic. As the discussion of terranes in ▶ Chap. 9 implied, terrane analysis is full of complexity and controversy. Such complexity can be difficult to resolve where ribbon terranes are doubled or tripled; did such duplication happen before the terranes docked and therefore the duplication represents separate collisional orogenies, or did the duplication occur after docking through transform motions? In the later case, only one orogenic event may have occurred. Like many sciences, geology has both lumpers and splitters. One geologists' major orogeny with several phases is another geologists' multiple orogenies. The collision of microcontinents must involve at least one subduction zone, possibly more. As discussed in ▶ Chap. 7, arcs undergo changes, some of which result in deformation—orogeny. Therefore, arcs can complicate the orogenic picture.

The Caledonian orogeny presents an example of multiple orogenies being miscorrelated and incorrectly linked to the "type Caledonian" orogeny. A classic study by McKerrow et al. (2000) illustrated this folly and traced the history of study of the Caledonian orogeny and used this history to refine and limit a term that was being used incorrectly. Their basic conclusion was that the term "Caledonian orogeny" should only be applied to those events directly related to the closing of the Iapatus Ocean—the collision of Baltica, Avalonia, and, Laurentia and associated island arcs.

So how many orogenies occurred prior to and during the assembly of Pangaea in the present North Atlantic region? There may never be an answer to this question. ◘ Figure 12.8 is a compilation of many of the terms that have been used to define Caledonian–Appalachian–Variscan orogenies and their subdivisions. Each term is related to the area and time frame that it represents, thus resulting in duplication of terms. Some represent relatively minor events such as island arc collapse or collision and others stand for major continent–continent collisions. One thing seems certain—there appears to be nearly continuous orogeny at some place in the Paleozoic as the Iapetus, Rheic, and other oceans closed and Pangaea was assembled.

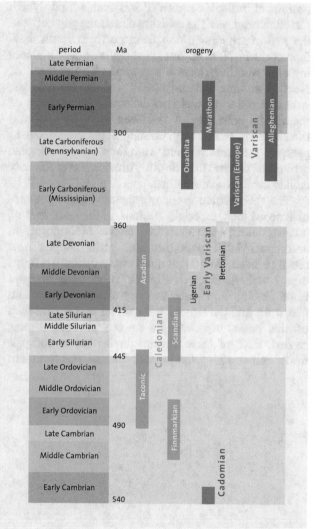

◘ **Fig. 12.8** Ages of orogenies and orogenic phases related to the assembly of western Pangaea (compiled from Ziegler 1990; McKerrow et al. 2000; and Hatcher 2002). Red shows the Cadomian orogeny, the basement to many later events. Green marks the classic Caledonian orogeny and the closing of Iapetus Ocean as defined by McKerrow et al. (2000). Pale blue marks early phases of the Variscan orogeny and deep blue marks the events that assembled Pangaea and closed the latest Paleozoic oceans, the Variscan and Alleghenian orogenies. Note that time scale bar is linear. Modified from Blakey (2007)

Regardless of how many terranes accreted in how many events to eastern Laurentia, the culminating *Alleghenian orogeny* involved the final collision between Gondwana and eastern Laurentia (◼ Figs. 12.6g and 12.7e) and the supercontinent Pangaea was born. But the Alleghenian orogeny was not everywhere equal. As the maps in ◼ Fig. 12.6 show, the eastern and southern margins of Laurentia were extremely irregular and consisted of various salients (large protruding areas) and reentrants (large embayments). The Reguibat promontory on the African portion of Gondwana struck the prominent salient between Pennsylvania and North Carolina head on. The collision closed the gap between the two in scissors-fashion from north to south. Farther north in New England, the collision was a glancing blow and deformation was minor. To the south of North Carolina, Laurentia took a sharp bend to the west and the force was also deflected. Here, pieces of the ribbon continent of the West Hun Terrane and their leading island arcs yielding a glancing blow to generate the Ouachita and Marathon orogenies; some geologists debate that South America never directly collided with North America.

The Alleghenian orogeny generated great volumes of igneous rock, especially plutons, in the main area of intense deformation described above. Metamorphism accompanied the plutonism and great nappes, now mostly eroded, and thrust sheets were driven westward over North America (◼ Fig. 12.7e). A huge foreland basin developed in which was deposited sandstone and mudstone and the largest coal reserves on Earth. Much of this basin, especially its eastern portion, is now eroded. Each of these features diminishes to both the north and south of the present Central and Southern Appalachians. The Alleghenian orogeny is the North American equivalent of the Variscan orogeny, although as explained below, much of the former is younger than its European counterpart.

The Ouachita and Marathon orogenies along southern Laurentia are of a completely different orogenic style and are commonly called "soft collisions. Only the forelands remain and in each, tremendous volumes of deep-water flysch were thrust over the continental shelf of Laurentia. The colliders are deeply buried under kilometers of Gulf Coastal Plain sediment and are believed to be oceanic arcs that fringed accreting terranes. The terranes are also buried, although some appear on deep seismic profiles. Permian deformation in the Marathon Mountains of West Texas marks the youngest datable event of the Appalachian history of eastern and southern North America, although undatable thrusting in the Central Appalachians likely lasted well into the Permian. In the Pennsylvanian, stresses on North America were apparently so great that parts of the craton were deformed by high-angle reverse faults to form the Ancestral Rocky Mountains (◼ Fig. 12.6g).

Not much is known about Appalachian events in Mexico and Central America and what is known is strongly debated. A few deep wells have penetrated Cadomian crust in places like Honduras, Oaxaca, and Yucatan. The presumed terranes that they represent have been correlated to the Armorician (West Hun) Terrane discussed above. In other places Paleozoic or Precambrian structures occur in the subsurface and on seismic records, but it is difficult to resolve Appalachian events from older Grenville events in such areas, especially where both may be present.

Young Orogens—The Earth's Loftiest Places

Contents

© The Author(s), under exclusive license to Springer Nature Switzerland AG 2022
W. Frisch et al., *Plate Tectonics*,
Springer Textbooks in Earth Sciences, Geography and Environment,
https://doi.org/10.1007/978-3-030-88999-9_13

In the preceding chapter various Proterozoic and Pale-ozoic orogens were described. Each of these ancient ranges is presently deeply eroded and typically the roots or cores of these mountains are exposed in up-lands or low mountains. Traces of some ancient moun-tains are now exposed in more recent orogenic regions where they have been extensively overprinted and sub-sequently uplifted. This chapter concerns more recent orogenic events and the mountains that were created in the Mesozoic and Cenozoic. Modern mountain ranges are the products of the same plate tectonic events that created older orogens. For example, the rocks in the late Proterozoic Arabian–Nubian Shield formed by plate processes now operating in the island arcs of eastern Asia that are characterized by large masses of subduc-tion related magmatism and long lasting subduction activity along the margin of a huge ocean.

The plate tectonic patterns of SE Asia clearly illus-trate the difficulties in reconstructing ancient orogenic events. A particularly complex pattern exists in the present region between Sulawesi (former Celebes) and the similarly shaped island Halmahera. Here, the small, purely oceanic Molucca Plate was completely sub-ducted within the last two million years. The Molucca Plate subducted towards both the west (down to 600 km depth) and to the east (down to 250 km depth) until it completely disappeared (◘ Fig. 13.1). As a con-sequence, the two island arcs that formed above the re-spective subduction zones are colliding. Further com-plicating matters, north of Halmahera, a third sub-duction zone was created, along which the Philippine

Sea Plate subducts towards the west. Such a situation will be impossible to correctly reconstruct in the geo-logic future and clearly illustrates how difficult a task it is to reconstruct ancient complex plate tectonic rela-tions. The Sulawesi–Halmahera setting demonstrates how complex intra-oceanic island arc systems with their various subduction polarities can be. Each island arc is generated by a subduction zone; arcs collide fol-lowing the disappearance of intervening oceanic realm thus contributing to the generation of continental crust and the growth of continents.

Young orogens are present on all continents and like their ancient counter-parts, they vary greatly in plate tectonic setting and orogenic style. The Andes, the mountains of East Australia, the New Zealand Alps, East Antarctica, and the North American Cordil-lera represent Cordilleran-style mountain belts that are characterized by an active continental margin that bor-ders a large ocean for long periods of time. In most of these ranges, a large number of terranes have been ac-creted to the continental margin during the Mesozoic and Cenozoic eras (see ▶ Chap. 9). Consequently, crus-tal thickening, deformation, metamorphism, and large volumes of magmatism have resulted. In the Andes, crustal thickening was mainly accomplished by the ad-dition of large volumes of magmatic rocks that caused isostatic uplift and the creation of a high mountain range. Here compressional forces also played an impor-tant role as large parts of the orogen are under com-pressive stress (▶ Chap. 7). Crustal thickness in the An-des is up to 70 km thus explaining the high mean eleva-tion of the mountain range. However, the highest peaks are all volcanoes. The North American Cordillera ex-perienced enormous crustal thickening as a conse-quence of extreme crustal shortening during thrusting events as well as through the addition of large volumes of magmatic rocks. However, the Cordilleran orogen has undergone extreme extension over the last 30 mil-lion years that has greatly modified its original config-uration.

A contrasting orogenic pattern developed from the Pyrenees, through the Alps and Caucasus, and east-ward into the Himalayas and Indochina where exten-sive continent–continent collision has occurred in the Mesozoic and Cenozoic. In its western part, the evo-lution is characterized by a Wilson cycle in which the Penninic–Ligurian Ocean opened in the Early Juras-sic as an extension of the Atlantic Ocean and closed in the Paleogene (see ◘ Fig. 4.9). Its closure was respon-sible for the formation of the Alps and the Apennines. Further east, the orogens represent several Wilson cy-cles related to the subduction of the Tethys Ocean, the formation of several microcontinents and interven-ing oceans, and the culminating intense collision of In-dia and Arabia. As India rapidly drifted northward, its collision with mainland central Asia caused the char-

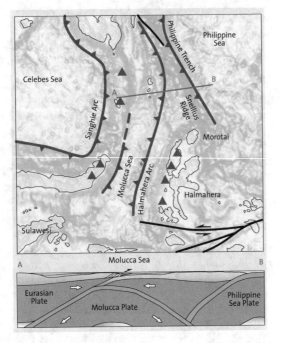

◘ **Fig. 13.1** Complex plate tectonic setting in the region of Sulaw-esi and Halmahera (Indonesia). Note that the Molucca Plate is com-pletely subducted (Hall 2000)

acteristic flexure in the trends of the mountain ranges that drape to the west and east of the indenter. The resulting Himalayas form the highest mountain range on Earth. The remainder of this chapter will describe in more detail several of the young orogens and their plate tectonic setting.

13.1 The Himalayas—A Mountain Range with Superlatives

The Himalayas not only comprise the highest mountain on Earth, but 10 of 14 mountains higher than 8,000 m (the other 4 being found in the neighboring Karakoram). The region forms the greatest relief on the continents and is characterized by areas of rapid uplift. Located on the back side of the mountains, the Tibetan Plateau represents the largest and highest plateau on Earth. The 2300 km-long Himalaya range was formed by the collision of India with Eurasia, but two smaller intervening continental pieces rifted from Gondwana and drifted northward and collided prior to the main Himalayan orogeny (■ Fig. 13.2). The suture zones of these earlier collisions are found in the Tibetan Plateau and occur between the Qiangtang (Changtang) block that collided in the Late Triassic and the Lhasa block that collided in the Early Cretaceous. These collisions were part of the extensive Cimmerian orogeny that closed older parts of the Tethys Ocean, the "Paleotethys" and opened behind them a new oceanic branch, the "Neotethys". The older sutures between these blocks and India are well documented in the Tibetan Plateau. Interestingly, the Neotethys suture zone is not found in the Higher Himalayas, but rather further north between the Transhimalaya or Gangdese belt, that forms the southern margin of the Tibetan Plateau (■ Fig. 13.3), and the Himalayas. The

headwaters of the Indus and Brahmaputra rivers, the latter called Yarlung Tsangpo in Tibet (*Tsangpo, Tibetan* river), follow this suture zone that is thus named Indus–Yarlung suture zone.

The Transhimalaya represents the pre-collision Cretaceous to Paleogene active continental margin that formed above a north-dipping subduction zone and is characterized by subduction-related magmatism; the volcanic chain, now eroded to its plutonic roots, is called Gangdese belt (■ Figs. 13.2 and 13.3). South of the Transhimalaya but north of the Indus–Yarlung suture zone, a sedimentary belt containing the Shigatse Flysch and the Qiuwu Molasse represents the forearc basin that formed south of the magmatic front (see ■ Fig. 7.18). The ophiolites aligned along the suture zone are parts of the oceanic basement of the forearc basin and thus are part of the upper plate along the southern margin of Asia. The Himalayas are situated entirely south of the suture zone, and therefore belong entirely to the stacked continental margin of India that was subsequently shortened by roughly 1000 km. Such suture zones, which represent the position of the former subducted ocean between the colliding continents, are usually positioned within the mountain range as is the case in the Alps. In this respect, the Himalayas are an asymmetric orogen with the suture flanking their margin; the stacked nappe package was generated entirely by thrusting and deformation of the Indian continental margin (■ Fig. 13.2).

13.2 Tectonic History of the Himalayas

The overall structure of the Himalayas is relatively straight forward. The orogen consists of three tectonic mega-units: (1) the nappe systems of the Higher Himalayas or "Tethys Himalayas", (2) the Lesser Himalayas,

■ **Fig. 13.2** Profile showing the plate tectonic evolution of the Tibetan Plateau and the Himalayas during the early Cretaceous (140 Ma), middle Cretaceous (100 Ma), and Paleogene (40 Ma) (Allègre et al. 1984)

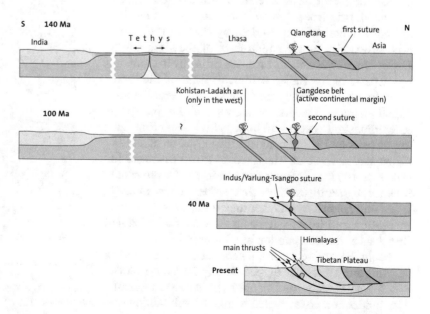

Fig. 13.3 Geological sketch of the tectonic components of the Himalayas. The indention of the northwestern and northeastern tips of the Indian continent formed the syntaxes of Nanga Parbat and Namche Barwa. Note the abrupt changes in structural trends across these syntaxes

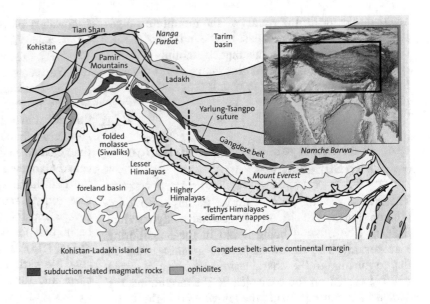

13

and (3) the Siwalik fold and thrust belt (■ Fig. 13.3). The nappe system contains sedimentary rocks, for instance limestones, that were deposited in the Tethys Sea as geologically recently as the Paleogene and now reside at the highest elevation on Earth. The Higher and Lesser Himalayas consist of ancient metamorphic basement that formed at the end of the Precambrian era during the Panafrican–Cadomian orogeny. It is overlain by a rather continuous sedimentary sequence that was deposited from Cambrian through Lower Tertiary on the continental passive margin of India. It embraces large areas on the northern slopes of the mountain range. During Paleogene (Eocene) the main nappe stacking took place and parts of the sedimentary sequence also experienced metamorphism. According to the northward dip of the subduction zone, thrusting was southvergent (■ Fig. 13.2). The lowest structural unit, the Siwaliks, represents the Neogene molasse that infilled the foredeep or foreland basin that developed on the downwarped Indian continent. The Siwaliks are both underlain and overlain by thrusts; they have been overridden by the nappe stack of the Higher and Lesser Himalayas and, in turn, are thrust over more interior parts of the Indian continent. Each of the three mega-units is internally imbricated into several individual nappes. Fensters (windows) and klippen provide important structural information regarding the thrust belts and help document the existence of broad thrust sheets, some of which record thrust distances in excess of 100 km. A fenster or window is an erosional hole through a thrust sheet that exposes a tectonically lower unit framed by a higher unit; a klippe is detached by erosion and forms a remnant of a nappe or higher thrust sheet that rests on top of a lower unit.

In the Early Cretaceous, a subduction zone formed in southern Tibet that dipped below the Lhasa block which was already welded onto the Asian continent. The subduction zone created a zone of subduction-re-

lated magmatism that formed the active continental margin of the Gangdese belt in the Transhimalaya (■ Figs. 13.2 and 13.3). The Kohistan–Ladakh island arc existed farther to the west on oceanic basement above an intra-oceanic subduction zone (■ Figs. 13.2 and 13.4). This subduction zone was the continuation of the subduction zone beneath the Gangdese belt on the one side and the subduction zone that was responsible for the obduction of the Semail ophiolite complex in Oman (see ■ Fig. 5.16) on the other. Another subduction zone formed north of the Kohistan–Ladakh island arc and dipped beneath the Eurasian continental margin (■ Fig. 13.4). This northerly zone subducted the oceanic basin north of the island arc, closed the intervening ocean, and generated the collision of the is-

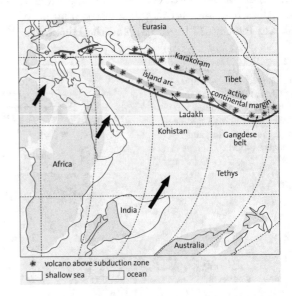

Fig. 13.4 Paleogeographic map of parts of the Tethys Ocean and SE Asia during the Cretaceous. This map shows the pre-Himalayan tectonic setting of the region prior to the collision between India and Tibet (Rolland 2002)

land arc with Eurasia in Late Cretaceous. These events preceded the collision of India with Eurasia and transformed southern Asia into an active continental margin on which the Gangdese belt was built. Subsequently, the approximately 5000 km-wide Neotethys Ocean was subducted. This was the tectonic setting that faced India as it drifted northward on the Neotethys-Indian plate. The ensuing collision of India with Eurasia generated the Himalayas and formed the Indus–Yarlung suture zone in Early Paleogene time. The suture zone is marked by ophiolites and a thick mélange succession.

Nanga Parbat and Namche Barwa Syntaxis

During collision, two sharp salients located on either side of the northern flank of the Indian continent protruded deeply into Asia and generated areas of extremely strong deformation. Such a point, where mountain ranges with different trends converge, is called a *syntaxis* (*Greek* convergence). The two syntaxes, Nanga Parbat and Namche Barwa, on the northwestern and northeastern corners of the Indian continent respectively (□ Fig. 13.3), are characterized by extremely rapid uplift and the exhumation of deeply buried rocks. The regions are characterized by some of the highest mountain peaks in the world, 8125 and 7756 m, respectively, and also display the greatest relief on Earth in the interior of a continent, approximately 7000 m over a horizontal distance of 20–30 km.

A narrow, north-plunging zone that trends perpendicular to the strike of the orogen characterizes the Nanga Parbat area. Here, once deeply buried rocks that experienced high-grade metamorphism and local melting in the last few millions of years, were rapidly uplifted to the surface. The rate of uplift of these rocks was determined to have been as high as to 7 mm/year or 7 km per million years (Zeitler et al. 1993). This very fast exhumation was enabled by tectonic denudation characteristic of metamorphic domes (see □ Fig. 3.19); the rocks were squeezed along steep faults as a vertical wedge. The tectonic setting in the eastern syntaxis of Namche Barwa probably resembles that around Nanga Parbat; however the region is less well studied because the area is extremely remote and difficult to access. Interestingly, Nanga Parbat and Namche Barwa appear to be related terms—*Nanga Parbat* meaning Naked Mountain (from *Sanskrit*).

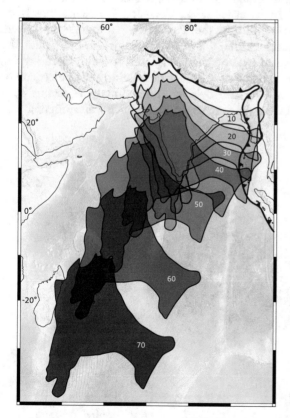

□ **Fig. 13.5** Map showing the drift locations of India relative to Asia since 70 Ma. The rapid northward drift of India during the Late Cretaceous and Early Paleogene abruptly slowed as India collided with Central Asia approximately 50 Ma (Patriat and Achache 1984). India penetrated more than 2000 km into Asia during the remainder of the Cenozoic

India rapidly marched northward towards Asia with a velocity of ca. 20 cm/year, a plate velocity that exceeds any modern example. This velocity considerably slowed to ca. 5 cm/year following the collision, yet India continued to protrude into Asia for more than 2000 km (□ Fig. 13.5). The collision resulted in a broad zone of deformation as the northern portion of India was subducted under Asia; this event contributed to the formation of the thickest crust on Earth and therefore formed the highest mountains. The irregular northern margin of the Indian continental crust first came into contact with Eurasia along its northwestern corner approximately 55 Ma. As a consequence, India underwent a counterclockwise rotation to close the remaining part of the Neotethys in scissor-like fashion from west to east. The closure of the Neotethys was completed approximately 40 Ma.

The effects of the collision between India and Asia not only dominated southern Asia, but also were global in nature. The drift rate of India considerably slowed at 50 Ma (□ Fig. 13.5) and the Indian Plate was reorganized at this time. The spreading of the mid-ocean ridge that lay between India and Australia and forced the latter continent northward decelerated and eventually stopped; this fused India and Australia into a single plate. The spreading axis of the Indian Ocean became established to the south to form its current location between the Antarctic Plate and the Indo-Australian Plate and consequently the drift direction of India changed by approximately 30° at 43 Ma. By then, the collision with Eurasia was almost completed. The forceful collision also had consequences for the worldwide plate drift pattern that had become reorganized in the Eocene. Many orogens around the globe reflect strong deformational phases during the Eocene. The sharp bend in the seamount line of the Hawaii–Emperor chain (see ◼ Fig. 6.8) that records a change in the direction of motion of the Pacific Plate of 60° at 42 Ma, coincides in time with the change in drift and geometry of the Indo-Australian Plate. Not surprisingly, the chain of events across the Pacific led to the opening of the Tasman and Coral Seas along eastern Australia and the formation of the Philippine Sea Plate to the north.

13.3   The Alps—An Untypical but Classic Orogen

The Alps can rightly be termed the classic orogen because they have been intensely investigated for more than two hundred years and many fundamental concepts and basic facts on mountain building came from these studies. However, serious problems concerning the understanding of the evolution of this orogen continued to arise; many geologists considered the Alps to be the most complicated orogen on Earth, especially when compared to the Himalayas with its much clearer evolutionary history. Some researchers believe the complexity of the Alps is related to the intense level of exploration that has been devoted to the range and that

problems have been disclosed that are not yet recognized in other orogens. But in recent years, researchers have discovered that the true complexity of the Alps is due to the fact that there are two orogens that generated the present mountains, an older Mesozoic event and a younger Cenozoic event. And this is what makes the Alps an untypical orogen.

The initial orogenic event occurred during the Middle Jurassic to Early Cretaceous and has been coined the Eoalpine ("early Alpine") orogeny. It encompassed the eastern part of the Alps and later formed the Austroalpine nappe system, the structurally highest of the three mega-units in the Alps (◼ Fig. 13.6). During this time, the Austroalpine realm was part of Adriatic Spur of Africa (Gondwana) and the lower or subducting plate. The upper plate comprised the Tethys realm with numerous microplates imbedded within it. Part of the upper plate was overthrust upon the Austroalpine unit and ophiolites derived from the ocean floor of the Tethys Ocean were obducted (◼ Fig. 13.7: 145 and 110 Ma). The Eoalpine orogeny continued into the Carpathians, Dinarides, and other mountain ranges further east where it is known as the Cimmerian orogeny; in central Asia it marks the collision of the Lhasa Terrane with Eurasia in Tibet. These events mark the partial closure of the Tethys Ocean as blocks rifted from Gondwana drifted northward to collide with Eurasia.

In its strict sense, the Alpine orogeny mostly occurred during the Paleogene. It marks the closure of the Penninic Ocean, a narrow oceanic realm that opened as an extension of the Atlantic Ocean during the Jurassic and Cretaceous. During the Paleogene orgeny, the Austroalpine and Southalpine realms occupied the position as the upper plate while different units

◼ **Fig. 13.6** Tectonic patterns of the Alps showing the three mega-units that comprise the nappe systems. Basement and overlying stratigraphic sequences of the Alpine cycle are differentiated for each mega-unit. The Austroalpine unit was thrust over the Penninic unit, which in turn was thrust over the Helvetic unit. The Peridadriatic Lineament is a large-scale fault zone that borders the northern margin of the Southern Alps

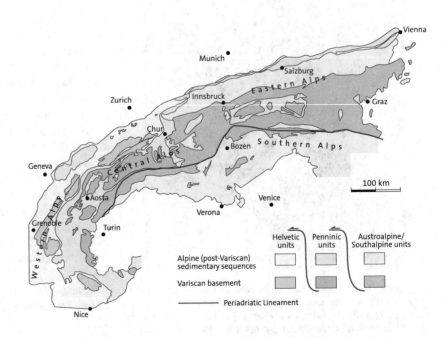

Fig. 13.7 Plate tectonic evolution of the Alps and adjacent parts of Europe and North Africa. Time slices from Late Triassic to Eocene show evolution in both map and cross-section views (after Frisch 1979). Geological units: AA = Austroalpine, SP = South Penninic (oceanic), MP = Middle Penninic, NP = North Penninic (partly oceanic), H = Helvetic; NCA = Northern Calcareous Alps (part of AA), Mol = Molasse zone

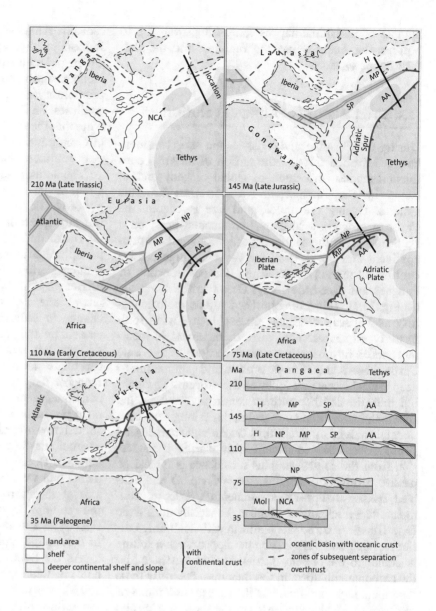

of the Penninic realm became subducted and underplated. Finally, the nappe stack was thrust over the European passive continental margin; imbrication of the nappe stack caused the Helvetic mega-unit to become the structurally lowest unit (■ Fig. 13.7). The Austroalpine unit, the dominant component of the Eastern Alps, was tectonically shaped separate from the Penninic and Helvetic units, the components that dominate the Central Alps and the arc of the Western Alps (■ Fig. 13.6). This interpretation of Alpine evolution clearly documents the differences, both in terms of tectonic behavior and tectonic history, of the different parts of the range; it also emphasizes the double role of the Austroalpine unit first as a lower plate, then as an upper plate. When these points are made clear, much of the confusion concerning Alpine evolution disappears.

The Penninic Ocean was a continuation of the Atlantic Ocean; the two were offset by a transform system that ran in two branches south and north of Iberia (■ Figs. 4.8 and 13.7). During the Middle Jurassic, the Atlantic–Penninic system generated the initial breakup of Pangaea and caused the separation of Laurasia and Gondwana. This system ultimately linked the Tethys Ocean with the Pacific Ocean via the Penninic–Ligurian, Atlantic, Gulf of Mexico, and Caribbean. Thus, the Penninic–Ligurian Ocean between Europe and Africa (remnants of the Ligurian Ocean are found in the Apennines) was a part of the early Atlantic and not a protrusion of the Tethys Ocean. The Penninic Ocean closed during the Paleogene orogeny as the former African continental margin (Austroalpine unit) was thrust over the European continental margin (Helvetic unit). The Penninic nappes including a microcontinent seamed by two oceanic branches became sandwiched in between. The result was a stack composed of a large number of individual nappes comprising, from top to bottom, the Austroalpine, the Penninic, and the Helvetic nappe systems.

13.4 Brief History of Alpine Evolution

During the Permian and Triassic, the supercontinent Pangaea formed through the unison of all large continental masses as a result of global orogeny orchestrated by the Appalachian–Variscan–Ural orogenic systems. During the middle and late Paleozoic, the Variscan orogeny welded new continental crust onto Baltica to form the southern portion of the European continent and its Variscan basement. A large passive continental margin bordered the European portion of Pangaea and faced the Tethys Ocean to the southeast (■ Fig. 13.7). This setting initiated Alpine geologic history as Variscan basement underpinned Alpine sedimentary sequences. The Variscan basement is metamorphosed in most Alpine areas because of the Late Carboniferous and Permian erosion and exhumation to deep crustal levels that followed the orogeny. Throughout the Triassic, the southern continental margin, the Austroalpine–Southalpine realm, lay along the continental margin of the Tethys Ocean and experienced strong subsidence coincident with marine transgression from the southeast. This subsidence is typical of passive continental margins (see ▶ Chap. 4) and enabled Triassic shallow-water sedimentary rocks to accumulate to a thickness of ~3000 m. Especially during the Late Triassic, a carbonate platform with a flanking barrier reef belt and large lagoon developed; such a setting is similar to the present Barrier Reef in Australia and the carbonate platform in the Bahamas. Presently these carbonate rocks are found in the rugged mountains of the Northern Calcareous Alps in the Eastern Alps and in the Dolomites in the Southern Alps. In deeper water towards the Tethys on the outer shelf and continental slope, the reef belt was framed by pelagic carbonates.

Towards the European continent, the Austroalpine realm was bordered by areas that later became part of the Penninic (Middle Penninic representing a later continental fragment split off from Europe) and Helvetic realms. Because they were more distant to the Tethys Ocean, these areas experienced much less subsidence in the Triassic when they were only partly flooded by the sea. Accordingly, the sediments deposited in this region are much thinner.

The break-up of Pangaea initiated early in the Jurassic following crustal extension that had occurred in the Triassic. The South Penninic Ocean formed as part of the overall rift system as discussed above. Slow spreading rates meant that the ocean remained fairly narrow and at its maximum width during the Creta-

ceous, it remained less than 1000 km wide. The Late Jurassic paleogeographic map shows that the Penninic Ocean played a similar role to that of the Central Atlantic and generated a separation between the African promontory/Adriatic Spur to the SE and Iberia and southwestern Europe to the NW (■ Fig. 13.7: 145 Ma). Note that the Austroalpine–Southalpine realm was part of Gondwana and was clearly separated from the European continent (Laurasia) by the narrow ocean.

The Jurassic sediments of the Austroalpine–Southalpine realm were deposited in an open sea; the supply of clastic material from the continental hinterland was largely cut off. The connection between the Atlantic Ocean and the Tethys Ocean created bottom currents that partly impeded sedimentation and thus were responsible for low sedimentation rates and even submarine gaps in the sequence; such a break in the rock record is termed a submarine unconformity. Dramatic change in sedimentation patterns had already been caused by the mass extinction at the Triassic/Jurassic boundary that wiped out marine organisms and the carbonate platform that they had built thereby strongly reducing the production of biogenic carbonates. As a consequence of continuous subsidence but less sediment production and supply, the sedimentation pattern shifted towards increasingly deep-water facies in the Austoalpine–Southalpine realm. Meanwhile, the South Penninic Ocean received mainly distal turbiditic sediments that were conveyed for long distances down the axis of the narrow oceanic trough. The shoulders of the Penninic Ocean became the sites of increasing subsidence rates as reflected in the passive margin deposits of the Middle Penninic and Helvetic realms.

As the South Penninic Ocean generated new oceanic crust during the Early to Middle Jurassic, ca. 180–170 Ma, intra-oceanic subduction and obduction in the Tethys Ocean initiated the Eoalpine orogenic cycle. Compressional tectonics and nappe formation ranged from the Austroalpine realm on the west to the Carpathians, Dinarides, Hellenides, and Turkey on the east. From Turkey and eastward, the orogenic event, typically called the Cimmerian orogeny, involved major microcontinent collision with southern Asia as part of the Tethys Ocean closed. The Adriatic Spur, which included the Austroalpine realm, was trapped between the expanding Penninic Ocean to the west and complex, poorly understood, compression associated with terrane accretion to the east. In the Austroalpine realm, the Eoalpine event began with the obduction of oceanic crust upon the Adriatic spur (■ Fig. 13.7: 145 Ma). Thrusting progressed from the Tethyan margin towards the continent, decelerated in the Late Jurassic, and accelerated again during the Early Cretaceous (Frisch and Gawlick 2003). The nappe fronts migrated northwestward until the early Late Cretaceous

Fig. 13.8 Panel cross sections and photograph showing tectonic elements of the Alps. Upper left—profile through the eastern Swiss Alps at ca. 30 Ma showing Alpine evolution immediately after slab breakoff (Schmid et al. 1996). Lower—cross section of the western Swiss Alps showing present relations of major tectonic elements (Trümpy 1980); note that location of photo is indicated. Upper right—photograph of the Matterhorn. The pyramidal peak consists of a piece of the Austroalpine unit, formerly part of Africa. The Austroalpine unit overlies the South Penninic unit that is composed of ocean floor material. The Middle Penninic unit, a continental fragment rifted from Europe is in the foreground. Note that the orientation of the photograph is reversed relative to that of the lower cross section

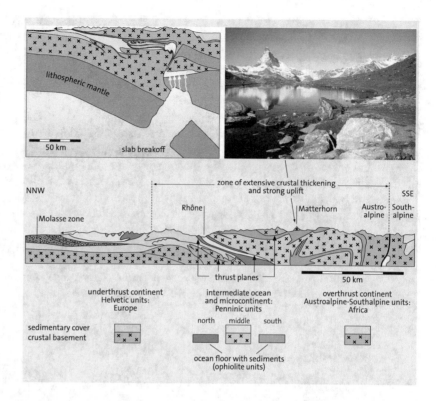

when parts of the Austroalpine basement, the passive continental margin, were subducted to depths of more than 50 km and transformed into eclogites. These high-pressure metamorphic rocks are part of the eclogite type locality, first described by Hauy (1822) from areas along the eastern margin of the Alps.

Beginning in the Middle Jurassic, crustal extension between the Middle Penninic and Helvetic realms led to the formation of the Middle Penninic microcontinent. Separation from Europe by the North Penninic Ocean occurred approximately at the Jurassic–Cretaceous boundary. The newly generated oceanic crust reached a limited width in the course of the Cretaceous and was flanked by thinned continental crust. The Middle Penninic microcontinent thus became isolated, bordered on both sides by deep oceanic troughs (South and North Penninic). The breakoff of continental fragments or microcontinents from larger continental mass is not uncommon; several examples were discussed in ▶ Chap. 12. The greater Alpine–Mediterranean mountain belt contains a number of such individual continental pieces. Their separation, mostly during the Cretaceous, reflects the long-term opening of the Atlantic Ocean and the resulting eastward movement of Africa relative to Europe that caused counterclockwise rotation of intermediate continental fragments. Both the Iberian microcontinent and the Adriatic Spur separated from Europe and underwent counterclockwise rotations (▪ Fig. 13.7: 110 and 75 Ma).

Lateral Tectonic Extrusion in the Alps

During the Alpine orogeny, the Southern Alps did not experience metamorphism. As a result, the tectonic block was relatively cool and strong and acted as an indenter that pushed against the main body of the Alps to the north. In map view, the Southern Alps consisted of two blocks with a triangular shape, each of which presented a sharp apex or salient towards the north (▪ Fig. 13.9). The salients, named the Insubric and the Dolomites indenters, acted somewhat independently through their respective tectonic histories. As the strong, brittle masses pushed northward, they generated crustal stacks north of the Periadriatic Lineament. The result was pattern referred to as lateral tectonic extrusion (▶ Chap. 11; Ratschbacher et al. 1991b). During the Early to Middle Miocene, ca. 22–12 Ma, metamorphic domes and conjugate strike-slip faults were formed as the lateral extrusion process operated across the Central and Eastern Alps.

In the metamorphic dome structures, Penninic rocks buried beneath the Austroalpine nappe system became rapidly exhumed and formed tectonic windows (▪ Fig. 13.9b). The most spectacular is the Tauern Window in the Eastern Alps, a 160 km-long and 30 km-wide structure that led to the synthesis of the nappe theory more than 100 years ago. Here, geologists postulated that far-traveled nappes are a consequence of lateral compression and an important attribute of orogens. The Tauern Window region consists of a brittle upper plate that comprises basement rocks that cooled following Cretaceous Eoalpine metamorphism and a lower plate of ductile, metamorphic Penninic rocks. Exhumation resulted as a consequence of the rapid extension and unroofing in the upper brittle plate; in response, the lower ductile plate was deformed and extended along an east–west trend during the rapid uplift. The pull-apart occurred across the entire length of the window (◻ Fig. 13.9c) as crustal extension in the Central and Eastern Alps nearly doubled the east–west dimension of the region.

Farther east of the Alps beneath the Neogene sedimentary cover of the Pannonian Basin, the Austroalpine unit is dissected by N–S-trending graben and horst structures, another indication of E–W stretching. But the rigid Austroalpine lid also accommodated E–W stretching through the formation of large strike-slip faults that enabled crustal wedges to escape in an eastward direction. The NE- to ENE-trending faults have left-lateral displacement and the SE- to SSE-trending faults document right-lateral displacement (◻ Fig. 13.9b). Along the ESE-trending Periadriatic Lineament, the right-lateral movement was more than 100 km.

The eastward extrusion in the Alps was coupled with subduction in the Carpathian arc to the east of the Pannonian Basin. Crustal thinning related to E–W extension in the Pannonian Basin was balanced by the convergence along a W-dipping subduction zone beneath the Carpathian arc.

13

◻ **Fig. 13.9** Oligocene and present tectonic maps of the Central and Eastern Alps. **a** Oligocene(ca.30 Ma) paleotectonicmap. **b** Present tectonic map (Frisch et al. 2000). The present elongated shape of this part of the Alps is the result of post-collisional lateral tectonic extrusion that led to considerable east–west extension of the orogen. In the Penninic realm, axial culminations of metamorphic domes were exhumed. **c** Structural cross section through the Tauern Window. The Austroalpine unit is characterized by crustal wedges that escaped eastward. The Ötztal (Ö) and Gurktal (G) blocks were originally continuous and adjacent (see **a**), but crustal extension has caused a separation of 160 km (see **b**)

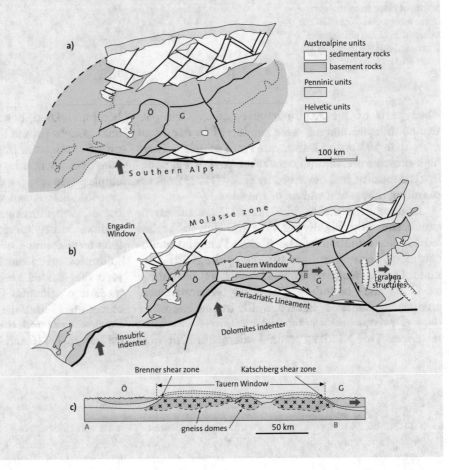

Although exact timing remains nebulous, Late Cretaceous southeast-directed subduction initiated along the southeastern margin of the South Penninic Ocean and the Alpine orogenic cycle was underway. The main orogenic phase lasted from Eocene to Oligocene. It led to the collision of the Austroalpine realm with the Middle Penninic continental mass, and, subsequently, with the European continental margin (the Helvetic realm). Penninic oceanic sediments and ophiolite fragments became sandwiched between the colliding continental masses. Deeply subducted regions experienced high-pressure metamorphism, mainly in the Penninic units where such rocks are widespread. Middle Penninic gneisses in the arc of the Western Alps experienced ultrahigh-pressure metamorphism with the formation of coesite during burial to a depth of ~100 km. It was in these rocks that subduction-related coesite was described for the first time (Chopin 1984). Thrusting of the Helvetic nappes continued in the Oligocene and Miocene and reflects the migration of the tectonic processes towards the European foreland.

A deep-marine trough formed in front of the advancing nappe stack and thick flysch sequences were deposited. During the Paleogene, this trough migrated across the downwarped European continental margin from the Penninic realm northward into the Helvetic realm. Due to underthrusting of the European margin, the convergence rate slowed and isostatic readjustment caused uplift of the nappe stack. In the Early Oligocene, ca. 30 Ma, increasing volumes of sediment were transported from the uplifting orogen into the foredeep, which therefore, became rapidly filled. This marks the change from flysch stage to molasse stage along the northern margin of the Alps. Rivers that drained the uplifted mountainous area transported large amounts of gravel that was discharged as conglomerate fans along the flanks of the Molasse zone. The balance between sedimentation rates, subsidence rates, and sea level change fluctuated widely. The Molasse zone was overfilled by sediments during the Late Oligocene and became land, flooded by a shallow sea for a short time in the Early Miocene, and then returned to a terrestrial setting. Finally, during the Late Miocene, between 10 and 8 Ma, the Molasse zone was uplifted and became an area of erosion.

The rapid uplift process in the Early Oligocene, that filled the foredeep with sediments was triggered by two major processes: (1) rapid stacking of numerous crustal sheets, and (2) slab breakoff of the oceanic part of the subducting plate (◘ Fig. 13.8). In the latter process, the heavy slab pull in the subduction zone lost its tug on the nappe stack when the lower plate slab broke off; the overlying stack responded with rapid uplift. Simultaneously, magmatic bodies ascended along the Periadriatic Lineament, the large fault zone that separates the Southern Alps from the main body of the Alps (◘ Figs. 13.6 and 13.8). The melts were extracted from an upward flow of the asthenospheric mantle that rose to fill the gap behind the broken slab; there it was mixed with lower crustal material.

13.5 The North American Cordillera—A Different Style of Orogen

Cordilleran-style mountain belts are characteristic of the regions adjacent to the Pacific Ocean—the Pacific ring of fire. The Pacific–Panthalassa system has been the dominant ocean, nearly half of Earth's surface area, since at least the Late Precambrian. Instead of contracting and expanding in classic Wilson cycles as the Iapetus, Rheic, Tethys, Indian, and Atlantic oceans have done, the Pacific Ocean has remained relatively unchanged in size for over one billion years. This is not to say that the ocean basin has not changed over this period of time; but rather, the ocean has expanded and subducted at approximately coeval rates but without the drifting of major continental blocks across the ocean basins. Like the Atlantic and Indian oceans, the Pacific Ocean has crust that has a maximum age of Early Jurassic. Consequently, continental regions that have bordered the Pacific since the Precambrian have experienced long-term, continuous subduction. As this is true of western North America and South America—regions with strong Hispanic heritage—and because cordillera means mountain range in Spanish, the term Cordilleran has been applied to such long-lasting active ocean–continent margins.

Cordilleran-style orogens border the modern Pacific and include the Great Dividing Range and New England Range of eastern Australia, the New Zealand Alps, the Antarctic Peninsula, the Andes, the Cordillera region of Central and North America, and the arc systems of eastern Asia that lie in the western Pacific. Two Cordilleran-like arcs form major bulges into the Atlantic, the Caribbean and Scotia. The remainder of this discussion will focus on the Cordillera of western North America.

The North American Cordillera comprises a collage of terranes that have accreted since the middle Paleozoic (◘ Fig. 13.10); prior to that, the region was a continental passive margin and adjacent oceanic crust. Many of these terranes were described in ▶ Chapter 9 so their descriptions are not repeated here. In detail, the Cordillera consists of many tectonic subdivisions that reflect local to regional tectonic history and has been significantly altered and overprinted by late Cenozoic extension and extensive volcanism. A simplified classification system that holds up for portions but not all of the Cordillera includes from east to west (1) foreland fold and thrust belt, (2) eastern terrane belt, (3) plutonic belt, and (4) western terrane belt that includes significant late Cenozoic transform fault systems. However, it should be pointed out that all of the subdivisions have some plutons and that the boundaries between these regions are not necessarily sharply defined.

The foreland fold and thrust belt consists of numerous imbricate low-angle thrust sheets of Late Precambrian, Paleozoic, and Mesozoic sedimentary rocks that were deposited mostly in a continental passive margin setting but also in both continental and oceanic back-arc settings. The depositional thicknesses of these rocks is appreciable and examples include 5–7 km of Cambrian strata in western Utah and eastern Nevada, and 7 km of Pennsylvanian–Permian strata in NW Utah. Rock thicknesses in most of the remaining Paleozoic

13

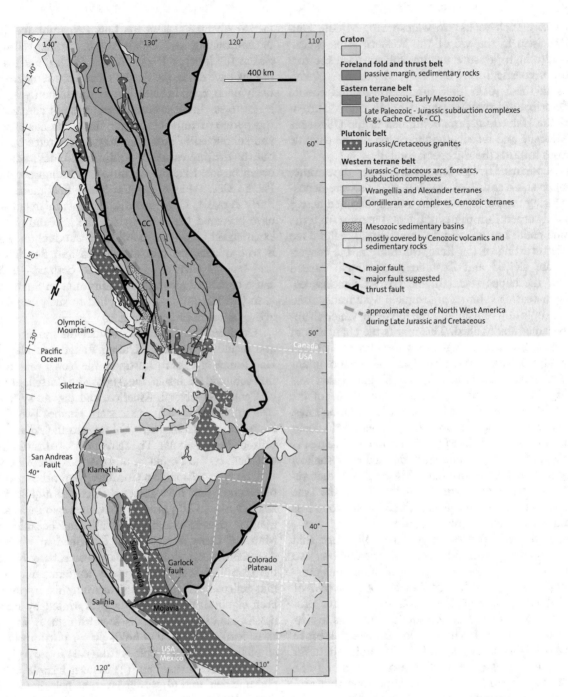

■ **Fig. 13.10** Simplified terrane map showing tectonic subdivisions used in text. Note that boundaries between subdivisions range from sharp to gradational and arbitrary

systems equals or exceeds 2 km, at least locally. The geometry and facies patterns of Late Precambrian through Devonian deposits define a classic example of a continental passive margin. A lower siliciclastic interval grades upwards into a monotonous, thick marine carbonate interval. Presently, these rocks are exposed in many of the ranges of the central Basin and Range Province of western Utah and Nevada. The main foreland fold and thrust belt developed in the Jurassic and Cretaceous during the Elko, Nevadan, and Sevier orog-

enies. Crustal shortening of up to 50% stacked thrust sheets on top of each other and onto the margin of the North American craton.

The eastern terrane belt comprises a complex assemblage of Late Precambrian, Paleozoic, and early Mesozoic terranes that represent arc, trench, back-arc, and open oceanic settings as well as local microcontinent fragments. In Canada, the eastern terrane belt is termed the Intermontane Superterrane; equivalents are presently found farther south from NW Mexico to Washington. These terranes were accreted in two or more orogenic events including the Late Devonian-Early Mississippian Antler orogeny and the Triassic Sonoman orogeny. As such, they document an active continental margin along western North America during this time. Some of the terranes are exotic to North America, although many show some North American affinities through paleontologic and detrital zircon studies and likely originated either as blocks that rifted from the parent continent and later returned, or as fringing, Japan-style arc systems. From central California to Yukon Territory in Canada, a terrane near the middle of the eastern terrane belt poses an interesting problem. The terrane, the Cache Creek Terrane of Canada and similar terranes in the USA, consists of ophiolites and associated oceanic materials with a Tethyan faunal affinity; yet this terrane is sandwiched between two terranes that mostly have North American affinities. Not surprisingly, several hypotheses have been proposed to explain this conundrum including transform offset of arc-microcontinents to generate the repetitive sequence, oroclinal folding of a ribbon-shaped microcontinent (◘ Fig. 9.4c, d), and an interpretation that one or both of the bounding terranes are exotic to North America and accreted from across the Panthalassa Ocean.

The plutonic belt stretches from Central America to Alaska and forms one of the most extensive batholithic systems on Earth. From NW Mexico to SE Alaska, these plutonic complexes include the Peninsular, Sierra Nevada, and Idaho batholiths in the USA, and the Coast Plutonic complex of western Canada. Together, they comprise the bulk of the Cordilleran

Mesozoic arc. The Sierra Nevada is probably the most studied of these and forms the basis for the following discussion. Although scattered Triassic plutons are present, Jurassic granites mark the maturity of the Cordilleran arc and Cretaceous plutons comprise the bulk of the magmatic complex. Plutonism and accompanying volcanism were not continuous throughout the Jurassic and Cretaceous, but rather, magmatism waxed and waned several times throughout the interval. During much of the Jurassic, the southern arc in Arizona and Nevada was built on continental North America while farther north, the arc was marine in style and was separated from the continent by an oceanic backarc basin. By the Cretaceous, the entire system was firmly pinned on the continent as an Andean arc. For many years, the Sierra Nevada complex has been interpreted to encompass a trinity of tectonic components, the Sierra Nevada magmatic arc, the Great Valley forearc basin complex, and the Franciscan trench-mélange complex (◘ Fig. 13.11a). Although this view is still widely embraced, it is not without problems; perhaps because the Sierra Nevada region is one of the most studied arc-forearc complexes on Earth, it may suffer the same problems as the Alps—too much knowledge raises more questions.

The Sierra Nevada and coeval plutonic complexes consist of multiple plutonic units, mostly ranging from granite to granodiorite in composition. The plutons were generated as the giant Farallon Plate, a plate as large as the present Pacific Plate, was subducted. The arc structure was built on a myriad of older materials that ranged from the North American craton to various previously accreted materials of the east-

◘ **Fig. 13.11** Restored cross sections showing **a** normal Mesozoic Cordilleran subduction, and **b** subduction of thickened, flat slab during Paleogene. Also shown on the Mesozoic cross section are the settings of key components of the arc complex

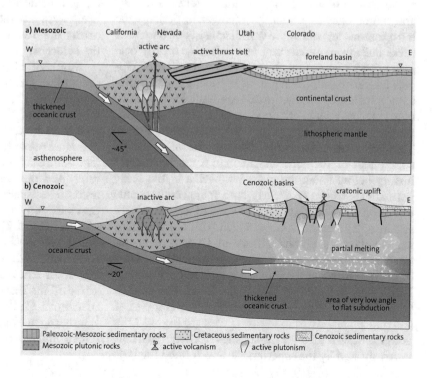

ern terrane belt. The Coast Plutonic complex of Canada was built on both the eastern terrane complex and the subsequently accreted exotic Insular Superterrane that contains the famous Wrangellia Terrane. In the forearc region, marine turbidites dominated the Great Valley forearc basin; detritus was mostly derived from the adjacent arc although forearc deposits are never found in stratigraphic contact with the arc; instead, the forearc basin was built on previously accreted terranes and other material, usually ophiolites. The Franciscan complex comprises one of the most complicated and well-studied trench-mélange deposits on Earth. Interestingly, the complex is never in stratigraphic contact with the supposedly adjacent Great Valley forearc, nor does the Franciscan have a known basement; perhaps neither fact is surprising given the complex tectonics associated with the dynamic trench system. However, some geologists have used this lack of pinning to adjacent terranes and tectonic elements to suggest that all or most of the Franciscan is exotic and originally formed perhaps thousands of kilometers to the south of its present location. During the maximum development of the Franciscan–Great Valley–Sierra Nevada arc trinity in the Middle Jurassic through much of the Cretaceous, the exotic Insular Superterrane was accreted. In spite of possible interference with normal subduction processes, there may have been little change in arc-forearc behavior, a problem that is discussed further below.

The western terrane belt includes diverse terranes and tectonic elements now west of the Cordilleran arc complexes. The Insular Superterrane was displaced by the late Cenozoic San Andreas Fault. The Insular Superterrane consists of two major terranes and several minor ones (▶ Chap. 9). Wrangellia, perhaps the most famous terrane in the world, consists of basement rocks built from middle and late Paleozoic mostly oceanic arc complexes and an overlying succession of Upper Triassic basalts thought to represent oceanic plateau volcanism. Younger arcs were built on this complex both before and after accretion to North America, which is generally interpreted to be Middle Jurassic. The far-traveled Alexander Terrane, the other major component of the superterrane, had a much different history before its accretion to Wrangellia in the Early Permian (▶ Chap. 9). Clearly both major components of the Insular Superterrane are exotic to western North America. The superterrane collided with North America at approximately the latitude of Oregon or northern California and then was transported southward due to oblique convergence between the Farallon Plate and North America (◨ Fig. 13.12). As the terrane docked with North America, it ploughed through a complex of fringing Japan- or Mariana-style arcs (◨ Fig. 13.12a). Some of these arcs and associated tectonic elements were captured by the superterrane and formed the Baja

BC block (▶ Chap. 9). How far south Baja BC transformed is a matter of great contention with hypotheses ranging from hundreds of kilometers to greater than 2000 km. The latter distance would have placed the block adjacent to Baja California, Mexico; it is now found in British Columbia (BC), hence Baja BC. Any distance south of central California would have placed the Baja BC block directly in the path of Farallon subduction under the Sierra Nevada (◨ Fig. 13.12b, c). Two recent studies place the southern-most margin of the block at the latitude of central California and southern California, respectively. Such a positioning would have meant that the Great Valley forearc basin complex was at least partly an interarc basin deposit (◨ Figs. 9.4d and 13.12c). Because the Franciscan complex is completely fault-bounded, it could have been displaced northward to its current position following the Late Cretaceous and Early Paleogene migration of Baja BC northward.

South of the current location of Baja BC on Vancouver Island, BC, the western terrane belt consists of two contrasting tectonic elements. North of the transform margin boundary in Northern California at Cape Mendocino, the forearc region remains in an active forearc setting. Terranes accreted during the Paleogene now comprise the coastal mountains of Oregon and Washington and are onlapped by Neogene forearc basin deposits and partly buried by plateau basalts. Siletzia, an accreted terrane mostly in Oregon, and portions of the Olympic Mountains of Washington are part of an obducted seamount and oceanic plateau assemblage (◨ Fig. 13.12d). Much of the former now lays buried beneath the extensive Neogene basalts of the Columbia River Plateau. The Columbia River Plateau basalts likely were generated as western North America drifted across the Yellowstone hotspot or a related feature in the Miocene.

South of Cape Mendocino, the tectonic setting is dramatically different. In fact, Southern California and the Baja California Peninsula of Mexico are no longer part of the North American Plate. The present transform system has completely shut down the Cordilleran arc southward to southern Baja, Mexico. The San Andreas Fault and similar right-lateral transform faults and the Gulf of California (Sea of Cortez) oceanic ridge system have caused a transfer of part of the previous North American Plate to the Pacific Plate (◨ Fig. 13.13). The transposed block consists of various prior arc and forearc components, and miscellaneous materials including locally a piece of cratonic North America. Pieces of the Cordilleran arc include parts of the Sierra Nevada now in the Salinian block of central California and the Penninsular Ranges of Southern California and adjacent Mexico. Both Franciscan and Great Valley equivalents are present west

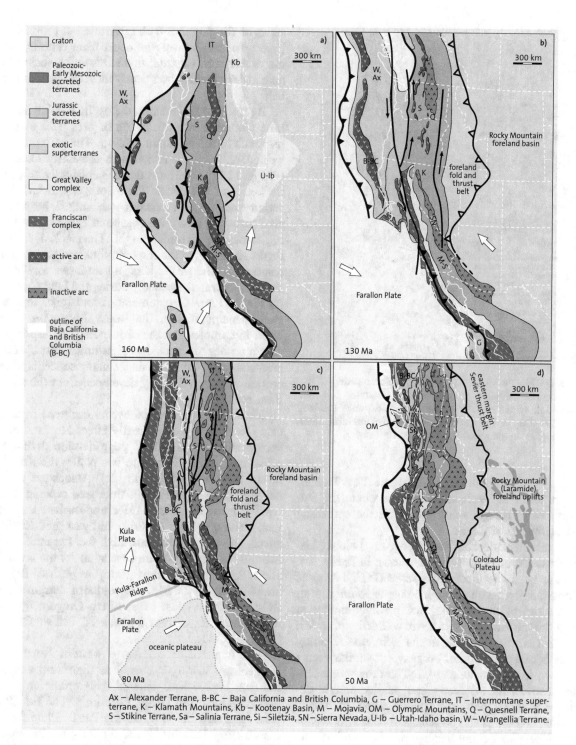

Ax – Alexander Terrane, B-BC – Baja California and British Columbia, G – Guerrero Terrane, IT – Intermontane super-terrane, K – Klamath Mountains, Kb – Kootenay Basin, M – Mojavia, OM – Olympic Mountains, Q – Quesnell Terrane, S – Stikine Terrane, Sa – Salinia Terrane, Si – Siletzia, SN – Sierra Nevada, U-Ib – Utah-Idaho basin, W – Wrangellia Terrane.

■ **Fig. 13.12** Paleogeographic/paleotectonic maps showing evolution of Cordillera from Late Jurassic to Eocene. **a** Late Jurassic showing initial accretion of Wrangellia–Alexander. **b** Early Cretaceous showing initial southward transport of Baja BC. **c** Late Cretaceous showing initial subduction of thickened oceanic plateau. Note separation of Farallon Plate into two plates. **d** Eocene showing accretion of seamount-plateau complex (OM). Note that southern margin of Baja BC is near its present position

of the San Andreas Fault from San Diego to San Francisco although the forearc basin deposits are absent along the Salinian block, a point that will be emphasized in the Laramide section below. Precambrian North American cratonic pieces are present in the Transverse Ranges east of Los Angeles. The Transverse Ranges got their name from the fact that their trend is E–W, transverse to the normal SE–NW orientation of the tectonic grain of the region. The Transverse Ranges formed when blocks of the Pacific Plate were caught in a large bend of the San Andreas system and rotated 120° clockwise, all in the last 7 million years. Ad-

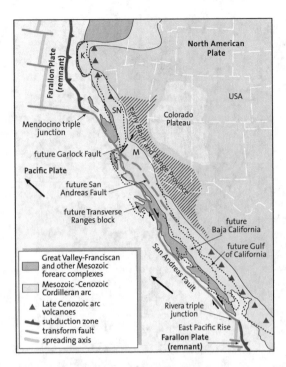

Fig. 13.13 Middle Miocene (15–12 Ma) tectonic setting of SW North American Cordillera showing major elements of early San Andreas Fault system and developing Basin and Range Province. Large arrows show motion of Pacific Plate relative to North America; small arrows show motion along San Andreas Fault

series of fringing arcs with uncertain but probably multiple polarities lay off the coast from central California to northern Canada; these arcs and associated oceanic material will subsequently form the eastern terrane belt.

By the Middle Jurassic, the exotic Insular Superterrane ploughed into this group of arcs. This event corresponds to the Nevadan orogeny. The entire mass was welded onto North America in the Early Cretaceous (■ Fig. 13.12b). Perhaps not coincidentally, this marks the onset of the classic Sevier orogeny. Both the Nevadan and Sevier orogenies are marked by west-vergent thrusting along the coast and east-vergent thrusting across the cratonic margin. The stacked, imbricate thrust sheets form the foreland fold and thrust belt. In response to the thrust load, an extensive foreland basin was generated especially during the Late Cretaceous (■ Fig. 13.12c). Conglomerate, sandstone, and mudstone accumulated in the Rocky Mountain foreland basin in complex stacked sequences that responded to varying rates of subsidence, sedimentation, and sea level changes. Not surprisingly, many concepts of modern sequence stratigraphy developed from the study of these rocks.

Meanwhile, a series of events occurred in the Pacific Ocean that would greatly affect North American geologic history. An E–W ridge developed that separated the Farallon Plate into two plates; the Kula Plate lay to the north (■ Fig. 13.12c). Associated with the newly formed Kula Plate, a thickened oceanic slab derived from a mid-ocean ridge approached the southwest coast of North America and was subducted. The effects of that event generated the Laramide Rocky Mountains—but more on that later. As the Kula–Farallon Ridge moved northward, it pushed Baja BC ahead of it. By 60 Ma, the southern margin of Baja BC lay off the coast of southern Oregon; by 50 Ma it was near its current latitude of British Columbia (■ Fig. 13.12d).

The final major change to western North America geologic history occurred as the mid-ocean ridge, the East Pacific Rise, entered the Cordilleran subduction zone near San Diego, California. The East Pacific Rise separated the NW-drifting Pacific Plate from the east-drifting Farallon Plate. As long as the Farallon Plate was adjacent to North America, normal subduction occurred. However, when the Pacific Plate encountered the continent, the rapidly drifting plate "pulled away" from the slowly westward drifting North America. A "space problem" developed and the tectonic regime responded by shutting down the Cordilleran arc and extending the western margin of North America (■ Fig. 13.13). A transform fault system developed and the continent was extended as much as 100% or more as the brittle crust thinned and broke as it was pulled

jacent to the 3 km-high ranges lies the Salton Trough, a region below sea level, and the area where the San Andreas Fault merges southward with the spreading center of the Sea of Cortez.

The middle Paleozoic to Paleogene geologic history of the Cordilleran Region is shown in ■ Fig. 9.4. The series begins in the Mississippian (■ Fig. 9.4a) as the long-ranging passive margin stage is coming to an end. A series of fast-moving arcs, possibly similar in style to the modern Caribbean Arc collided with much of western North America. In the USA, the resulting orogeny is called the Antler orogeny but similar events are known from Canada as well. The orogeny accomplished two events, (1) deep-water passive margin deposits were thrust onto the shallow-water carbonate shelf and (2) the arcs and possible associated microcontinents lingered off shore throughout the remainder of the Paleozoic and formed a Pacific-style marine arc system. During the Pennsylvanian and Permian, the arcs drifted away from North America and a large oceanic backarc was formed (■ Fig. 9.4b, c). During the Late Permian and into the Triassic (■ Fig. 9.4d), the arcs collapsed back against the continent to generate the Sonoman orogeny. It was along this coastal margin that the initial Cordilleran arc was constructed. By Early Jurassic, the arc was built on cratonic North America to the south but remained a marine arc to the north. A

apart. Commencing in southern Arizona and later expanding NW, areas of greatest extension formed metamorphic core complexes as the rapidly spreading brittle upper plates unroofed deep crustal rocks that responded through extremely rapid uplift. The extensive horizontal normal faults associated with regional extension are some of the largest-such features on Earth and as brittle rock broke and rotated domino-style, the Basin and Range Province was formed. What had been a high mountain system formed by compressional tectonics in the Cretaceous, had become an extensional, collapsed orogen in the Neogene. To the east and north of the Basin and Range Province, a large block of cratonic North America now rises nearly undeformed but greatly elevated above the collapsed orogen, the Colorado Plateau.

13.6 Laramide Rocky Mountains—An Orogenic Mystery Solved

The Central and Southern Rocky Mountains bend around the eastern margin of the Colorado Plateau from New Mexico to Wyoming and SW Montana; and somewhat inexplicitly, they lie more than 1500 km from the eastern margin of the Pacific Ocean. The Southern and Central Rocky Mountains discussed here are collectively part of a much greater range called the Rocky Mountains that stretch from Alaska to New Mexico. The Central and Southern Rockies are also referred to as the Laramide Rocky Mountains after one of the individual uplifts, the Laramide Uplift near Laramie, Wyoming. The structure of the Rockies has long been known, but their origin has been shrouded in mystery. Although original studies believed that the Rockies were primarily the product of vertical uplift, for over fifty years, a compressional origin has been documented. Unlike most other orogens that comprise elongate, continuous zones of deformation (including the remainder of the Rocky Mountains), the Rockies consist of individual, elongate uplifts separated by structural basins (◻ Fig. 13.14). Although all ranges share common structural elements, each range also possesses individual characters. The most unusual characteristic of the Rockies is their position firmly within cratonic North America (◻ Fig. 13.12d).

Laramide, unfortunately, has several different meanings. In its strictest sense, which is not used much anymore, it refers only to the Paleocene–Eocene uplift near Laramie, Wyoming. More broadly, but still somewhat restricted, it refers to the area of the Late Cretaceous–Eocene orogen of specific style that comprises the Central and Southern Rockies and Colorado Plateau; this is how we will use it here. It is also used as a time term to describe broad orogenic events of western

North America, and unfortunately, even broader areas. And finally, it has been used to describe a specific style of orogen, which except for a relatively small area of the Andes in South America, is essentially identical in scope to our preferred usage above.

The Laramide Rocky Mountains consist of Proterozoic, basement-cored uplifts with distinctive mushroom shape (◻ Fig. 13.15). Most ranges have the following characteristics: (1) Precambrian basement exposed in center of range, (2) asymmetric flanks with thrust faults producing mushroom shaped flare of basement rocks over flanking, deformed, locally overturned Paleozoic and Mesozoic sedimentary rocks, (3) flanking sedimentary basins with several-kilometer-thick deposits of Paleocene and Eocene continental deposits that were derived from the adjacent uplifts, (4) pediment-carved multiple high-level erosion surfaces that document episodic uplift-erosion cycles, (5) local Late Cretaceous–Eocene granitic plutonic intrusions and associated volcanism, (6) widespread Oligocene and Miocene volcanic fields, and (7) ragged, glaciated central peaks and ridges. Most ranges are at least locally onlapped by Paleogene sedimentary rocks that unconformably overlie deformed rocks as young as Maastrichtian (latest Cretaceous) thereby indicating Late Cretaceous to Eocene (Laramide) age of uplift and deformation. Some uplifts are more difficult to date directly, but all have rock sequences that suggest or permit a Late Cretaceous to Eocene age of uplift and deformation. The Colorado Plateau has a number of monoclines that are also believed to be Laramide in age. Of the uplifts on the Colorado Plateau, only the San Rafael Swell (◻ Fig. 13.14) can be directly dated as the adjacent Uinta Basin contains Eocene deposits derived from the adjacent uplift. However, the structural style, asymmetric uplifts with basement thrust faults, and abundant indirect evidence suggest that Colorado Plateau uplifts are Laramide in age. Why the Colorado Plateau escaped more serious deformation, it is surrounded on all flanks by deformation, is still a question that remains largely unanswered.

The style and composition of Laramide plutons and volcanics suggest that they are subduction related. The significance of this will be discussed below. Several subduction-related (?) Eocene–Oligocene laccoliths occur across the Colorado Plateau and similar igneous structures occur in the Rockies. Oligocene and Miocene volcanic rocks are post-orogenic and in some cases overlie Laramide uplifts. The San Juan Volcanic Field unconformably overlies an erosion surface at over 2500 m elevation that is carved across the Laramide San Juan Uplift (◻ Fig. 13.14). Similarly, the Neogene volcanics of the Yellowstone hot spot overlie several Laramide uplifts. Many of the high-level erosion surface carved across the Rockies are Oligocene in age; those on the eastern side of the Rockies form sloping

13

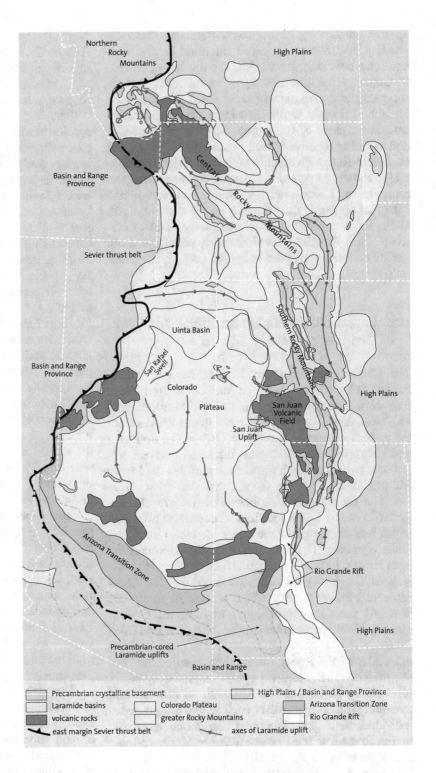

surfaces that can be projected eastwards to match with similar erosion surfaces on the High Plains.

The above descriptions of the Rockies suggest a straight-forward interpretation of geologic history: uplift and compressional deformation of Laramide age accompanied by foreland basin sedimentation, subduction-related igneous activity, post-orogenic erosion and subsequent multiple uplift events, post-orogenic volcanism, and Pleistocene glaciation. But several aspects remain puzzling: why are the ranges so far inland from Cordilleran tectonics? Why are the ranges scattered and separated by basins? Why is there an absence of regional metamorphism? What are subduction-related igneous rocks doing 1500 km from the Cordilleran subduction zone? The partial answer came in the early 1980's. From the large volume of data and evidence, Dickinson and Snyder (1978) suggested flat-slab subduction. The flat slab did not reach the depth required

◻ Fig. 13.15 Schematic cross sections showing Laramide tectonic style. **a** Front range uplift near Denver, Colorado; **b** Uinta uplift in northern Utah

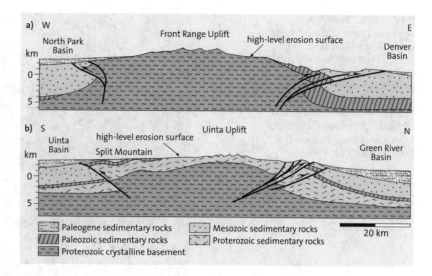

to partially melt the overlying asthenosphere until it reached the location of the Rockies (◻ Fig. 13.11b). An area in the Andes was used as a modern example of such slab behavior. A number of papers followed that either presented more data or modified the flat-slab hypothesis. Several workers suggested that the migration of Baja BC (see above) triggered the flat-slab subduction. But flat-slab subduction per se, did not fully explain the characteristics of the Rockies, especially their arcuate shape around the Colorado Plateau.

An important piece of the puzzle came not from studies in the Rockies or Colorado Plateau, but rather from studies in the Mojave and Sierra Nevada regions of California and in the NW Pacific Ocean. Based on both detailed and broad, regional data, Saleeby (2003) proposed a model that answered many questions concerning the tectonic setting and significance of the Rockies. The model proposes that a piece of a thickened mid-ocean ridge drifted eastward on the Farallon Plate and entered the Cordilleran subduction zone ca. 85 Ma. Coincidentally, this event may have triggered the breakup of the Farallon Plate into a northern Kula Plate (◻ Fig. 13.12c). The thickened slab resisted normal subduction at the edge of the North American Plate along the Southern California coast (◻ Fig. 13.11b). The forearc of Salinia, which at the time was at the southern margin of the Sierra Nevada, was tectonically removed, either by transform transport to the north, subduction with the thickened oceanic slab, or both. The thickened oceanic slab was almost completely subducted by the beginning of the Paleogene and the locus of its frontal margin progress is well marked across Mojavia, under the Colorado Plateau, and into the Rockies.

The subduction generated a Late Cretaceous regional metamorphic event across Mojavia. As it continued eastward under the Colorado Plateau, it triggered regional uplift of over 2 km and formed monoclines.

As the front penetrated the Rockies region, partial melting generated plutons and associated volcanics and local, sharp, compressive uplifts and associated foreland basins. The pattern of the collective uplifts marks the penetration of the slab at depth into the region; note how the ranges curve concave to the SW in response to the pressure applied at depth (◻ Fig. 13.14). The scattered distribution of the ranges reflects inherited crustal strengths and weaknesses of the Proterozoic crust. Perhaps the repeated uplift-erosion cycles were generated by irregularities in the slab or by currents set in motion in the asthenosphere as the slab moved through. Currently, deep seismic records show the slab at depth under the central USA. The slab left behind a series of dramatic mountain ranges in what otherwise would have been a low-lying temperate desert.

13.7 Epilog

For an epilog we come back to the Alps, the classical orogen, and the surrounding areas. In the Alps, the process of lateral tectonic extrusion (see box above) generated the present E–W elongation. Before the final orogeny, the Alpine trend was oriented NE–SW instead of E–W. The extremely complex arcuate structure of the Apennines–Alps–Carpathians mountain belt is due to indenter tectonics of the Adriatic Plate and the irregular margins of the colliding continents. B. Székely (2001) pursued another thought that was already instigated decades ago. He compared the complex pattern of mountain ranges with the atmospheric currents in major storms or cyclones. The course of the mountain ranges and their motion pattern are partly similar to those of cyclonic storms (◻ Fig. 13.16). If the Alps are considered analogous to a warm front and the Apennines to a cold front, astonishing comparisons arise.

■ **Fig. 13.16**   The Alps-
Apennine vortex (digital elevation
model; Globe Task Team 1999) as
a possible analogy to atmospheric
cyclones (Székely 2001). The
weather cyclone (false color
satellite image) is from Project
Weather World 2010, University
of Illinois at Urbana–Campaign,
USA, and shows a storm in the
USA on March 13, 1993

13

The Apennines have rotated counterclockwise since the Miocene and thus they approached the Alps, whose position has remained relatively stable. Such scenarios are typical in the relations between cold and warm fronts. Of course, lithospheric plates possess completely different physical properties compared to atmospheric masses. However, things that appear so differently over the short-term scale, can lead to quite similar processes and phenomena over long periods. Plate tectonics teaches this with many of its phenomena. Although the fundamental concept of comparing cyclones with mountain belts had been proposed by a few authors since the 1970's, the comparison of vortices between the two different mediums has had little positive resonance among experts—but this is often the case with unconventional ideas: one only has to recall Wegener's theory of continental drift in the early twentieth century.

Supplementary Information

© The Editor(s) (if applicable) and The Author(s), under exclusive license to Springer Nature Switzerland AG
2022
W. Frisch et al., *Plate Tectonics*, Springer Textbooks in Earth Sciences, Geography and Environment,
https://doi.org/10.1007/978-3-030-88999-9

Glossary

* Reference to back inside cover

Italics Term also in glossary

Abukuma-type metamorphism* Low-pressure regional metamorphism, typically occurring in the magmatic belt above subduction zones

Acidic → *Magma* and → *magmatic rock* with more than 65 weight-% SiO_2

Active continental margin Continental margin with a → *subduction zone* dipping beneath. Represents a destructive → *plate boundary*. See also → *passive continental margin*

Alkali feldspar Mineral, potassium–sodium → *feldspar*

Alkaline Term for magma or magmatic rock rich in alkalis (Na_2O, K_2O) with reference to the content on silica (SiO_2) or alumina (Al_2O_3)

Alkaline basalt Basalt with high content of alkalis (Na, K) and lower content of silica as compared to → *tholeiitic basalt*. Derived from mantle → *peridotite* with low degree of partial melting (<10%), typically occurring in → *graben* structures and above → *hot spots*

Amphibole Mineral group similar to *pyroxene* but containing hydroxyl ions. An important member of this mineral group is hornblende

Amphibolite Rock derived from basalt or gabbro by regional metamorphism; main mineral constituents are amphibole and plagioclase

Amphibolite facies* Medium- to high-grade regional metamorphism

Anatexis* Melting of rocks in continental crust during high-grade regional metamorphism

Andesite* Volcanic rock, typically formed above subduction zones

Anhydrite Mineral, calcium sulfate; *cf* → *gypsum*

Aragonite Mineral, calcium carbonate of same chemical composition as → *calcite* but crystallizing in a different crystal system. Occurring in shells of certain organisms and in high-pressure metamorphic rocks

Arkose → *Sandstone* containing a considerable amount of feldspar (>20–25%) besides quartz

Assimilation Incorporation of wall rocks into a magma chamber by melting

Asthenosphere Shell of the upper mantle directly below the base of the → *lithosphere,* mostly in depths between 100 and 250 km. Contains low amounts of melt

Aulacogen Failed → *graben structure* fi lied with extremely thick sediments

Barrow-type metamorphism* Medium-pressure regional metamorphism, typically occurring in collision zones of continents (see → *continent–continent collision*)

Barite Mineral of high density, barium sulfate

Basalt* Volcanic rock, very common at the ocean floor and in → *–flood basalts*. Basaltic magma evolves by partial melting of → *peridotite* in the → *mantle*

Basement Metamorphic and igneous base on which a younger sedimentary cover sequence was deposited; e.g., in the Alps the Variscan basement of the Alpine sequences

Basic → *Magma* and → *magmatic rock* with 45–53 weight-% SiO_2

Batholith Large body of → *plutonic* rock mainly composed of → *granodiorite* and → *granite,* typically occurring in the → *magmatic belt* above → *subduction zones* or in collision zones of continents (see → *continent–continent collision*)

Benioff zone Seismically active zone at convergent plate boundaries which obliquely plunges down to a maximum depth of 700 km indicating a → *subduction zone*

Blueschist High-pressure metamorphic rock formed in → *blueschist facies*. Contains → *glau-cophane*

Blueschist facies* Lower grade of → *high-pressure metamorphism; cf* → *eclogite facies*

Breccia Rock mainly composed of angular rock fragments. May evolve tectonically (fault breccia) or by sedimentary processes

Brittle Rigid behavior of rock with fracturing during deformation (as opposed to → *ductile*)

Calc-alkaline Magmas and magmatic rocks mainly formed above *subduction zones* and in collisional belts

Calc-alkaline basalt → *Basalt* with substantial contents of calcium, alkalis, and aluminum. Typically formed above → *subduction zones*. Typically a → *high-alumina basalt*

Calcite Mineral, calcium carbonate. Main constituent of limestone and chalk, formed primarily by organic processes in shallow, warm seas

Calcite compensation depth Oceanic boundary layer below which no calcitic sediments are deposited because calcite dissolution is higher than calcite supply

Carbonatite Magmatic rock mainly composed of → *calcite* or → *dolomite*

CCD See → *calcite compensation depth*

Chlorite Greenish mineral similar to mica, typical of low-grade metamorphism

Chromite Mineral of the → *spinel* group, chrome ore

Clastic sediment Sediment formed by transportation and deposition of mineral and rock fragments

Claystone → *Clastic sedimentary rock* composed of quartz and clay minerals with grain-sizes below 0.06 mm or, → *sensu stricto*, below 0.004 mm; *cf* → *siltstone*

Clinopyroxene See → *pyroxene*

Coccolithophorids Carbonate-secreting algae

Coesite High-pressure modification of → *quartz*

Collision Forceful interaction between plates, e.g., → *continent–continent collision*

Concordant Strata that show parallel bedding without break in sedimentation or contain interlayered rocks (e.g., → *sills*) that are parallel to bedding; *cf* → *discordant*

Conglomerate Sedimentary rock mainly composed of grains larger than sand-size, usually pebbles and cobbles

Conjugate faults Connected system of two differently oriented *fault* sets that form during the same tectonic event. Their sense of movement is interconnected

Conservative plate boundary See *plate boundary*

Constructive plate boundary See *plate boundary*

Contact metamorphism* Metamorphism in the thermal aureole of *intrusions*

Continent–continent collision Collision of continental parts of plates as a consequence of subduction of intervening oceanic parts of plates. Leads to *orogeny*

Continental crust Outermost shell of the solid Earth, ca. 35 km thick on average and forming the continents and shelf areas. Average composition is andesitic with about 60% SiO_2. Beneath mountain ranges thickness increases up to ca. 70 km

Continental margin Edge of continental crust, usually bordered by slope-rise; two broad types, → *active continental margin* and → *passive continental margin*

Core Innermost zone of the Earth below ca. 2900 km depth, mainly composed of iron and nickel. The outer core is liquid, the inner core (below ca. 5100 km) is solid

Craton Old, cooled and therefore relatively stiff parts of continents. Most cratons underlain by Precambrian → *basement*

Crust See → *continental crust*, → *oceanic crust*

Curie temperature Temperature below which a mineral acquires and preserves magnetic properties (after Pierre Curie). 680 °C for hematite, 580 °C for magnetite

D″-layer Lowermost layer of the Earth's mantle, thickness between 100 and 500 km (mostly 200–250 km). Believed to be area where most mantle plumes originate

Dacite* Volcanic rock of → *acidic*/→ *intermediate* composition, typically formed above → *subduction zones*

Deep earthquake Earthquake with focus between 350 and 700 km depth

Deep sea trench Depression at the margins of oceans with water depths up to 11 km. Indicates a convergent –a*plate boundary* where a subducting plate plunges down into the mantle

Depleted mantle Mantle impoverished in → *incompatible elements* during partial melting and extraction of basaltic melts

Destructive plate boundary See → *plate boundary*

Devitrification Growth of fine minerals from a precursor of volcanic glass

Diagenesis* Compaction and alteration of sediments at temperatures below 200 °C (grades into lowest ranks of → *metamorphism*)

Diamond High-pressure modification of graphite, formed in the Earth's mantle and in → *subduction zones* at depths of more than 100 km as well as by meteorite impacts

Diapir Rising, mostly tube-shaped rock body of highly variable diameter. The ascent is a result of density inversion, e.g., → *salt* beneath limestone, → *serpentinite* beneath oceanic crust, hot → *peridotite* in the mantle beneath cooler peridotite (→ *hot spot*). The rising rock material must be easily deformable; the ability of deformation is mainly dependent on the temperature

Diatoms Siliceous algae capable of forming major sedimentary rock sequences rich in chert. Although algae live and die in shallow water, they usually accumulate in deep-water deposits

Differentiation Modification of the composition of a magma by → *assimilation* of wall rock or crystallization and separation of minerals (e.g., by gravity). Basic rocks typically differentiate to more acidic rocks

Dike Magmatic tabular, → *discordant* rock body forming a mostly subvertical sheet, typically several decimeters or meters thick. Commonly a feeder channel under a volcanic edifice. See also → *sill*

Diopside A pyroxene (clinopyroxene) mineral enriched in magnesium and calcium

Diorite* *Intermediate* composition plutonic rock typically formed in the magmatic belt above *subduction zones*. Equivalent to *andesite*

Discordant Strata or rocks that are non-parallel or cross cutting; *cf* → *concordant*, → *unconformity*

Dolerite Subvolcanic basaltic rock occurring in → *dikes* and → *sills*. Intermediate in texture between basalt and gabbro

Dolomite Mineral, calcium–magnesium carbonate. Also sedimentary rock mainly composed of the mineral dolomite; *cf* → *limestone*

Ductile Plastic behavior of rock without fracturing during deformation (as opposed to → *brittle*)

Dunite* Kind of → *peridotite* formed in the uppermost mantle directly below the oceanic crust

Eclogite Rock formed by → *high-pressure metamorphism* in → *subduction zones*, mostly from rocks of the → *oceanic crust*

Eclogite facies* Higher grade of → *high-pressure metamorphism*; *cf* → *blueschist facies*

Ensialic island arc → *Island arc* with basement of → *continental crust*

Ensimatic island arc → *Island arc* with basement of → *oceanic crust*

Enstatite Mineral, magnesium → *pyroxene* (orthopyroxene)

Earth's core See → *core*

Earth's crust See → *crust*

Earth's mantle See → *mantle*

Facies Sum of rock characteristics as compared to other facies. Sedimentary facies are acquired in sediment according to place and conditions of formation (e.g., marine facies, deep-water facies, sandy facies). Metamorphic facies reflect the pressure and temperature conditions during metamorphism

Fault Displacement of two blocks along a fracture. Normal fault: Inclined fault plane with the hanging-wall block (block above the fault plane) moving downwards relative to the footwall block (expression of horizontal extension, e.g., during formation of a graben). Reverse fault: Hanging-wall block moving upwards relative to the footwall block (expression of horizontal compression). Thrust fault: reverse fault with a shallow-dipping fault plane (expression of strong horizontal contraction, formation of → *nappes*). Strike-slip fault: Horizontal movement of two blocks past each other along a steep-dipping fault; may be dextral (right-lateral; clockwise ball-bearing effect) or sinistral (left-lateral; anticlockwise ball-bearing effect)

Feldspar Most common aluminum-rich mineral in the Earth's → *crust*. → *Plagioclase* (Na–Ca feldspar) and → *alkali feldspar* (K–Na feldspar) represent two different mineral groups

Fenster See → *teconic window*

Flood basalt Thick, widespread sheets of flat-lying → *basalts* formed above a → *hot spot*. Horizontal layers of lava and vertical fissures cause a staircase-like morphology, therefore also termed trap basalt (*trappa, Swedish* stairs). May cover areas more than 1,000,000 km^2

Flower structure Subvertical fault fanning upwards towards the surface

Fluid, fluid phase Highly mobile gas or liquid phase in a rock, circulating along grain boundaries and fissures or by volume diffusion. Mostly "water" as hydroxyl (OH^-) ions, or CO_2. Serves as transport medium of elements during metamorphism. Decreases the melting temperature of rocks substantially

Fluviatile sediments → *Terrestrial* through a boundary layer, e.g., the Earth's surface. Measured in milliwatt per square meter (mW/m^2)

Hematite Mineral, iron oxide (Fe_2O_3) with iron in trivalent state. Common state of iron at Earth's surface since presence of oxygen in atmosphere (ca. 2000 Ma)

High-alumina basalt → *Calc-alkaline basalt* rich in aluminum ($Al_2O_3 > 16$ weight-%), occurs above → *subduction zones*

High-pressure metamorphism* Metamorphism typical of → *subduction zones*

Hornblende Mineral of the → *amphibole* group

Hornfels Rock formed at conditions of high-grade → *contact metamorphism*

Horst Elevated block between two → *grabens*. Horst-graben structure typical of rift zones and dominate American Basin and Range Province

Hot spot Place in the Earth's → *crust* under which a mantle → *diapir* rises. Characterized by volcanism (e.g., Hawaii)

Hyaloclastite Broken → *volcanic glass* in the spandrels (between bulges) of pillow lavas

Hydrothermal activity Hot water and steam circulating in cracks and pores of rocks. M

I-type ("igneous") granite Granitic rock evolved from more → *basic* magmatic rocks by → *differentiation*, typically occurring above → *subduction zones* (compare → *S-type granite*)

Illite A clay mineral, similar to white mica and derived by weathering of feldspars. Typical clay in many marine sedimentary rocks; *cf* → *kaolinite*

Incompatible elements Chemical elements (e.g., alkalis) incompatible with mantle rocks, i.e., they cannot be integrated in the main mineral phases of mantle → *peridotite*. During partial melting they preferably enter the melt. Also termed "lithophile elements"

Indenter Stiff crustal block that is pushed into deformable crust (e.g., Indian indenter in the Himalayas, South-Alpine indenter in the Alps)

Intermediate → *Magma* and → *magmatic rock* with 53–65 weight-% SiO_2

Intermediate earthquake Earthquake with focus between 70 and 400 km depth

Intraplate magmatism, Intraplate volcanism Magmatism (volcanism) in the interior of plates above → *hot spots* or along → *graben structures*

Intrusion → *Magma* penetrating older rocks and crystallizing at depth

Island arc Arcuate chain of islands with active volcanoes above → *subduction zones*

Isostasy Buoyancy equilibrium of crustal blocks of different thickness and density

Kaolinite A clay mineral derived from weathering of feldspars. Typical clay in many continental sedimentary rocks; *cf* → *illite*

Komatiite Peridotitic or basaltic volcanic rock formed by high percentage of partial melting of → *peridotite;* rich in magnesium (MgO > 18 weight-%)

Kyanite Mineral, aluminum silicate typical of → *Barrow-type* and → *high-pressure metamorphism*

Lava Liquid form of rock (→ *magma*) extruded onto surface of Earth. Magma releases dissolved gas when approaching the surface, because of pressure release. The gas may remain enclosed as bubbles in the solidifying lava

Leucite Mineral similar to potassium feldspar but undersaturated in SiO_2

Lherzolite* Kind of → *–peridotite* widely distributed in the upper mantle

Limestone Sedimentary rock mainly composed of → *calcite* and mainly formed in the shelf areas and the open ocean above the → *calcite compensation depth* by the accumulation of shells and skeletal particles of organisms

Listric fault Curved → *fault*, subvertical near the surface and becoming flatter at depth. Causes tilting of the hanging-wall block (block above the fault)

Lithophile elements See → *incompatible elements*

Lithosphere Outer solid layer of rocks that form the plates. Layer includes the → *crust* (continental or oceanic) and the → *lithospheric mantle*. Sicknesses typically range between 70 and 150 km but swell to more than 200 km beneath mountain ranges

Lithospheric mantle Uppermost rigid part of the Earth's → *mantle* that belongs to the → *lithsphere*. Together with the → *crust* it forms the lithospheric → *plates*

Lysocline Boundary layer in the sea water below which the dissolution (e.g., of → *calcite*) increases substantially. For calcite mostly 1500–2000 m above the → *calcite compensation depth*

Magma Molten rock, generally composed of melt, dissolved gas, and crystallized minerals

Magmatic arc, ~ belt Magmatic (volcanic) zone above a → *subduction zone*

Magmatic (volcanic) front Line of abrupt onset of magmatism above a → *subduction zone*

Magmatic rock Rock formed by cooling from a melt (see → *magma*)

Magnesiowuestite Mineral, magnesium oxide (MgO) that prevails in the lower → *mantle* together with → *perovskite*

Magnetite Mineral, iron oxide (Fe_3O_4) with iron in both bivalent and trivalent state

Mantle Shell of the Earth between crust and → ■ *core*. The boundary between upper and lower mantle is at a depth of ca. 660 km

Marine sediment Sediment deposited in the ocean realm. Includes shallow-water deposits of the shelf (on → –*continental crust*) and abyssal deposits of the deep ocean (on → *oceanic crust*)

Mélange Mixture of different rocks formed by sedimentary and/or tectonic processes. Characterized by a block-in-matrix structure: more rigid rock types form blocks in a matrix of softer and strongly deformed rock. The blocks range from meters to kilometers in size. Typically form in trenches and some foredeeps

Metamorphic core complex See → *metamorphic dome*

Metamorphic dome Dome-shaped bulge of highly metamorphosed rocks formed by substantial crustal extension. Synonym

Metamorphic core complex

Metamorphism* Transformation of rocks, typically (except → *contact* and → *ocean–floor metamorphism*) achieved by burial to greater depths which causes an increase in pressure and temperature. The mineral association (→ *paragenesis*) is adapted to the changing pressure and temperature conditions because minerals and paragenesis have different fields of stability. The presence of a → *fluid phase* plays an important role because it transports ions and thus enables mineral reactions

Mid-ocean(ic) ridge Elongate ridges in the oceans where new oceanic → *lithosphere* is formed. Constructive plate → *boundary*

Migmatite "Mixed" rock—highly metamorphosed schists or gneisses that were partly melted (see → *anatexis*). The melt phase, typically of granitic composition, solidified within the gneisses and schists thus giving the mixed appearance to the rock

Moho Common acronym of Mohorovicic discontinuity. Boundary layer between → *crust* and → *mantle*

Molasse Sequence of → *marine* and → *terrestrial* sediments deposited in the foredeep of a rising mountain range. Largely composed of the erosional detritus of the mountain range. Usually syn- to post-orogenic

Montmorillonite A clay mineral of the → *smectite* group. A typical weathering product of volcanic ashes

Mudstone, mudrock General sedimentary rock term for any combination of silt and clay; usually lacks parting or fissility; *cf* → *shale*, → *siltstone*

Mylonite Strongly deformed rock in a shear zone and recrystallized to a fine-grained mineral association. Deformation occurs in a *ductile* manner

Nappe Tectonic unit thrust upon another unit at km-scale. Result of strong horizontal shortening during orogenesis (see → *fault*)

Nepheline Mineral similar to sodium feldspar but undersaturated in silica (SiO_2)

Normal fault See → *fault*

Obduction Thrusting of an → *ophiolite* and/or associated accretionary prism and/or accreted terrane over a continental margin or island arc. Opposite to → *subduction*

Ocean-floor metamorphism → *Metamorphism* in the area of mid-ocean ridges that transforms the mineral assemblages of the still hot rocks of the → *oceanic* crust by circulating waters and aqueous → *fluids*

Oceanic crust Outermost shell of the solid Earth forming the ocean floors. Average thickness 6–8 km. The composition of the oceanic crust is basaltic (ca. 50% SiO_2)

Oceanic plateau Submarine plateaus with thickened → *oceanic crust* standing above the abyssal plains

Olivine Mineral, magnesium(–iron) silicate, undersaturated in silica (SiO_2). Most important constituent of → *peridotite* in the → *mantle* to a depth of ca. 400 km

Omphacite Mineral, sodium → *pyroxene,* as a high-pressure mineral and important constituent of → *eclogite*

Ophiolite Rock association of the oceanic → *lithosphere,* directly accessible due to tectonic processes (→ *obduction* or scrape-off during → *subduction*). Belts of ophiolites mark → *suture zones* in orogens

Orogen Mountain range formed by → *orogenesis*

Orogenesis, orogeny Mountain building process, initiated by the collision of continents or island arcs and characterized by crustal thickening, → *nappe* formation, rock deformation, → *metamorphism* and → *anatexis*

Orthophyroxene See → *pyroxene*

Paired metamorphic belt Parallel arrangement of a → *high-pressure metamorphic* belt and an → *Abukuma-type metamorphic* belt. Indicates former subduction activity

Pangaea Supercontinent in the Late Paleozoic and Early Mesozoic (ca. 300–175 Ma)

Panthalassa Giant ocean opposing → *Pangaea*

Paragenesis Metamorphic mineral association depending on the chemical composition of the rock as well as the pressure and temperature during → *metamorphism*. Used to estimate the pressure–temperature conditions during metamorphism

Passive continental margin Continental margin with the → *continental crust* connected to the adjacent → *oceanic crust*. Does not represent a → *plate boundary*. See also → *active continental margin*

Pelagic sediment Sediment formed in the open ocean (outer shelf, deep sea) and only slightly influenced by terrigenous input (*cf* → *terrigenous sediment*)

Peridotite * Rock forming the Earth's → *mantle,* mainly composed of → *olivine* and —*pyroxene*

Perovskite Mineral group, most frequent as Mg–Si spinel ($MgSiO_3$) in the lower → *mantle* (between a depth of ca. 660 and 2900 km)

Phonolite* → *Alkaline* volcanic rock occurring in → *graben structures*

Phreatic, phreatomagmatic Characterization of explosive volcanic events as a consequence of access of surface waters into the magma chamber and the subsequently developing gas pressure

Pillow lava Basaltic lava with pillow-like structures formed during subaquatic extrusion

Plagioclase Sodium–calcium → *feldspar*

Plagiogranite* → *Plutonic rock* rich in → *plagioclase,* typically occurring in sections of → *oceanic crust.* Similar to → *tonalite*

Plate Part of the rigid outer shell of the solid Earth, consisting of → *crust* (oceanic, continental) and → *lithospheric mantle.* The Earth is divided into approximately a dozen plates, each with individual movements

Plate boundary Constructive (divergent) plate boundary: Plate boundary along which new oceanic → *lithosphere* is formed (→ *mid-ocean ridge*). Destructive (convergent) plate boundary: Plate boundary along which one plate subducts beneath the other (→ *subduction zone*). Conservative plate boundary: Plate boundary along which plates slide past each other (→ *transform fault*)

Pluton Body of → *plutonic rocks* solidified from a → *magma* chamber

Plutonic rock → *Magmatic rock* that solidified at depth; usually coarse-grained due to slow cooling

Pyrite Mineral, iron sulfide

Pyrope Mineral of the → *garnet* group formed during → *high-pressure metamorphism;* constituent of → *eclogite*

Pyroxene Mineral, magnesium(–iron) silicate (e.g., the orthopyroxene enstatite), similar to → *olivine* but saturated in silica. Diopside and augite (clinopyroxenes) contain relevant amounts of calcium. Important constituent of → *peridotite,* → *gabbro,* and → *basalt*

Quartz SiO_2, common mineral in continental crust. Important constituent of most → *sandstones*

Quartzite Metamorphic rock derived from → *sandstone,* mainly composed of → *quartz*

Radiolarian Protozoan with siliceous shell

Radiolarite Rock derived from the remains of → *radiolarians,* typically composed of very fine grained quartz. Common in deep-sea deposits; commonly associated with arc terranes

Radiometric age determination Dating of minerals and rocks using the decay of radioactive isotopes (e.g., uranium–lead method)

Regional metamorphism* → *Metamorphism* affecting large crustal bodies in the → *magmatic belt* above → *subduction zones* and during mountain building processes. Subdivided into → *Abukuma-type* and → *Barrow-type metamorphism*

Regression Seaward shift of the coastline due to sea-level drop, sediment influx, or uplift of the shelf area. Opposite to → *transgression*

Rhyolite* Volcanic rock rich in silica (SiO_2). Equivalent to → *granite*

Rift Graben structure (see → *graben*)

S-type granite Granitic rock formed by → *anatexis* of → *continental crust,* commonly through melting if sediments ("S") (compare → *S-type granite*)

Salt Sedimentary rock formed by evaporation of saline water. Strong indicator of paleo aridity

Sandstone → *Clastic* sedimentary rock originating from sand (grain size 0.06–2 mm). → *Quartz* is typically the predominating constituent

Sapropel → *Claystone* or → ▪ *siltstone* rich in organic (bituminous) matter. Lack of oxygen in the sediment or in the water body directly above prevents decomposition of organic matter. Petroleum source rocks (black shales) evolve from sapropel by → *diagenesis*

Sea-floor spreading Spreading of newly formed ocean floor (oceanic → *lithosphere*) at the → *mid-ocean ridge*

Seamount Submarine volcano, frequently arranged in chains that formed above → *hot spots*

Serpentine Metamorphic mineral transformed from → *olivine* or orthopyroxene (see → *pyroxene*) by absorption of water

Serpentinite Metamorphic rock mainly composed of → *serpentine*

Shale Sedimentary rock composed of silt and clay that displays parting or fissility. Typical of low-energy sedimentary environments; cf → *mudstone, siltstone*

Shallow earthquake Earthquake with focus in depths of less than 100 km

Shoshonite → *Basalt* with high potassium content

Shield Stable continental crust that consolidated in the Precambrian

Sial Acronym (silicium and aluminum) coined by Wegener to characterize → *continental crust*

Siliceous sediment Sediment composed of siliceous shells of organisms (→ *radiolaria* or → *diatoms*). The original opal is transformed into → *quartz* during → *diagenesis* of the sediment

Sill Tabular, → *concordant* magmatic rock body that forms a subhorizontal sheet, typically several decimeters or meters thick. Often deflected feeder channel under a volcanic edifice. See also → *dike*

Siltstone Sedimentary rock with grain size between 0.004 and 0.06 mm. Main constituent is quartz; cf → *shale, mudstone*

Sima Acronym (silicium and magnesium) coined by Wegener to characterize → *oceanic crust* and mantle

Skarn Carbonatic rock transformed by → *contact metamorphism*

Slab breakoff Breakaway of the subducted oceanic part of a lithospheric plate after → *continent–continent collision.* Causes isostatic (see → *isostasy*) uplift of a mountain range in consequence of the loss of the counterweight

Smectite A group of clay minerals typically formed from weathering of volcanic ash

Spinel Mineral group comprising magnetite (Fe^{+2}–Fe^{+3} spinel), chromite (Fe^{+2}–Cr^{+3} spinel), Mg–Si spinel and others (occurs as Mg–Si spinel in the deeper upper → *mantle* between 400 and 660 km depth)

Stishovite Highest-pressure modification of → *quartz*

Stratovolcano Volcano with steep slopes (up to 40° inclination) and alternating layers of lava and volcanic ash

Strike-slip fault See → *fault*

Subduction, subduction zone Sinking of mostly oceanic parts of litho-spheric plates into the depth of the upper → *mantle* at convergent → *plate boundaries*

Suture, suture zone Elongate belt along which two colliding continents were welded together during → *continent–continent collision,* typically char-acterized by tectonized remnants of oceanic lithosphere (see → *ophiolites*) squeezed in between ("ophiolitic suture")

Syntaxis Region where large tectonic structures with different regional trend converge, mostly caused by the indentation of the spur of an → *indenter*

Tectonics Study of large-scale structures of the lithosphere, stress acting on rock bodies and their reaction by movement and deformation (strain). Deformation can be → *ductile* or → *brittle* and acts in all dimensions from sub-microscopic scale to plate scale (plate tectonics)

Terrane Far-traveled crustal block accreted to a continent. Due to its remote origin, the terrane shows a different geological evolution compared to adja-cent parts of the continent

Terrestrial sediment Sediment deposited on land—in continental settings (e.g., river or lake deposits); *cf* → *terrigenous sediment*

Terrigenous sediment → *Clastic* sediment deposited in the ocean but composed of fragments derived from a continent or island and generally transported into the ocean by rivers or wind

Tholeiite, tholeiitic basalt → *Basalt* poor in potassium but slightly richer in silica (SiO_2) than → *alkaline basalt*. Mostly formed at → *mid-ocean ridges* and above highly productive → *hot spots*. Tholeiites indicate a relatively high percentage (15–25%) of partial melting of the peridotitic source rock in the upper → *mantle*

Thrust See → *fault*

Tilted block Block above a → *listric fault* tilted by normal displacement along the fault

Tonalite* → *Plutonic rock* rich in → *plagioclase*, typically occurring in sub-duction-related magmatic belts and continental crust. Similar to → *plagiog-ranite*

Trachyte* Intermediate volcanic rock that commonly occurs in → *graben* structures

Transform fault Strike-slip → *fault* cutting through oceanic and continen-tal → *lithosphere* and connecting segments of → *mid-ocean ridges* or → *sub-duction zones*. Conservative → *plate boundary*

Transgression Continentward shift of the coastline due to sea-level rise, erosion, or subsidence of the coastal area. Opposite to → *regression*

Trap basalt See → *flood* basalt

Trondhjemite Magmatic rock similar to → *plagiogranite* but contain-ing → *plagioclase* richer in sodium

Turbidity current Suspension of sediment in water that glides down a subaquatic (mostly submarine) slope. Common at continental margins. The deposited layer typically shows → *graded bedding*

Tuff Volcanic ashes solidified to rocks

Turbidite Deposit from a → *turbidity current,* characterized by → *graded bedding*. Common in deep water adjacent to uplifted regions

Ultrabasic *Magma* and *magmatic rock* with less than 45 weight-% SiO_2

Ultrahigh-pressure metamorphism* → *Metamorphism* in → *subduction zones* with extremely high pressure (burial depth more than 80 km)

Unconformity A break in deposition within a stratigraphic package in which time is missing. Three common types: disconformity—time missing in → *con-cordant* rocks; non-conformity—sedimentary deposits on crystalline rocks; angular unconformity—angle (→ *discordance*) between rock layers

Viscosity Resistance of a liquid to flow. Generally, → *acidic* magmas and those with lower amounts of dissolved → *fluids* have higher viscosities

Volcanic arc, volcanic belt See → *magmatic arc*

Volcanic front See → *magmatic front*

Volcanic glass Non-crystallized rock material formed from a melt by fast chilling (typically in contact with water). The atoms are not arranged in a crystal lattice but rather have an arrangement similar to liquids

Volcanic rock → *Magmatic rock* (→ *lava* or 2219 → *tuff*) formed at the Earth's surface (subaerial or submarine)

Wadati-Benioff zone See → *Benioff zone*

Wilson cycle Tectonic cycle that operates at the scale of 10^8 years. After break-up of a continent an ocean is formed and then subducted. The cycle ends with the complete consumption of the oceanic realm, → *continent–con-tinent collision* and mountain building

Window (or fenster) In a (tectonic) window a lower tectonic unit which was overthrust by a higher tectonic unit (→ *nappe*) becomes visible when the stack of nappes forms a bulge and the higher unit becomes partly eroded. The higher unit forms a frame around the exposed lower unit (the window)

Zeolite Mineral group that forms at low temperatures (*diagenesis* and very low-grade *metamorphism*)

Zeolite facies* Very low-grade → *regional metamorphism*

Zircon Mineral, zirconium silicate. A tough mineral that resists weathering and is valuable in → *radiometric age determination* and → *terrane* analysis

References

1. Allègre CJ et al (1984) Structure and evolution of the Himalaya–Tibet orogenic belt. Nature 307:17–22

2. Ampferer O (1906) Über das Bewegungsbild von Faltengebirgen. Jahrb k k Geol Reichsanstalt 56:539–622

3. Amstutz A (1951) Sur l'évolution des structures alpines. Arch Sci (Genève) 4:323–329

4. Anderson DL (1982) Hotspots, polar wander, Mesozoic convection and the geoid. Nature 297:391–393

5. Anderson DL, Dziewonski AM (1984) Seismic tomography. Sci Am 251(4):60–68

6. Anhaeusser CR (1978) The geological evolution of the primitive earth—evidence from the Barberton Mountain Land. In: Tarling DH (ed) Evolution of the earth's crust. Academic Press, London New York San Francisco, pp 71–106

7. Armstrong R, Wilson AH (2000) A SHRIMP U-Pb study of zircons from the layered sequence of the Great Dyke, Zimbabwe, and a granitoid anatectic dyke. Earth Planet Sci Lett 180:1–12

8. Auzende JM, Bideau D, Bonatti E, Can–nat E, Honnorez J, Lagabrielle Y, Malavielle J, Mamaloukas–Frangoulis V, Mevel C (1989) Direct observation of a section through slow–spreading oceanic crust. Nature 337: 726–729

9. Baker BH, Wohlenberg J (1971) Structure and evolution of the Kenya Rift Valley. Nature 229:538–542

10. Baker HB (1911) The origin of the moon. Free Press, Detroit

11. Ballard RD, Moore JG (1977) Photographic atlas of the mid-Atlantic rift valley. Springer, New York, p 114

12. Barrow G (1893) On an intrusion of muscovite–biotite gneiss in the southeastern highlands of Scotland, and its accompnying metamorphism. Quart J Geol Soc Lond 49:330–358

13. Bearth P (1959) Über Eklogite, Glaukophanschiefer und metamorphe Pillowlaven. Schweiz Mineral Petrogr Mitteilungen 39:267–286

14. Benioff H (1954) Orogenesis and deep crustal structure: additional evidence from seismology. Geol Soc Amer Bull 65:385–400

15. Berger WH (1974) Deep–sea sedimentation. In: Burk CA, Drake CL (eds) The geology of continental margins. Springer, New York–Heidelberg–Berlin, pp 213–241

16. Bertrand MA (1884) Rapports de structure des Alpes de Glaris et du bassin houiller du Nord. Bull Soc Geol France 3eme Ser 12:318–330

17. Bickle MJ (1978) Heat loss from the earth: a constraint on Archaean tectonics from the relation between geothermal gradients and the rate of plate production. Earth Planet Sci Lett 40:301–315

18. Bickle MJ, Nisbet EG, Martin A (1994) Achean greenstone belts are not oceanic crust. J Geology 102:121–138

19. Blakey RC (2007) Carboniferous–Permian paleogeography of the Assembly of Pangaea: In: Wong TE (ed) Proceedings of the XVth international congress Carboniferous Permian Stratigraphy Utrecht 2003. Royal Dutch Academy Arts Sciences, Amsterdam, pp 443–456

20. Blakey RC (2008) Gondwana paleogeography from assembly to breakup—a 500 million year odyssey: In: Fielding CR, Frank TD, Isbell JL (eds) Resolving the late paleozoic ice age in time and space: Geological Society of America Special Paper, vol 441, pp 1–28

21. Blakey RC (2016) Global paleogeography and tectonics in deep time series. Colorado Plateau Geosystems Inc., World Wide Web. ▶ https://deeptimemaps.com/map-room/

22. Blundell D, Freeman R, Mueller S (1992) A continent revealed. The European Geotraverse. Cambridge University Press, Cambridge, 73 + 275 pp, 14 maps

23. Bonatti E (1994) The Earth's mantle below the oceans. Sci Am 270(3):26–33

24. Bott MPH (1982) The interior of the earth: its structure, constitution and evolution, 2nd edn. Edward Arnold, London, p 403

25. Boudier F, Nicolas A (1985) Harzburgite and lherzolite subtypes in ophiolitic and oceanic environments. Earth Planet Sci Lett 76:84–92

26. Boudier F, Ceuleneer G, Nicolas A (1988) Shear zones, thrusts and related magmatism in the Oman ophiolite: initiation of thrusting on an oceanic ridge. Tectonophysics 151:275–296

27. Bowring SA, Williams IS (1999) Priscoan (4.00–4.03) orthogneisses from northwestern Canada. Contrib Mineral Petrol 134:3–16

28. Braun A, Marquardt G (2001) Die bewegte Geschichte des Nordatlantiks. Spektrum der Wissenschaft 6(2001):50–59

29. Brewer PG, Densmore CD, Munns R, Stanley RJ (1969) Hydrography of the Red Sea brines. In: Degens ET, Ross DA (eds) Hot brines and recent heavy metal deposits in the Red Sea. Springer, Berlin, pp 138–147

30. von Buch LC (1824) Über geognostische Erscheinungen im Fassathal. Von Leonard's Mineral Taschenbuch 1824:396–437

31. Bullard EC, Everett JE, Smith AG (1965) The fit of the continents around the Atlantic. Phil Trans R Soc London A258:41–51

32. Bullen KE (1942) The ellipticities of surfaces of equal density in the Earth's interior. Royal Soc New Zealand Trans Proceed 72:141–143

33. Burke KC, Wilson JT (1976) Hot spots on the Earth's surface. In: American S (ed) Continents adrift and continents aground. Freeman, San Francisco, pp 58–75

34. Burke K (2011) Plate tectonics, the Wilson Cycle, and mantle plumes: geodynamics from the top. Annu Rev Earth Planet Sci 39:1–29

35. Campbell BJ, Jeanthon C, Luther GW, Cary SC (2001) Growth and phylogenetic properties of novel epsilon proteobacteria enriched from Alvinella pompejana and deep–sea hydrothermal vents. Appl Environ Microbiol 67:4566–4572

36. Cann JR (1981) Ore deposits of the ocean crust. In: Tarling DH (ed) Economic geology and geotectonics. Blackwell, Oxford, pp 119–134

37. Chase CG (1979) Subduction, the geoid, and lower mantle convection. Nature 282:464–468

38. Chen F, Todt W, Hann HP (2003) Zircon and garnet geochronology of eclogites from the Moldanubian Zone of the Black Forest, Germany. J Geol 111:207–222

39. Chopin C (1984) Coesite and pure pyrope in high–grade blueschists of the Western Alps: a first record and some consequences. Contrib Mineral Petrol 86:107–118

40. Christiansen RL, Foulger GR, Evans JR (2002) Upper–mantle origin of the Yellowstone hotspot. Geol Soc Amer Bull 114:1245–1256

41. Christie-Blick N, Biddle KT (1985) Deformation and basin formation along strike–slip faults. Soc Econ Paleont Mineral Spec Publ 37:1–34

42. Cocks CRM, Fortey RA (1982) Faunal evidence for oceanic separations in the Palaeozoic of Britain. J Geol Soc London 139:465–478

43. Coffin MF, Eldholm O (1993) Large igneous provinces. Sci Am 269(4):42–49

44. Condie KC (1973) Archaean magmatism and crustal thickening. Geol Soc Amer Bull 84:2981–2992

45. Condie KC (1997) Plate tectonics and crust–al evolution. Butterworth-Heinemann, Oxford, p 282

46. Condie KC, Moore JM (1977) Geochemistry of Proterozoic volcanic rocks from the Grenville Province, eastern Ontario. Geol Ass Canada Spec Paper 16:149–168

47. Cowan DS (1985) Structural styles in Mesozoic and Cenozoic mélanges in the western Cordillera of North America. Geol Soc Amer Bull 96:451–462

48. Crough ST, Morgan WJ, Hargraves RB (1980) Kimberlites: their relation to mantle hotspots. Earth Planet Sci Lett 50:260–274

49. Curray JR, Moore DG (1974) Sedimentary and tectonic processes in the Bengal deep–sea fan and geosyncline. In: Burk CA, Drake CL (eds) the geology of continental margins. Springer, New York–Heidelberg–Berlin, pp 617–627

50. Dana JD (1873) On some results on the Earth's contraction from cooling, including a discussion of the origin of mountains and the nature of the Earth's interior. Am J Sci Ser 3(5):423–443

51. DeMets C, Gordon R, Argus D, Stein S (1990) Current plate motions. Geophys J Int 101:425–478

52. DeMets C, Gordon RG, Argus DF, Stein S (1994) Effect of recent revisions to the geomagnetic reversal time scale on estimate of current plate motions. Geophys Res Lett 21:2191–2194

53. Descartes R (1644) Principia philosophiae

54. Dewey JF (1972) Plate tectonics. In: Planet Earth (Scientific American), 124–135, San Francisco 1974

55. Dickinson WR, Synder WS (1978) Plate tectonics of the Laramide Orogeny. Geol Soc Amer Memoir 151:355–366

56. Dingle RV (1980) Large allochthonous sediment masses and their role in the construction of the continental slope and rise of southwestern Africa. Mar Geol 37:333–354

57. Dobrzhinetskaya LF, Eide EA, Larsen RB, Sturt BA, Tronnes RG, Taylor WR, Poshukhova TV (1995) Microdiamonds in high–grade metamorphic rocks from the Western Gneiss region, Norway. Geology 23:597–600

58. Doglioni C, Harabaglia P, Merlini S, Mongelli F, Peccerillo A, Piromallo C (1999) Orogens and slabs vs. their direction of subduction. Earth Sci Rev 45:167–208

59. Drake R, Vergar M, Munizaga F, Vincente JC (1982) Geochronology of Mesozoic-Cenozoic magmatism in Central Chile, lat 31°–36°S. Earth Sci Rev 18:353–363

60. Drury SA (1981) The history of the Earth's crust: the Archaean Eon and before. In: Smith DG (ed) The Cambrige encyclopedia of earth sciences. Cambridge University Press, Cambridge, pp 250–261

61. Einsele G (1992) Sedimentary basins. Springer, Berlin-Heidelberg, p 628

62. Einsele G, Liu B, Dürr W, Frisch W, Liu G, Luterbacher HP, Ratschbacher L, Ricken W, Wendt J, Wetzel A, Yu G, Zheng H (1994) The Xigaze forearc basin: evolution and facies architecture (Cretaceous, Tibet). Sed Geol 90:1–32

63. Élie de Beaumont JB (1841) Les Vosges. Explication de la Carte Géologique de la France, vol I, Paris

64. Élie de Beaumont JB (1852) Notice sur les systèmes des montagnes. 3 vol, Paris

65. England PC (1981) Metamorphic pressure estimates and sediment volumes for the Alpine orogeny: an independent control on geobarometers? Earth Planet Sci Lett 56:387–397

66. de Saint–Fond BF (1788) Essai sur l'histoire naturelle des roches de Trapp. Rue et Hôtel Serpente, Paris, 159 pp

67. Foley SF, Buhre S, Jacob DE (2003) Evolution of the Archaean crust by delamination and shallow subduction. Nature 421:249–252

68. Franke W (2002) Die Vereinigten Platten von Europa. In: Wefer G (ed) Expedition Erde. Alfred Wegener Foundation, Berlin, pp 30–35

69. Frisch W (1979) Tectonic progradation and plate tectonic evolution of the Alps. Tectonophysics 60:121–139

70. Frisch W, Al Shanti A (1977) Ophiolite belts and the collision of island arcs in the Arabian Shield. Tectonophysics 43:293–306

71. Frisch W, Gawlick H-J (2003) The nappe structure of the central Northern Calcareous Alps and its disintegration during Miocene tectonic extrusion—a contribution to understanding the orogenic evolution of the Eastern Alps. Int J Earth Sci (Geol Rundschau) 92:712–727

72. Frisch W, Loesche J (1986) Plattentektonik. Wissenschaftliche Buchgesellschaft, Darmstadt, 190 pp

73. Frisch W, Ménot R–P, Neubauer F, von Raumer JF (1990) Correlation and evolution of the Alpine basement. Schweizerische Mineralogische Petrogra–phische Mitteilungen 70: 265–285

74. Frisch W, Meschede M (2005) Plattentektonik. Kontinentverschiebung und Gebirgsbildung. Primus and Wissenschaftliche Buchgesellschaft, Darmstadt, 196 pp

75. Frisch W, Neubauer F (1989) Pre-Alpine terranes and tectonic zoning in the eastern Alps. Geol Soc Amer Spec Paper 230:91–100

76. Frisch W, Meschede M, Sick M (1992) Origin of the Central American ophiolites: evidence from paleomagnetic results. Geol Soc Amer Bull 104:1301–1314

77. Frisch W, Dunkl I, Kuhlemann J (2000) Post-collisional orogen–parallel large–scale extension in the Eastern Alps. Tectonophysics 327:239–265

78. Gebauer D, Schertl HP, Brix M, Schreyer W (1997) 35 Ma old ultrahigh–pressure metamorphism and evidence for very rapid exhumation in the Dora Maira massif, Western Alps. Lithos 41:5–24

79. Grand SP, van der Hilst RD, Widiyantoro S (1997) Global seismic tomography: A snapshot of convection in the Earth. GSA (Geol Soc Amer) Today 7/4: 1–7

80. Granet M, Achauer U, Wilson M (1995) Nachweis eines Mantelplumes unter Frankreichs Zentralmassiv durch seismische Sondierung. Spektrum der Wissenschaft 10(1995):28–33

81. Green HW (1994) Solving the paradox of deep earthquakes. Sci Am 271(3):50–57

82. Griffiths RW, Campbell IH (1990) Stirring and structure in mantle starting plumes. Earth Planet Sci Lett 99:66–78

83. Hall J (1815) On the vertical position and the convolutions of certain strata and their relation with granite. Trans R Soc Edinburgh 7:79–85

84. Hall R (2000) Neogene history of collision in the Halmahera region, Indonesia. In: Proceedings of 27th annual convention on Indonesian Petroleum Association, pp 487–493

85. Hann HP, Chen F, Zedler H, Frisch W, Loeschke J (2003) The Rand Granite in the southern Schwarzwald and its geodynamic significance in the Variscan belt of SW Germany. Int J Earth Sci 92:821–842

86. Hatcher RD (2002) Alleghanian (Appalachian) orogeny, a product of zipper tectonics: rotational transpressive continent–continent collision and closing of ancient oceans along irregular margins. In: Martinez Catalan JR, Hatcher RD, Arenas R, Diaz Garcia F (eds) Variscan–Appalachian dynamics: the building of the late Paleozoic basement. Geol Soc Amer Spec Paper 364:199–208

87. Hauy RJ (1822) Traité de Minéralogie, 2nd edn. Bachelier, Paris

88. Hawkins JW (1974) Geology of the Lau basin, a marginal sea behind the Tonga arc. In: Burk CA, Drake CL (eds) The geology of continental margins. Springer, New York, pp 505–524

89. Heaman LM, Kjarsgaard BA (2000) Timing of eastern North American kimberlite magmatism: continental extension of the Great Meteor hospot strack? Earth Planet Sci Lett 178:253–268

90. Heezen BC, Hollister CD (1971) The face of the deep. Oxford University Press, New York, p 659

91. Heirtzler JR, LePichon X, Baron JG (1966) Magnetic anomalies over the Reykjanes ridge. Deep-Sea Res 13:427–443

92. Herzig PM, Hannington MD (2000) Input from the deep: hot vents and cold seeps. In: Schulz HD, Zabel M (eds) Marine geochemistry. Springer, Berlin, Heidelberg, New York, pp 397–416

93. Hess HH (1946) Drowned ancient islands of the Pacific basin. Am J Sci 244:772–791

94. Hoffman PF (1980) Wopmay orogen: a Wilson cycle of Early Proterozoic age in the northwest of the Canadian Shield. Geol Assoc Canada Spec Paper 20:523–549

95. Hoffman PF (1991) Did the breakout of Laurentia turn Gondwanaland insideout? Science 252:1409–1412

96. Holmes A (1931) Radioactivity and earth movements. Geol Soc Glasgow Trans 18:559–606

97. Holmes A (1944) The Machinery of continental drift: the search for a mechanism. Principles of physical geology. Nelson and Sons, London, pp 505–509

98. Howell DG (ed) (1985) Tectonostrati–graphic terranes of the circum–Pacific region. Circum–Pacific Council for energy and mineral resources, Houston. Earth Science Series, vol 1, 581 pp

99. Hutton J (1795) Theory of the Earth with proofs and illustrations, vol 2. Cadell, Davies and Creech, Edinburgh, London

100. Hyndman RD (1996) Schwere Erdbeben nach langer seismischer Stille. Spektrum der Wissenschaft 10(1996):64–72

101. Illies JH (1974) Taphrogenesis: Rhine–graben, continental rift systems, and Martian rifts. In: Illies J, Fuchs K (eds) Approaches to taphrogenesis. Schweizerbart, Stuttgart, pp 1–13

102. Isacks B, Molnar P (1969) Mantle earthquake mechanisms and the sinking of the lithosphere. Nature 223:1121–1124

103. Jahn B, Caby R, Monie P (2001) The oldest UHP eclogites of the World: age of UHP metamorphism, nature of protoliths and tectonic implications. Chem Geol 178:143–158

104. Jones DL, Cox A, Coney P, Beck M (1982) The growth of western North America. Sci Am 247(5):70–84

105. Jordan JL (1803) Mineralogische, berg– und hüttenmännische Reisebemerkungen, vorzüglich in Hessen, Thüringen, am Rheine und im Sege–Altkirchner Gebiet

106. Karig DE (1971) Origin and development of marginal basins in the western Pacific. J Geophys Res 76:2542–2561

107. Kearey P, Vine FJ (1990) Global Tectonics. Blackwell, Oxford, 302 pp

108. Kontinen A (1987) An early Proterozoic ophiolite—the Jormua mafic–ultra–mafic complex, northeastern Finland. Precambr Res 35:313–341

109. Kopf A (2003) Schlote, die Schlamm statt Feuer speien. Spektrum der Wissenschaft 1(2003):38–47

110. Kossmat F (1927) Gliederung des varis–tischen Gebirgsbaues. Abhandlungen sächsisches geol Landesamt 1:3–40

111. Kraml M, Bull A (2000) Sodaseen im Ostafrikanischen Grabenihre Entstehung und Bedeutung. Verein Naturforschende Gesellschaft Freiburg im Breisgau 88(89):85–118

112. Kröner A (ed) (1981) Precambrian plate tectonics. Elsevier, Amsterdam, p 781

113. Kroner U, Mansy J-L, Mazur S, Aleksandrowski P, Hann H-P, Huckriede H, Laquement F, Lamarche J, Ledru P, Pharaoh TC, Zedler H, Zeh A, Zulauf G (2008) Variscan tectonics. In: McCann T (Hrsg.): The geology of Central Europe. Vol. 1: Precambrian and Paleozoic. The Geological Society, London, pp 599–664

114. Kuhlemann J, Spiegel C, Dunkl I, Frisch W (1999) A contribution to the middle Oligocene paleogeography of central Europe: new evidence from fission track ages of the southern Rhine-Graben. Neues Jahrbuch Geol Paläont Abhandlungen 214:415–432

115. Kusky MT, Li J-H, Tucker RD (2001) The Archean Dongwanzi ophiolite complex, North China Craton: 2.505–billion–year–old oceanic crust and mantle. Science 292:1142–1145

116. Lago BL, Rabinowicz M, Nicolas A (1982) Podiform chromite ore bodies: a genetic model. J Petrol 23:103–125

117. Larson RL (1995) The mid-Cretaceous superplume episode. Sci Am 272(2):82–86

118. LePichon X (1968) Sea floor spreading and continental drift. J Geophys Res 73:3661–3697

119. Lister GS, Davis A (1989) The origin of metamorphic core complexes and detachment faults formed during Tertiary continental extension in the northern Colorado River region, USA. J Struct Geol 11:65–94

120. Lugeon M (1902) Les grandes nappes de recouvrement des Alpes du Chablais et de la Suisse. Bull Soc Géol France, 4ème Sér, 1:723–823

121. Lyell C (1833) Principles of Geology. Bd 3, John Murray, London, 398 pp

122. Macdonald KC, Fox PJ, Perram PJ, Eisen MF, Haymon RM, Miller SP, Carbotte SM, Cormier M-H, Shor AN (1988) A new view of the mid-ocean ridge from the behaviour of ridge–axis discontinuities. Nature 335:217–225

123. Makris J, Menzel H, Zimmermann J, Gouin P (1975) Gravity field and crustal structure of north Ethiopia. In: Pilger A, Rösler A (eds) Afar depression in Ethiopia. Schweizerbart, Stuttgart, pp 135–144

124. Matte P (2001) The Variscan collage and orogeny (480–290 Ma) and the tectonic definition of the Armorica microplate. Terra Nova 13:122–128

125. McKenzie DP (1978) Some remarks on the development of sedimentary basins. Earth Planet Sci Lett 40:25–32

126. McKenzie DP, Parker RL (1967) The North Pacific: an example of tectonics on a sphere. Nature 216:1276–1280

127. McKerrow WS, Mac Niocaill C, Dewey JF (2000) The Caledonian orogeny redefined. J Geol Soc 157:1149–1154

128. Meschede M (1986) A method of discriminating between different types of mid-ocean ridge basalts and continental tholeiites with the Nb–Zr–Y diagram. Chem Geol 56:207–218

129. Meschede M, Pelletier B (1994) Structural style of the accretionary wedge in front of the North d'Entrecasteaux Ridge (ODP Leg 134). In: Proceedings of ocean drilling program. Scientific Results, College Station, Texas, pp 417–429

130. Meschede M, Frisch W, Herrmann UR, Ratschbacher L (1997) Stress transmission across an active plate boundary: an example from southern Mexico. Tectonophysics 266:81–100

131. Meschede M, Zweigel P, Frisch W, Völker D (1999) Mélange formation by subduction erosion: the case of the Osa mélange in southern Costa Rica. Terra Nova 11:141–148

132. Meschede M, Schmiedl G, Weiss R, Hemleben C (2002) Benthic foraminiferal distribution and sedimentary structures suggest tectonic erosion at the Costa Rica convergent plate margin. Terra Nova 14:1–12

133. Miyashiro A (1973) Metamorphism and metamorphic belts. Allen and Unwin, London, p 492

134. Moore GF, Billman HG, Hehanussa PE, Karig DE (1980) Sedimentology and paleobathymetry of trench–slope deposits, Nias Island, Indonesia. J Geol 88:161–180

135. Morelli A, Dziewonski AM (1987) Topography of the core–mantle boundary and lateral homogeneity of the liquid core. Nature 325:678–683

136. Morgan WJ (1968) Rises, trenches, great faults, and crustal blocks. J Geophys Res 73:1959–1982

137. Neumann GA, Forsyth DW (1993) The paradox of the axial profile: isostatic compensation along the mid-Atlantic Ridge? J Geophys Res 98:17891–17910

138. Nicolas A (1995) The mid-Oceanic ridges. Mountains below sea level. Springer, Berlin–Heidelberg, 217 pp

139. O'Connor JM, Stoffers P, Wijbrans JR, Shannon PM, Morrissey T (2000) Evidence from episodic seamount volcanism for pulsing of the Iceland plume in the past 70 Myr. Nature 408:954–958

140. Okomura K, Yoshioka T, Kuscu I (1993) Surface faulting on the North Anatolian Fault in these two millennia. US Geol Survey Open–file Rep 94–568: 143–144

141. O'Neil J, Carlson RW, Francis D, Stevenson RK (2008) Neodymium–142 evidence for Hadean mafic crust. Science 321:1828–1831

142. Pallas PS (1777) Observation sur la Formation des Montagnes. St Petersburg

143. Parsons B, Sclater JG (1977) An analysis of the variation of the ocean floor bathymetry and heat flow with age. J Geophys Res 82:803–827

144. Patriat P, Achache J (1984) India-Eurasia collision chronology has implications for crustal shortening and driving mechanism of plates. Nature 311:615–621

145. Pearce JA (1983) The role of sub–continental lithosphere in magma genesis at destructive plate margins. In: Hawkes–worth CJ, Norry MJ (eds) Continental Basalts and Mantle Xenoliths, Shiva Publications, Nantwich, pp 230–249

146. Pearce JA, Cann JR (1973) Tectonic setting of basic volcanic rocks determined using trace element analysis. Earth Planet Sci Lett 19:290–300

147. Polat A, Herzberg C, Münker C, Rodgers R, Kusky T, Li J, Fryer B, Delaney J (2006) Geochemical and petrological evidence for a subduction zone origin of Neoarchean (ca. 2.5 Ga) peridotites, central orogenic belt, North China craton. Geological Society of America Bulletin 118: 771–784

148. Pflug R (1982) Bau und Entwicklung des Oberrheingrabens. Wissenschaftliche Buchgesellschaft, Darmstadt, 145 pp

149. Platt JP (1986) Dynamics of orogenic wedges and the uplift of high pressure metamorphic rocks. Geol Soc Amer Bull 97:1037–1053

150. Ratschbacher L, Riller U, Meschede M, Herrmann U, Frisch W (1991) Second look at suspect terranes in southern Mexico. Geology 19:1233–1236

151. Ratschbacher L, Frisch W, Linzer H-G, Merle O (1991) Lateral extrusion in the Eastern Alps, Part 2: structural analysis. Tectonics 10:257–271

152. Reading HG (1986) Sedimentary environments and facies, 2nd edn. Blackwell, Oxford, p 615

153. Reigber C, Gendt G (1996) Geodätische Messung der Plattentektonik. Spektrum der Wissenschaft 1(1996):115–117

154. Richter FM (1985) Models for the Archean thermal regime. Earth Planet Sci Lett 73:350–360

155. Roll A (1979) Versuch einer Volumenbilanz des Oberrheintalgrabens und seiner Schultern. Geol Jahrbuch A 52:3–82

156. Rolland Y (2002) From intra–oceanic convergence to post–collisional evolution: example of the India-Asia convergence in NW Himalaya, from Cretaceous to present. J Virtual Explor 8:185–208

157. Saleeby JB (2003) Segmentation of the Laramide Slab—evidence from the southern Sierra Nevada region. Geol Soc Amer Bull 113:655–668

158. Sarnthein M, Diester–Haass L (1977) Eolian sand turbidites. J Sed Petrol 47:868–890

159. Saussure HB de (1796) Voyages dans les Alpes. Bd. 4, Fauche–Borel, Neuchâtel

160. Schermer ER, Howell DG, Jones DL (1984) The origin of allochthonous terranes. Ann Rev Earth Planet Sci 12:107–131

161. Schidlowski M (1983) Evolution of photoautotrophy and early atmospheric oxygen. Precambrian Res 20:319–335

162. Schmid SM, Pfiffner OA, Froitzheim N, Schönborn G, Kissling E (1996) Geophysical–geological transect and tectonic evolution of the Swiss-Italian Alps. Tectonics 15:1036–1064

163. Schmincke H-U (1981) Volcanic activity away from plate margins. In: Smith DG (ed) The Cambridge encyclopedia of Earth sciences. Cambridge University Press, Cambridge, pp 201–209

164. Schmincke H-U (2004) Volcanism. Springer, Heidelberg-Berlin, p 324

165. Scholl DW (1974) Sedimentary sequences in the North Pacific trenches. In: Burke CA, Drake CL (eds) The geology of continental margins. Springer, New York–Heidelberg–Berlin, pp 493–504

166. Scholz CH, Barazangi M, Sbar ML (1971) Late Cenozoic evolution of the Great Basin, western United States, as an ensialic inter-arc basin. Geol Soc Amer Bull 82:2979–2990

167. Schreiner (1984) Hegau und westlicher Bodensee. Sammlung Geol Führer vol 62, 2nd edn, Borntraeger, BerlinStuttgart, 103 pp

168. Schubert G, Yuen DA, Turcotte DL (1975) Role of phase transitions in a dynamic mantle. Geophys J Royal Astr Soc 42:705–735

169. Schubert G, Turcotte DL, Olson P (2001) Mantle convection in the earth and planets. Cambridge University Press, Cambridge (941 S)

170. Schwinner R (1920) Vulkanismus und Gebirgsbildung. Ein Versuch. Zeitschrift Vulkanologie 5:175–230

171. Scott DJ, Helmstaedt H, Bickle MJ (1992) Purtuniq ophiolite, Cape Smith belt, northern Quebec, Canada: a reconstructed section of Early Proterozoic oceanic crust. Geology 20:173–176

172. Seibold E, Berger WH (1982) The Sea Floor. Springer, Berlin, p 288

173. Sieh KE (1978) Prehistoric large earthquakes produced by slip on the San Andreas fault at Pallett Creek, California. J Geophys Res 83:3907–3939

174. Smith WHF, Sandwell DT (1997) Global seafloor topography from satellite altimetry and ship depth soundings. Science 277:1957–1962

175. Snider A (1859) Le Création et ses Mystères Devoilés. Franck and Dentu, Paris

176. Stein RS, Barka AA, Dietrich JH (1997) Progressive failure on the North Anatolian fault since 1939 by earthquake stress triggering. Geophys J Int 128:594–604

177. Steno N (Stensen N) (1669) De solido intra solidum naturaliter contento dissertationis prodromus. Florence

178. Stern RJ (2002) Subduction zones. Rev Geophys 40/4:3–1 to 3–38

179. Stille H (1913) Tektonische evolution und revolution in der Erdrinde. Veit, Leipzig

180. Suess E (1885–1909): Das Antlitz der Erde. 5 vol., Tempsky–Freytag, Vienna

181. Sykes LR (1967) Mechanism of earthquakes and nature of faulting on the mid-ocean ridges. J Geophys Res 72:2131–2153

182. Székely B (2001) On the surface of the Eastern Alps—a DEM study. Tübinger Geowissenschaftliche Arbeiten 60, 157 pp

183. Tackley PJ (2008) Geodynamics: layer cake or plum pudding? Nat Geosci 1:157–158

184. Tait JA, Bachtadse V, Franke W, Soffel HC (1997) Geodynamic evolution of the European Variscan fold belt: palaeomagnetic and geological constraints. Geol Rundschau 86:585–598

185. Talwani M, LePichon X, Ewing M (1965) Crustal structure of the mid-ocean ridges. J Geophys Res 70:341–352

186. Tapponnier P, Peltzer G, Armijo R (1986) On the mechanics of the collision between India and Asia. Geol Soc Amer Spec Publ 19:115–157

187. Tarney J (1976) Geochemistry of Archaean high–grade gneisses, with implications as to the origin and evolution of the Precambrian crust.

In: Windley BF (ed) The early history of the Earth. Wiley-Interscience, London, pp 405–417

188. Tarney J, Dalziel IWD, de Wit MJ (1976) Marginal basin "Rocas Verdes" complex from S Chile: a model for Archaean greenstone belt formation. In: Windley BF (ed) The early history of the Earth. Wiley-Interscience, London, pp 131–146

189. Taylor FB (1910) Bearing of the Tertiary mountain belt on the origin of the earth's plan. Geol Soc Amer Bull 21:179–226

190. Termier P (1904) Les nappes des Alpes orientales et la synthèse des Alpes. Bull Soc Géol France, 4ème Sér 3:711–765

191. Torsvik TH, Burke K, Steinberger B, Webb SJ, Ashwal LD (2010) Diamonds sampled by plumes from the core-mantle boundary. Nature 466:352–355

192. Torsvik TH, van der Voo R, Doubrovine PV, Burke K, Steinberger B, Ashwald LD, Trønnes RG, Webb SJ, Bulla AL (2014) Deep mantle structure as a reference frame for movements in and on the Earth. PNAS 111:8735–8740

193. Trümpy R (1980) An Outline of the Geology of Switzerland. Wepf and Co, Basel, 109 pp

194. Turcotte DL, Emerman SH (1983) Mechanisms of active and passive rifting. Tectonophysics 94:39–50

195. Umhoefer PJ, Blakey RC (2006) Moderate (1600 km) northward translation of Baja British Columbia from southern California: an attempt at reconciliation of paleomagnetism and geology. In: Haggart JW, Enkin RJ, Monger JWH (eds) Paleogeography of the North American Cordillera: evidence For and against large–scale displacements. Geol Assoc Canada Spec Paper 46:305–327

196. Uyeda S, Kanamori H (1979) Back–arc opening and the mode of subduction. J Geophys Res 84:1049–1061

197. Van Fossen MC, Kent DV (1992) Paleo–magnetism of 122 Ma plutons in New England and the mid-Cretaceous paleomagnetic field in North America. J Geophys Res 97:19651–19661

198. Vine FJ, Matthews DH (1963) Magnetic anomalies over oceanic ridges. Nature 199:947–949

199. Visser DJL et al. (1984) Geological map of the republics of South Africa. Dept Min Energy Affairs, South Africa

200. Vogel A (1994) Die Kern–Mantel–Grenze: Schaltstelle der Geodynamik. Spektrum der Wissenschaft 11(1994):64–72

201. von Huene R, Scholl DW (1991) Observations at convergent margins concerning sediment subduction, subduction erosion, and the growth of continental crust. Rev Geophys 29:279–316

202. von Huene R, Ranero CR, Weinrebe W (2000) Quaternary convergent margin tectonics off Costa Rica, segmentation of the Cocos Plate, and Central American volcanism. Tectonics 19:314–334

203. Wadati K (1935) On the activity of deep–focus earthquakes in the Japan Islands and neighbourhoods. Geophys Mag 8:305–325

204. Walker RJ, Morgan JW, Horan MF (1995) Osmium–187 enrichment in some plumes: evidence for core–mantle interaction? Science 269:819–821

205. Walther J (1894) Lithogenesis der Gegenwart. Einleitung in die Geologie als Historische Wissenschaft, vol 3, Fischer, Jena, pp 535–1055

206. Wegener A (1912) Die Entstehung der Kontinente. Geol Rundschau 3:276–292

207. Wegener A (1915) Die Entstehung der Kontinente und Ozeane. Vieweg, Braunschweig, p 94

208. Wegener A (1929) Die Entstehung der Kontinente und Ozeane, 4th edn. Vieweg, Braunschweig, p 231

209. Wegener A (1966) The Origin of Continents and Oceans. Dover Publications, New York, 246 pp (translated from the fourth revised German edition by John Biram)

210. Werner D, Doebl F (1974) Geothermal anomalies and consequences for diagenesis and thermal waters. In: Illies J, Fuchs K (eds) Approaches to taphrogenesis. Schweizerbart, Stuttgart, pp 182–191

211. Wernicke B (1981) Low–angle normal faults in the Basin and Range province: nappe tectonics in an extending orogen. Nature 291:645–648

212. White RS, McKenzie D (1995) Mantle plumes and flood basalts. J Geophys Res 100:17543–17585

213. Wiens DA (2001) Seismological constraints on the mechanism of deep earthquakes: temperature dependence on deep earthquake source properties. Phys Earth Planet Inter 127:145–163

214. Wilde SA, Valley JW, Peck WH, Grahams CM (2001) Evidence from detrital zircons for the existence of continental crust and oceans on the Earth 4.4 Gyr ago. Nature 409:175–178

215. Wilson JT (1963) Evidence from islands on the spreading of the ocean floor. Nature 197:536–538

216. Wilson JT (1965) A new class of faults and their bearing on continental drift. Nature 207:343–347

217. Wilson M (1989) Igneous petrogenesis. Unwin Hyman, London, p 466

218. World Data Center for Marine Geology & Geophysics (2003) Total Sediment Thickness of the World's Oceans & Marginal Seas. ▶ http://www.ngdc.noaa.gov/mgg/sedthick/sedthick.html, Boulder

219. Wyld SJ, Umhoefer PJ, Wright JE (2006) Reconstructing northern Cordilleran terranes along known Cretaceous and Cenozoic strike–slip faults: implications for the Baja British Columbia hypothesis and other models. In: Haggart JW, Enkin RJ, Monger JWH (eds) Paleogeography of Western North America: evidence For and Against Large–Scale Displacements. Geological Association of Canada Special Paper, vol 46, pp 277–298

220. Zeitler PK, Chamberlain CP, Smith HA (1993) Synchronous anatexis, metamorphism, and rapid denudation at Nanga Parbat (Pakistan Himalaya). Geology 21:347–350

221. Zhao X, Antretter M, Solheit P, Inokuchi H (2002) Identifying magnetic carriers from rock magnetic characterization of Leg 183 basement core. Proc Ocean Drilling Program Init Repts 183, ▶ http://www-odp.tamu.edu/publications/183_IR/183ir.htm

222. Ziegler PA (1990) Geological atlas of western and central Europe. Shell International Petroleum, Maatschappij, 239 pp

Index

Printed in the United States
by Baker & Taylor Publisher Services